Michel Hervé

Analyticity in Infinite Dimensional Spaces

Walter de Gruyter
Berlin · New York 1989

Author
Michel Hervé
Mathématiques
Université de Paris VI
F-75252 Paris Cedex 05

Library of Congress Cataloging-in-Publication Data

Hervé, Michel (Michel A.)
    Analyticity in infinite dimensional spaces / Michel Hervé.
    p.   cm.--(De Gruyter studies in mathematics : 10)
    Bibliography: p.
    Includes index.
    ISBN 0-89925-205-2 : 1. Analytic functions. 2. Harmonic functions.
3. Analytic mappings.   I. Title.   II. Series.
QA331.7.H47   1989   88-31955
515.7'3--dc 19   CIP

Deutsche Bibliothek Cataloging-in-Publication Data

Hervé, Michel:
Analyticity in infinite dimensional spaces / Michel Hervé. –
Berlin ; New York : de Gruyter, 1989
    (De Gruyter studies in mathematics ; 10)
    ISBN 3-11-010995-6 (Berlin) Gewebe
    ISBN 0-89925-205-2 (New York) Gewebe
NE: GT

Cover design: Rudolf Hübler, Berlin. Typesetting: Asco Trade Typesetting Ltd., Hong Kong. Printing: Druckerei Gerike GmbH, Berlin. – Binding: Dieter Mikolai, Berlin.

Foreword

During the last twenty years, the theory of analyticity in infinite dimensions has developed from its foundations into a structure which may be termed harmonious, provided that one accepts to do without some features of the finite dimensional case. This harmony is of course favoured by the choice of a unique setting – locally convex spaces over the complex field, analytic maps into sequentially complete spaces – and a central topic: plurisubharmonicity, where a multitude of results obtained by different authors – Pierre Lelong, the founder of the notion, Gérard Coeuré, Christer Kiselman and others – deserved to be brought together. The reader will find the precise contents of each chapter in the summary which opens it.

The concern for unity has inevitably led to the omission of several other topics in spite of their indisputable interest, and among these I insist on the local theory of analytic sets. But the methods used have little in common with those used in this book; the material consists essentially in two theses: "Sous-ensembles analytiques d'une variété analytique banachique" (Paris, 1969) by Jean-Pierre Ramis, "Ensembles analytiques complexes dans les espaces localement convexes" (Paris, 1969) by Pierre Mazet, and for the time being there is little more to say on the subject.

This book is a tribute to all the authors it mentions. The first chapters owe much to the paper [Boc Sic]; comparatively short sections are devoted to some topics extensively developed in the excellent monographs published as North Holland mathematics studies: "Pseudo-convexité, convexité polynomiale et domaines d'holomorphie" by Philippe Noverraz (n° 3); "Analytic functions and manifolds in infinite dimensional spaces" by Gérard Coeuré (n° 11); "Holomorphic maps and invariant distances" by Franzoni and Vesentini (n° 40); "Complex analysis in locally complex spaces" by Sean Dineen (n° 57); "Complex analysis in Banach spaces" by Jorge Mujica (n° 120). To bring a new contribution was no easy task, but it seemed to me that classical potential theory deserved a better place, as a special tribute to the recently deceased Marcel Brelot.

Finally I express my gratitude to Professors Heinz Bauer and Peter Gabriel who accepted this book in their renowned series.

November 1988   Michel Hervé

Contents

Chapter 1 Some topological preliminaries   1

Summary   1
1.1 Locally convex spaces   2
1.2 Vector valued infinite sums and integrals   6
1.3 Baire spaces   9
1.4 Barrelled spaces   11
1.5 Inductive limits   13

Chapter 2 Gâteaux-analyticity   19

Summary   19
2.1 Vector valued functions of several complex variables   20
2.2 Polynomials and polynomial maps   28
2.3 Gâteaux-analyticity   35
2.4 Boundedness and continuity of Gâteaux-analytic maps   43
Exercises   50

Chapter 3 Analyticity, or Fréchet-analyticity   51

Summary   51
3.1 Equivalent definitions   52
3.2 Separate analyticity   58
3.3 Entire maps and functions   65
3.4 Bounding sets   73
Exercises   79

Chapter 4 Plurisubharmonic functions   81

Summary   81
4.1 Plurisubharmonic functions on an open set Ω in a l.c.
     space X   82
4.2 The finite dimensional case   87
4.3 Back to the infinite dimensional case   94
4.4 Analytic maps and pluriharmonic functions   104

4.5 Polar subsets 107
4.6 A fine maximum principle 120
Exercises 127

Chapter 5 Problems involving plurisubharmonic functions 129

Summary 129
5.1 Pseudoconvexity in a l.c. space X 130
5.2 The Levi problem 135
5.3 Boundedness of p.s.h. functions and entire maps 144
5.4 The growth of p.s.h. functions and entire maps 146
5.5 The density number for a p.s.h. function 154
Exercises 162

Chapter 6 Analytic maps from a given domain to another one 163

Summary 163
6.1 A generalization of the Lindelöf principle 164
6.2 Intrinsic pseudodistances 168
6.3 Complex geodesics and complex extremal points 179
6.4 Automorphisms and fixed points 184
Exercises 194

Bibliography 195

Glossary of Notations 201

Subject Index 205

Chapter 1
Some topological preliminaries

Summary

The aim of this first chapter is only to present the topological tools and set the notation which will be used throughout the book.

1.1. A. Locally convex topologies; the metrizable case.
    B. Bounded sets. Complete and quasi-complete spaces.
    C. The adjoint space; the weakened topology.
    D. Strong and weak* topologies on the adjoint space.
       Reflexivity and semi-reflexivity.

1.2. A. Summable families.
    B. The integral of a vector valued function.

1.3. A. Meagre sets and Baire spaces.
    B. Topological properties involving the Baire property.
    C. Banach and Fréchet spaces.

1.4. Barrelled spaces.
Prop. 1.4.2. The bounded sets are the same for the weakened topology and the initial one.
Rem. 1.4.3. Semireflexive spaces.

1.5. A. The gauge of an absorbent balanced convex set.
    B. Inductive limits.
    C. Strict inductive limits.

Chapter 1
Some topological preliminaries

1.1 Locally convex spaces

All topologies considered throughout the book shall be *Hausdorff*; there-fore, any finite dimensional vector space will be endowed with its euclidean topology ([B0]$_2$, Chap. I, §2, Th. 2).

(A) All topological vector spaces considered will be over $\mathbb{C}$ and *locally convex*; in this chapter, (X, Π) will denote a vector space X with the Hausdorff topology defined by some family Π of seminorms such that $\sup\limits_{p \in \Pi} p(x) > 0$ for all $x \in X \setminus \{0\}$; a fundamental system of neighbour-hoods of x in this topology is made up of the finite intersections of sets $x + p^{-1}([0, \alpha]), p \in \Pi, 0 < \alpha < \infty$. Then all seminorms in Π are continuous, but the family $\text{csn}(X, \Pi)$ of all continuous seminorms on (X, Π) is generally larger than Π: $p \in \text{csn}(X, \Pi)$ if and only if p is majorized by a finite linear combination, with positive coefficients, of seminorms in Π; another fun-damental system of neighbourhoods of x is made up of the sets $x + p^{-1}([0, 1]), p \in \text{csn}(X, \Pi)$.

If Π is *countable*, namely $\Pi = \{p_n : n \in \mathbb{N}^*\}$, then (X, Π) is *metrizable* since the formula

$$d(x, y) = \sum_{n=1}^{\infty} \frac{1}{n^2} \frac{p_n(x - y)}{1 + p_n(x - y)}$$

provides a distance defining the same topology as Π, and obviously in-variant under translations: $d(x + z, y + z) = d(x, y)$. Conversely, if (X, Π) is metrizable, then its topology can be defined by a countable family of seminorms (since the origin has a countable fundamental system of neigh-bourhoods) and also by an increasing sequence of seminorms.

A linear map $\lambda \colon X \to Y$ is a continuous map $(X, \Pi) \to (Y, \Pi')$ if and only if $q \circ \lambda \in \text{csn}(X, \Pi)$ for all $q \in \Pi'$; in particular, the topology of (X, Π) is finer than the topology of (X, Π') if and only if $\Pi' \subset \text{csn}(X, \Pi)$, and both topologies coincide if and only if $\text{csn}(X, \Pi') = \text{csn}(X, \Pi)$.

(B) A set B in the locally convex space (X, Π) is *bounded* if $p(B)$ is bounded, i.e. $\sup_{x \in B} p(x) < +\infty$, for all $p \in \Pi$ or for all $p \in \text{csn}(X, \Pi)$. Then the closure $\bar{B}$ of B is bounded too.

Whenever useful in order to avoid unessential difficulties (See for example 2.1.8 or 2.2.2.B), analytic maps will be assumed to have their values in *complete* (in some sense) locally convex (l.c.) spaces. A filter $\mathscr{F}$ on X is a Cauchy filter in (X, Π) if for each $p \in \Pi$ and each $\varepsilon > 0$ there is an $A \in \mathscr{F}$ such that $p(x - y) \leqslant \varepsilon$ for all $x, y \in A$; (X, Π) is complete if any Cauchy filter converges, and *quasi-complete* if any Cauchy filter on a closed bounded subset of (X, Π) converges. In other words, (X, Π) is quasi-complete if any Cauchy filter $\mathscr{F}$, for which there is a bounded subset B of (X, Π) such that $A \cap B \neq \emptyset$ for all $A \in \mathscr{F}$, converges (to some point $\in \bar{B}$).

Since any Cauchy sequence is bounded, quasi-complete (q.c.) implies sequentially complete (s.c.); if (X, Π) is metrizable, then complete, quasi-complete, sequentially complete, are equivalent properties. A *Fréchet space* is a metrizable space (X, Π) in which they hold.

The topology of (X, Π) can be defined by *one norm* if and only if the origin has bounded neighbourhoods. In fact, if Π contains $p_1, \ldots, p_n$ such that the set $\{x \in X : p_i(x) \leqslant 1, i = 1, \ldots, n\}$ is bounded, then each $p \in \Pi$ satisfies $p \leqslant \text{const.} \sup(p_1, \ldots, p_n)$; therefore $\sup(p_1, \ldots, p_n)$ is a norm defining the same topology as Π.

The space (X, Π) is *bornological* if all seminorms on X which are bounded on each bounded set are continuous. Given any l.c. space (X, Π), there always exists a unique topology on X for which the bounded sets are the same and which makes the space bornological: it is defined by the family of all seminorms on X which are bounded on each bounded set, and of course finer than the given one.

If a linear map $\lambda: X \to Y$ is continuous from (X, Π) to (Y, Π'), B bounded in (X, Π) implies $\lambda(B)$ bounded in (Y, Π'). The converse statement is true only if (X, Π) is bornological.

(C) Given a locally convex topology on the space X, X' or $\mathscr{L}(X, \mathbb{C})$ will denote the *adjoint space*, the set of all linear maps $x': X \to \mathbb{C}$ which are continuous, i.e. $|x'|$ is a continuous seminorm, and $x'(x)$ or $\langle x, x' \rangle$ the image of x under x'. The family of seminorms $|x'|$ defines the so-called *weakened topology* of X, which is coarser than the initial one, but related to it as follows.

Proposition 1.1.1. a) *The weakened topology is Hausdorff.*
b) *A convex set in X, if closed for the initial topology, is also closed for the weakened one.*

c) *The continuous linear maps $X \to \mathbb{C}$ are the same for the initial topology and for the weakened one.*

Proof. a) Given $x_0 \in X \setminus \{0\}$, there is a $p \in \Pi$ such that $p(x_0) > 0$; then, by the Hahn-Banach theorem ([Bo]$_2$, Chap. II, §6, Th. 1), the linear map $\mathbb{C}x_0 \ni \zeta x_0 \mapsto p(x_0)\zeta$ can be extended into $x' \in X'$ such that $|\langle x, x' \rangle| \leq p(x)$ for all $x \in X$ and $\langle x_0, x' \rangle = p(x_0) > 0$.

b) As a consequence of the same theorem (ibid., §3, n° 3), a convex set which is closed for the initial topology is the intersection of some sets $\{x \in X : \mathrm{Re}\langle x, x' \rangle \leq \alpha\}$, $x' \in X'$, which are closed for the weakened topology. $\square$

(D) Several locally convex topologies on X' are natural: the *weak* topology* defined by the seminorms $x' \mapsto |\langle x, x' \rangle|$; the *topology of compact convergence*, a finer one, by the seminorms $x' \mapsto \sup_{x \in K} |\langle x, x' \rangle|$ where K is any compact set in X; the *strong topology*, a still finer one, defined by the seminorms $x' \mapsto \sup_{x \in C} |\langle x, x' \rangle|$, where $C = \{x \in X : |\langle x, x' \rangle| \leq c(x')$ for all $x' \in X'\}$ and the $c(x')$ are arbitrarily chosen real positive numbers. In other words: the strong topology is defined by the seminorms $x' \mapsto \sup_{x \in B} |\langle x, x' \rangle|$, where B is bounded for the weakened topology, and it will turn out (Prop. 1.4.2) that the bounded sets are the same for this topology and the initial one.

The sets C will be useful because of their algebraic properties: balanced and convex. The corresponding sets $C' = \{x' \in X' : |\langle x, x' \rangle| \leq 1$ for all $x \in C\}$ are a fundamental system of strong neighbourhoods of the origin in X'.

The three above defined topologies on X' give birth to three adjoint spaces, each of which is larger than the previous one.

Theorem 1.1.2. a) *The adjoint space to X' for the weak* topology is always X.*

b) *The adjoint space to X' for the strong topology is also X if and only if the above defined sets*

$$C = \{x \in X : |\langle x, x' \rangle| \leq c(x') \text{ for all } x' \in X'\}$$

are compact for the weakened topology. This property, which is known as semireflexivity (s.r.) of X, reflexivity if X is a normed space, has the following consequences:

c) *any closed subspace of X is s.r. too;*

d) *X' with the strong topology is s.r. too;*

e) *X is quasi-complete for the weakened and initial topologies.*
  See also Remark 1.4.3 below.

Proof. a) If a linear map $x^*: X' \to \mathbb{C}$ is continuous for the weak* topology, one can find a finite number of $x_j \in X$ and real $\alpha_j > 0$ such that

$$|\langle x^*, x' \rangle| \leqslant \sum_j \alpha_j |\langle x_j, x' \rangle| \quad \text{for all } x' \in X'.$$

Hence follows that x^* vanishes on $\bigcap_j \ker(x' \mapsto \langle x_j, x' \rangle)$ and therefore, by an algebraic argument ([Ta]$_2$, Th. 3.5.C), x^* is a linear combination of the x_j.
b) First assume that X is the adjoint space to X' for the strong topology. Given a set C, let $c_0(x') = \sup_{x \in C} |\langle x, x' \rangle| \leqslant c(x')$ for all $x' \in X'$, so that we also have

$$C = \{x \in X : |\langle x, x' \rangle| \leqslant c_0(x') \text{ for all } x' \in X'\}.$$

Then any linear map $f: X' \to \mathbb{C}$ satisfying $|f(x')| \leqslant c_0(x')$ for all $x' \in X'$ is strongly continuous and C (with the weakened topology) is exactly the set (with the topology of pointwise convergence) of all linear maps f satisfying $|f(x')| \leqslant c_0(x')$ for all $x' \in X'$, a closed subset (for the same topology) of the set of all maps $f: X' \to \mathbb{C}$ satisfying the same inequalities.
  The latter set (with the topology of pointwise convergence) can be identified with the product space $\prod_{x' \in X'} \bar{\Delta}[0, c_0(x')]$ of compact discs, which is compact by the Tychonov theorem ([Bo]$_1$, Chap. I, §10, Th. 3).
  Conversely, assume that each C is compact for the weakened topology, and let X'' be the adjoint space to X' for the strong topology. Given $x_0'' \in X''$, choose C so that $|\langle x_0'', x' \rangle| \leqslant 1$ for the x' in the corresponding C': we claim that x_0'' actually is an element of this C. In fact, the weak* topology on X'', defined by the seminorms $x'' \mapsto |\langle x'', x' \rangle|$, induces the weakened topology on the subspace X of X'', and for this topology the convex set C is compact. Then ([Bo]$_2$, Chap. II, §3, n° 3): if $x_0'' \notin C$, by a) above there is an $x' \in X'$ such that $\mathscr{Re}\langle x_0'', x' \rangle$ exceeds 1 and $1 \geqslant \sup_{x \in C} \mathscr{Re}\langle x, x' \rangle$ or $1 \geqslant \sup_{x \in C} |\langle x, x' \rangle|$ since C is balanced: this is a contradiction.
c) Let Y be a closed subspace of X (with the topology induced by the initial topology of X). By the Hahn-Banach theorem: Y is the set of common zeroes of some subfamily of X', hence is closed also for the weakened topology, and any $y' \in Y'$ is the restriction to Y of some $x' \in X'$; therefore the weakened topology of Y is also induced by the weakened topology of X, and each set $\{y \in Y : |\langle y, y' \rangle| \leqslant c(y') \text{ for all } y' \in Y'\}$ is the intersection with Y of some set C above.

d) If the adjoint space to X' for the strong topology is X, then the analogue in X' of the weakened topology is the weak* one, and the analogues in X' of the sets C are the sets $\{x' \in X': |\langle x, x' \rangle| \leqslant c(x) \text{ for all } x \in X\}$, which are compact for the weak* topology by the argument already used in part b) of the proof.

e) Let the property hold. If there is a bounded (for either topology) subset of X meeting each set in the filter $\mathscr{F}$ on X, there are also a set C such that $A \cap C \neq \emptyset$ for all $A \in \mathscr{F}$, and a point $x_0 \in C$ such that any weakened neighbourhood of x_0 meets any set $A \in \mathscr{F}$. If $\mathscr{F}$ is a Cauchy filter for the weakened topology, it follows at once that each weakened neighbourhood of x_0 contains a set $A \in \mathscr{F}$.

Now let $\mathscr{F}$ be also a Cauchy filter for the initial topology. Given $p \in \Pi$ and $\varepsilon > 0$, choose $A \in \mathscr{F}$ such that $p(x - y) \leqslant \dfrac{\varepsilon}{2}$ for all x, $y \in A$, then choose $x \in A$: since $A \subset x + p^{-1}\left(\left[0, \dfrac{\varepsilon}{2}\right]\right)$, x_0 in the closure of A and $x + p^{-1}\left(\left[0, \dfrac{\varepsilon}{2}\right]\right)$ closed (Prop. 1.1.1.b) for the weakened topology imply $x_0 \in x + p^{-1}\left(\left[0, \dfrac{\varepsilon}{2}\right]\right)$, hence $A \subset x_0 + p^{-1}([0, \varepsilon])$. $\square$

1.2 Vector valued infinite sums and integrals

Let (X, Π) be a vector space with a given locally convex topology.

(A) Let I be an infinite set of indeces i and $\mathscr{F}$ the family of finite subsets F of I; for each $i \in I$, let x_i be a vector in X [resp.: a map $T \ni t \mapsto x_i(t) \in X$]. The family $(x_i)_{i \in I}$ is *summable* to $S = \sum\limits_{i \in I} x_i$ [resp.: uniformly summable to $S(t) = \sum\limits_{i \in I} x_i(t)$] if for each $p \in \Pi$ and each $\varepsilon > 0$ there is an $F_0 \in \mathscr{F}$ such that $F_0 \subset F \in \mathscr{F}$ implies $p\left(S - \sum\limits_{i \in F} x_i\right) \leqslant \varepsilon$ [resp.: $p\left(S(t) - \sum\limits_{i \in F} x_i(t)\right) \leqslant \varepsilon$ for all $t \in T$]. If the latter property holds, and each map $t \mapsto x_i(t)$ is continuous, so is the map $t \mapsto S(t)$.

Proposition 1.2.1. a) *A necessary, but not sufficient (unless X has a finite dimension) condition for the summability of the family $(x_i)_{i \in I}$ is the boundedness of the set of all finite sums $\sum\limits_{i \in F} x_i$.*

b) *A necessary condition for the summability [resp.: uniform summability] of the family $(x_i)_{i \in I}$ is the Cauchy condition, namely the existence, for each $p \in \Pi$ and each $\varepsilon > 0$, of an $F_1 \in \mathscr{F}$ such that $F \in \mathscr{F}$, $F_1 \cap F = \emptyset$ imply*

$$p\left(\sum_{i \in F} x_i\right) \leqslant \varepsilon \left[\text{resp.: } p\left(\sum_{i \in F} x_i(t)\right) \leqslant \varepsilon \text{ for all } t \in T\right].$$ *Therefore $(x_i)_{i \in I}$ summable demands $\{i \in I: p(x_i) > 0\}$ at most countable for all $p \in \Pi$, and $\{i \in I: x_i \neq 0\}$ at most countable if X is metrizable.*

c) *If (X, Π) is q.c. (or s.c. with I countable): the above Cauchy condition is sufficient, and therefore the summability of the family of real numbers $p(x_i)$ [resp.: uniform summability of the family of real functions $p(x_i)$] for all $p \in \Pi$ is also a sufficient condition; another consequence is the summability [resp.: uniform summability] of any infinite subfamily of a summable [resp.: uniformly summable] family.*

Proof of a), c). Let the Cauchy condition hold for the family of vectors $(x_i)_{i \in I}$. Given $p \in \Pi$, there is a finite set of indeces $i_1, \ldots, i_n$ such that

$$\{i_1, \ldots, i_n\} \cap F = \emptyset \quad \text{implies } p\left(\sum_{i \in F} x_i\right) \leqslant 1.$$

Then $p\left(\sum_{i \in F} x_i\right) \leqslant M(p) = 1 + p(x_{i_1}) + \cdots + p(x_{i_n})$ for all $F \in \mathscr{F}$: all finite sums $\sum_{i \in F} x_i$ lie in the bounded set $\{x \in X: p(x) \leqslant M(p) \text{ for all } p \in \Pi\}$, and the sets $\left\{\sum_{i \in F} x_i: F \supset F_1\right\}$, where F_1 runs through $\mathscr{F}$, make up the basis of a Cauchy filter converging to $S \in X$. If I is countable, i.e. $I = \{i_n: n \in \mathbb{N}^*\}$, the sums $y_n = \sum_{k=1}^{n} x_{i_k}$ make up a Cauchy sequence converging to S.

If $F_1 \cap F = \emptyset$ implies $p\left(\sum_{i \in F} x_i\right) \leqslant \varepsilon$, then $F' \supset F_1$ implies $p\left(S - \sum_{i \in F'} x_i\right) \leqslant \varepsilon$: hence follows the sufficiency of the Cauchy condition for the uniform summability of a family of maps. $\square$

Example 1.2.2. [Coe]. Let (X, Π) be q.c. (or s.c. with I countable), $(x_i)_{i \in I}$ a summable family of vectors in X and the α_i complex variables, i.e. $\alpha = (\alpha_i)_{i \in I} \in \mathbb{C}^I$.

a) The family $(\alpha_i x_i)_{i \in I}$ is summable for $\|\alpha\| = \sup_{i \in I} |\alpha_i| < \infty$ and uniformly summable on $A = \{\alpha \in \mathbb{C}^I: \|\alpha\| \leqslant 1\}$.
b) The map $A \ni \alpha \mapsto \sum_{i \in I} \alpha_i x_i$ is continuous for the product topology on A and therefore its image in X is compact.

Proof. a) Since it is enough to prove the uniform summability on A of the families $((\mathscr{R}e\,\alpha_i)x_i)$ and $((\mathscr{I}m\,\alpha_i)x_i)$, we may assume $\alpha_i \in [-1, 1]$ for all $i \in I$.

Let $F_1 \cap F = \emptyset$ imply $p\left(\sum_{i \in F} x_i\right) \leqslant \dfrac{\varepsilon}{3}$; any $F \in \mathscr{F}$ such that $F_1 \cap F = \emptyset$ can be written as $F = \{i_1, \ldots, i_n\}$ with $-1 \leqslant \alpha_{i_1} \leqslant \cdots \leqslant \alpha_{i_n} \leqslant 1$, and from $\sum_{i \in F} \alpha_i x_i = \alpha_{i_1}(x_{i_1} + \cdots + x_{i_n}) + (\alpha_{i_2} - \alpha_{i_1})(x_{i_2} + \cdots + x_{i_n}) + \cdots + (\alpha_{i_n} - \alpha_{i_{n-1}})x_{i_n}$ follows $p\left(\sum_{i \in F} \alpha_i x_i\right) \leqslant \dfrac{\varepsilon}{3}[|\alpha_{i_1}| + (\alpha_{i_n} - \alpha_{i_1})] \leqslant \varepsilon.$

b) $S(\alpha) = \sum_{i \in I} \alpha_i x_i$ defines a linear map $S\colon \mathbb{C}^I \to X$. With the same choice of F_1 as above, $F_1 \cap F = \emptyset$ implies, for all $\alpha \in \mathbb{C}^I$, $p\left(\sum_{i \in F} \alpha_i x_i\right) \leqslant 2\varepsilon\|\alpha\|$, hence $p\left[S(\alpha) - \sum_{i \in F_1} \alpha_i x_i\right] \leqslant 2\varepsilon\|\alpha\|$. Then, for $\alpha, \alpha' \in A$:

$$p[S(\alpha') - S(\alpha)] \leqslant \sum_{i \in F_1} |\alpha_i' - \alpha_i| p(x_i) + 2\varepsilon\|\alpha' - \alpha\|$$

$$\leqslant \sum_{i \in F_1} |\alpha_i' - \alpha_i| p(x_i) + 4\varepsilon. \qquad \square$$

(B) Let μ be a real or complex Radon measure on a compact metric space T and f a continuous map $T \to (X, \Pi)$, a sequentially complete locally convex (s.c.l.c.) space.

Proposition 1.2.3. a) *An element y of X, which by definition is the vector integral $\int f d\mu$, is uniquely determined by the relation $\langle y, x'\rangle = \int \langle f, x'\rangle\, d\mu$ for all $x' \in X'$.*
b) *for all $p \in \mathrm{csn}(X, \Pi)$:*

$$p(\textstyle\int f\, d\mu) \leqslant \|\mu\| \sup_T (p \circ f),$$

where $\|\mu\| = \sup\{|\int \varphi\, d\mu|\colon \varphi \text{ continuous } T \to \mathbb{C},\ |\varphi| \leqslant 1\}$, and more precisely $p(\int f d\mu) \leqslant \int (p \circ f)\, d\mu$ if $\mu \geqslant 0$.
c) *Let f be the pointwise limit of a sequence of continuous maps $f_k\colon T \to (X, \Pi)$: if the real functions $p \circ f_k$ are bounded together for each $p \in \Pi$, then $\int f d\mu = \lim \int f_k\, d\mu$.*

Proof. a) Since T is a compact metric space, for each $n \in \mathbb{N}^*$ we can find a finite partition of T into Borel sets T_n^j, $j = 1, \ldots, J(n)$, with diameters $\leqslant \dfrac{1}{n}$; if we also choose $t_n^j \in T_n^j$ and set

$$y_n = \sum_{j=1}^{J(n)} \mu(T_n^j)f(t_n^j),$$

then $\int \langle f, x' \rangle \, d\mu = \lim \langle y_n, x' \rangle$ for all $x' \in X'$ and, on account of Proposition 1.1.1a, it is enough to check that (y_n) is a Cauchy sequence.

In fact

$$y_m - y_n - \sum_{j=1}^{J(m)} \sum_{k=1}^{J(n)} \mu(T_m^j \cap T_n^k)[f(t_m^j) - f(t_n^k)],$$

where either $T_m^j \cap T_n^k = \emptyset$ or $\text{dist.}(t_m^j, t_n^k) \leqslant \dfrac{1}{m} + \dfrac{1}{n}$; therefore, given $p \in \Pi$ and $\varepsilon > 0$, if

$$\text{dist.}(t, t') \leqslant \frac{2}{n_0} \quad \text{implies} \quad p[f(t) - f(t')] \leqslant \frac{\varepsilon}{\|\mu\|},$$

then $m, n \geqslant n_0$ imply $p(y_m - y_n) \leqslant \varepsilon$ since $\sum_j \sum_k |\mu(T_m^j \cap T_n^k)| \leqslant \|\mu\|$.

b) Similarly, the first inequality in b) follows from $\sum_j |\mu(T_n^j)| \leqslant \|\mu\|$. The second one follows from

$$\int (p \circ f) \, d\mu = \lim \sum_j \mu(T_n^j) \, p \circ f(t_n^j).$$

c) If $\mu \geqslant 0$, the latter inequality may be used: for all $p \in \Pi$, $p(\int f_k \, d\mu - \int f \, d\mu) \leqslant \int p \circ (f_k - f) \, d\mu$, which tends to zero by the classical Lebesgue theorem. $\square$

1.3 Baire spaces

(A) In a topological space X: a set M is *meagre* if it is contained in a countable union of closed sets without interior point. Now let $A \subset Y \subset X$: if $\bar{A}$ contains an open nonempty set U, then $\bar{A} \cap Y$ contains $U \cap Y$ which is an open nonempty subset of Y; therefore any meagre subset of Y is also a meagre set in X.

X is a *Baire space* if any meagre set in X has an empty interior; then any open set in X is also a Baire space; for any meagre set M in X, $Y = X \backslash M$ is also a Baire space. In fact, if M' is a meagre subset of Y, then $M \cup M'$ is a meagre set in X; an open nonempty subset of Y contained in M' would be an intersection $U \cap Y$, with U open nonempty contained in $M \cup M'$.

Proposition 1.3.1. *Let X be a Baire space and Z a metric space: if a sequence (f_n) of continuous maps $X \to Z$ has a pointwise limit f, there is a meagre set M in X such that f is continuous at each point $\in X \backslash M$; therefore $f|_{X \backslash M}$ is continuous, which is a weaker statement.*

Proof. For all $\varepsilon > 0$, X is the union of the closed sets

$$A_{n,\varepsilon} = \{x \in X: \text{dist. } [f_n(x), f_{n+k}(x)] \leqslant \varepsilon \text{ for all } k \in \mathbb{N}\};$$

any open nonempty set in X is Baire and therefore meets $B_\varepsilon = \bigcup_n \overset{\circ}{A}_{n,\varepsilon}$, so this open set is dense and $\bigcap_{\varepsilon > 0} B_\varepsilon = \bigcap B_{1/n}$ is the complement of a meagre set M.

Let $a \in X \setminus M$: for each $\varepsilon > 0$ we have $a \in B_\varepsilon$ and can choose n so that $a \in \overset{\circ}{A}_{n,\varepsilon}$; then $x \in \overset{\circ}{A}_{n,\varepsilon}$ implies dist. $[f_n(x), f(x)] \leqslant \varepsilon$, hence

$$\text{dist. } [f(x), f(a)] \leqslant 2\varepsilon + \text{dist. } [f_n(x), f_n(a)] \leqslant 3\varepsilon$$

for x in a suitable neighbourhood of a.   □

Proposition 1.3.2. *For any set A in a topological space X, the set $\{x \in A: A \cap V$ meagre for some neighbourhood V of $x\}$ is meagre.*

Proof. First consider a family $(U_i)_{i \in I}$ of mutually disjoint open sets. If $R_i \subset U_i$ and $\overset{\circ}{R}_i = \emptyset$ for all $i \in I$, then $\overline{\bigcup_{i \in I} R_i}$ also has an empty interior: in fact, if the last set contained a nonempty open set ω, we could choose $i \in I$ and $a \in \omega \cap R_i \subset \omega \cap U_i$; since $R_j \cap U_i = \emptyset$ for all $j \neq i$, necessarily $\omega \cap U_i \subset \overline{R}_i$, which is impossible. Consequently, if $M_i \subset U_i$ and M_i is meagre for all $i \in I$, then $\bigcup_{i \in I} M_i$ is also meagre: in fact, $M_i = \bigcup_{n \in N} R_i^n$ with $\overline{R_i^n}^{\,\circ} = \emptyset$ for all $i \in I$, which implies $\overline{\bigcup_{i \in I} R_i^n}$ without any interior point.

Next consider a maximal family $(U_i)_{i \in I}$ (which exists by Zorn's lemma) of mutually disjoint open sets such that each $A \cap U_i$ is meagre; let $W = \bigcup_{i \in I} U_i$. If some nonempty open set U were contained in the closure of $A \cap \complement W$, hence in $\complement W$, $A \cap U$ could not be meagre, U would be contained in the complement of the open set

$$G(A) = \{x \in X: A \cap V \text{ meagre for some neighbourhood } V \text{ of } x\}.$$

So $G(A) = X$ implies that the closure of $A \cap \complement W$ has no interior point; but $A \cap W$ is meagre since each $M_i = A \cap U_i$ is meagre: $G(A) = X$ if and only if A is meagre.

Finally any point $x \in G(A)$ has a neighbourhood V in $G(A)$ such that $A \cap V$ is meagre, or $[A \cap G(A)] \cap V$ meagre; replacing X by the open set $G(A)$, we have $G[A \cap G(A)] = G(A)$, which by the last result means $A \cap G(A)$ meagre in $G(A)$, or meagre in X.   □

(B) Any complete metric space X is a Baire space. In fact, given a sequence $(A_n)_{n \in \mathbb{N}^*}$ of closed sets in X without any interior point, we can successively find in any open nonempty set U: a closed ball B_1 such that $A_1 \cap B_1 = \emptyset$, with a radius $\leqslant 1$; a closed ball $B_2 \subset B_1$ such that $A_2 \cap B_2 = \emptyset$, with a radius $\leqslant \frac{1}{2}, \ldots$, a closed ball $B_n \subset B_{n-1}$ such that $A_n \cap B_n = \emptyset$, with a radius $\leqslant \frac{1}{n}, \ldots$. The Cauchy sequence of centres of the balls B_n has a limit which lies in U and in the intersection of the $X \backslash A_n$, so that this intersection is dense in X.

Any locally compact space is also a Baire space, but this is of no use to us since, by the Frederic Riesz theorem ([Bo]$_2$, Chap. I, §2, Th. 3) locally compact vector spaces have finite dimensions. More interesting is the following fact.

If $(X_i)_{i \in I}$ is any infinite family of complete metric spaces, the product space $X = \prod_{i \in I} X_i$ is Baire. The above argument also yields this fact, if each closed ball B_n with a radius $\leqslant \frac{1}{n}$ is replaced by a set $\left(\prod_{i \in I \backslash F_n} X_i \right) \times \left(\prod_{i \in F_n} B_n^i \right)$, where F_n is a finite subset of I and each B_n^i a closed ball with a radius $\leqslant \frac{1}{n}$ in the corresponding space X_i. This product space X is metrizable if and only if I is countable.

(C) Now we consider locally convex spaces; by the results in **(B)**, the following ones are Baire spaces:

a) for any set X, the space $Y = \mathbb{C}^X$ of all functions $y: X \to \mathbb{C}$, endowed with the topology of pointwise convergence, which is defined by the seminorms $y \mapsto |y(x)|$, $x \in X$;
b) any *Banach space*, i.e. normed and complete;
c) any *Fréchet space*, i.e. locally convex, metrizable (or, equivalently by §1.1, with a topology defined by a countable family of seminorms) and complete.

1.4 Barrelled spaces

A locally convex space (X, Π) is *barrelled* if any seminorm on X which is lower semicontinuous is actually continuous.

From this definition follows at once that any vector space X is barrelled for the topology defined by all possible seminorms on X. The remark holds in particular for $X = c_{00}(\mathbb{N})$, the space of complex sequences with only a finite number of nonzero terms; but this space cannot be Baire, for any Hausdorff vector space topology, since any finite dimensional subspace is closed and its interior empty. Therefore the following proposition has no converse.

Proposition 1.4.1. *If a locally convex space* (X, Π) *is Baire, it is also barrelled.*

Proof. If the seminorm p is lower semicontinuous, X is the union of the closed sets $\{x \in X : p(x) \leqslant n\}$, $n \in \mathbb{N}^*$, which are the images under homeomorphisms of the first one of them. Since (X, Π) is Baire, each of them has a nonempty interior, there are $a \in X$ and a neighbourhood V of the origin such that $x \in V$ implies $p(a + x) \leqslant 1$, hence $p(x) \leqslant 1 + p(a)$; then, for all $\varepsilon > 0$:

$$x \in \frac{\varepsilon}{1 + p(a)} V \quad \text{implies } p(x) \leqslant \varepsilon,$$

and p continuous at the origin implies p continuous at any point x_0 since $|p(x) - p(x_0)| \leqslant p(x - x_0)$. $\quad\square$

We use this result to prove

Proposition 1.4.2. *In a locally convex space* (X, Π): *a set* B *is bounded if* (*and obviously only if*) $\sup_{x \in B} |\langle x, x' \rangle| < +\infty$ *for all* $x' \in X'$; *in other words, the bounded sets are the same for the weakened topology and the initial one, and may be termed "bounded sets".*

Proof. Given $p \in csn(X, \Pi)$, the subspace $E' = \{x' \in X' : |x'| \leqslant \alpha p$ for some $\alpha > 0\}$ of X' is a Banach space with the norm $\|x'\| = \inf\{\alpha > 0 : |x'| \leqslant \alpha p\}$. Since $|\langle x, x' \rangle| \leqslant p(x)\|x'\|$ for all $x \in X$, $x' \in E'$, this norm on E' makes each linear map $E' \ni x' \mapsto \langle x, x' \rangle$ continuous; then it makes the (finite by the assumption) seminorm $E' \ni x' \mapsto \sup_{x \in B} |\langle x, x' \rangle|$ lower semicontinuous, therefore continuous: for a suitable $\beta > 0$, we have $\sup_{x \in B} |\langle x, x' \rangle| \leqslant \beta \|x'\|$ for all $x' \in E'$, and we now prove $\sup_{x \in B} p(x) \leqslant \beta$.

Take any $a \in B$ with $p(a) > 0$, and let ζ be a complex variable: by the Hahn-Banach theorem ([Bo]$_2$, Chap. II, §6, Th. 1), the linear map $\zeta a \mapsto p(a)$. ζ can be extended into an $a' \in X'$ such that $|a'| \leqslant p$, hence $a' \in E'$, $\|a'\| \leqslant 1$, $\sup_{x \in B} |\langle x, a' \rangle| \leqslant \beta$, and $p(a) = \langle a, a' \rangle \leqslant \beta$. $\quad\square$

Remark 1.4.3. By the above proposition: for the weakened topology on X, the sets C considered in Theorem 1.1.2.b are compact (i.e. X semireflexive, s.r.) if and only if the bounded sets are relatively compact.

Theorem 1.4.4. *Given a locally convex space (X, Π), let M' be any subset of the adjoint space X'.*

a) *M' equicontinuous implies*
(i) *M' weak* relatively compact and* (ii) *M' strongly bounded;*
either property (i) *or* (ii) *implies*
(iii) *M' weak* bounded.*
b) *If (X, Π) is barrelled: the above four properties of M' are equivalent, and X' quasi-complete for the weak* and the strong topologies.*

Proof. a) M' equicontinuous means the existence of $p \in \operatorname{csn}(X, \Pi)$ such that $|x'| \leqslant p$ for all $x' \in M'$. Then $\sup_{\substack{x \in B \\ x' \in M'}} |\langle x, x' \rangle| < +\infty$ for each bounded set $B \subset X$, i.e. (ii); and, with the weak* topology, the set of all $x' \in X'$ such that $|x'| \leqslant p$ is compact, as a closed subset of the set of all maps $f: X \to \mathbb{C}$ satisfying $|f(x)| \leqslant p(x)$ for all $x \in X$.
b) Let (X, Π) be barrelled. If M' has property (iii), then $\sup_{x' \in M'} |x'|$ is a lower semicontinuous, hence continuous, seminorm on X.

Now let M' be bounded (for both topologies considered) and meet each set in the filter $\mathscr{F}'$ on X': since M' is weak* relatively compact, there is $x'_0 \in X'$ such that any weak* neighbourhood of x'_0 meets any set in $\mathscr{F}'$. If $\mathscr{F}'$ is weak* Cauchy, it follows at once that $\mathscr{F}'$ converges to x'_0 for the weak* topology. If $\mathscr{F}'$ is also strongly Cauchy: for each bounded set $B \subset X$ and each $\varepsilon > 0$, there is $A' \in \mathscr{F}'$ such that $\sup_{x \in B}|\langle x, x' - y' \rangle| \leqslant \varepsilon$ for all x', $y' \in A'$, hence also $\sup_{x \in B}|\langle x, x' - x'_0 \rangle| \leqslant \varepsilon$ for all $x' \in A'$, since x'_0 belongs to the weak* closure of A'. □

1.5 Inductive limits

(A) For any set A, in the vector space X, which is *absorbent* (i.e. any $x \in X$ lies in αA for sufficiently large $|\alpha|$), the *gauge* q of A is defined on X by $q(x) = \inf\{\alpha > 0: x \in \alpha A\}$ and is positively 1-homogeneous: $q(tx) = tq(x)$ for all $t \in \mathbb{R}_+$; q is related to A by the inclusion $A \subset q^{-1}([0,1])$, and also by the inclusion $A \supset q^{-1}([0,1[)$ if A is *star-shaped* with respect to the origin (i.e. $x \in A$, $t \in [0,1]$ imply $tx \in A$).

If the absorbent set A is also *convex* (i.e. $x, x' \in A$ implies $\alpha x + \alpha' x' \in A$ for all $\alpha, \alpha' \in \mathbb{R}_+$ with $\alpha + \alpha' = 1$), then q is a Minkowski functional, i.e.

also satisfies $q(x + y) \leqslant q(x) + q(y)$: in fact $\alpha > q(x)$, $\beta > q(y)$ implies $\dfrac{x}{\alpha} \in A$, $\dfrac{y}{\beta} \in A$, hence $x + y \in (\alpha + \beta)A$. If A is moreover *balanced* (i.e. $x \in A$, $|\alpha| \leqslant 1$ imply $\alpha x \in A$), then q is a seminorm.

Conversely, any seminorm p on X is the gauge of the absorbent, balanced, convex (a.b.c.) sets $\{x \in X: p(x) < 1\}$ and $\{x \in X: p(x) \leqslant 1\}$. Therefore the semi-norms in a family π defining a locally convex topology on X can be given as the gauges of a family $\mathscr{A}$ of a.b.c. sets A, and this topology is Hausdorff if and only if $\bigcap\limits_{A \in \mathscr{A}} A$ does not contain any 1-dimensional subspace of X.

When the space X is endowed with a topology: a) if q is the gauge of A, A absorbent and star-shaped with respect to the origin: A closed [resp. open] implies $A = q^{-1}([0, 1])$, [resp. $q^{-1}([0, 1[)$] and q lower [resp. upper] semi-continuous;

b) if A is a.c.: A a neighbourhood of the origin implies q continuous, and conversely;

c) X is barrelled if and only if any closed a.b.c. set (i.e. *barrel*) is a neighbourhood of the origin.

(B) Let I be a preordered set in which any two elements i, i' have a common majorant j $(i \prec j, i' \prec j)$ and $(X_i, \Pi_i)_{i \in I}$ a family of l.c. spaces with the following properties: the X_i are linear subspaces of a vector space X which is their union; whenever $i \prec j$, $X_i \subset X_j$ and (X_j, Π_j) induces on X_i a topology which is coarser than (X_i, Π_i). Then the family Π of all seminorms p on X such that $p|_{X_i} \in \mathrm{csn}(X_i, \Pi_i)$ for all $i \in I$ or, equivalently, the family $\mathscr{A}$ of all a.b.c. sets A such that $A \cap X_i$ is a neighbourhood of the origin in (X_i, Π_i) for all $i \in I$, defines the finest l.c. topology on X making each injection $X_i \to X$ continuous: (X, Π) is the *inductive limit* of the family $(X_i, \Pi_i)_{i \in I}$.

By the above characterization of barrelled spaces: if each (X_i, Π_i) is barrelled, so is (X, Π).

Proposition 1.5.1. *Let* (X, Π) *be the inductive limit of an increasing sequence of Hausdorff subspaces* (X_n, Π_n) *in which the origin has a neighbourhood* V_n *relatively compact in* (X_{n+1}, Π_{n+1}) *(hence also a fundamental system of such neighbourhoods).*

a) *A set* $\Omega \subset X$ *is open in* (X, Π) *if and only if* $\Omega \cap X_n$ *is open in* (X_n, Π_n) *for each* n; *in particular,* (X, Π) *is Hausdorff too.*

b) *Let* Ω *be open in* (X, Π): *a map* h *of* Ω *into a topological space is continuous for the topology of* (X, Π) *if and only if, for each* n, $h|_{\Omega \cap X_n}$ *is continuous for the topology of* (X_n, Π_n).

Proof of a). V_n may be assumed b.c.; let $\Omega \cap X_n$ be open in (X_n, Π_n) for each n. Given $a \in \Omega \cap X_n$, since $\Omega \cap X_{n+1}$ is open in (X_{n+1}, Π_{n+1}), there is a $\lambda_n > 0$ such that $a + \lambda_n V_n$ is included in a compact subset $a + K_n$ of $\Omega \cap X_{n+1}$; since $a + K_n$ is also a compact subset of $\Omega \cap X_{n+2}$, there is a $\lambda_{n+1} > 0$ such that $a + \lambda_n V_n + \lambda_{n+1} V_{n+1}$ is included in a compact subset $a + K_n + K_{n+1}$ of $\Omega \cap X_{n+2}$, and so on; $\Omega - a$ is a neighbourhood of the origin in (X, Π) since it contains $\lambda_n V_n$, $\lambda_n V_n + \lambda_{n+1} V_{n+1}, \ldots$.

In particular $\{0\}$ is closed in (X, Π). $\square$

(C) Let (X_n, Π_n) be a sequence of l.c. Hausdorff spaces with the following properties: (X_n) is a strictly increasing sequence of linear subspaces of a vector space X which is their union; for each n, (X_{n+1}, Π_{n+1}) and (X_n, Π_n) induce the same topology on X_n. Then, again, the family Π of all seminorms p on X such that $p|_{X_n} \in csn(X_n, \Pi_n)$ for all n or, equivalently, the family $\mathscr{A}$ of all a.b.c. sets A such that $A \cap X_n$ is a neighbourhood of the origin in (X_n, Π_n) for all n, defines the finest l.c. topology on X making each injection $X_n \to X$ continuous: (X, Π) is the *strict inductive limit* of the sequence (X_n, Π_n).

Proposition 1.5.2. *Let (X, Π) be the strict inductive limit of a sequence of Hausdorff subspaces (X_n, Π_n).*

a) *Every seminorm $\in csn(X_n, \Pi_n)$ is the restriction to X_n of a seminorm $\in csn(X_{n+1}, \Pi_{n+1})$, hence also of a seminorm $\in csn(X, \Pi)$; therefore (X, Π) is Hausdorff too and induces on each X_n the topology (X_n, Π_n).*

b) *If moreover X_n is closed in (X_{n+1}, Π_{n+1}), then each seminorm $\in csn(X_n, \Pi_n)$ is the restriction to X_n of a seminorm $\in csn(X_{n+1}, \Pi_{n+1})$, which can be chosen taking an arbitrarily large value at a given point $y \in X_{n+1} \backslash X_n$.*

Proof. a) Let $p_0 \in csn(X_n, \Pi_n)$: $\{x \in X_n: p_0(x) \leqslant 1\}$, as a neighbourhood of the origin in (X_n, Π_n), is the intersection of X_n with some neighbourhood of the origin in (X_{n+1}, Π_{n+1}); so there is a $p' \in csn(X_{n+1}, \Pi_{n+1})$ such that $p_0 \leqslant p'|_{X_n}$, and the a.b.c. set in X_{n+1}, also a neighbourhood of the origin,

$$A = \{\alpha x + \alpha' x': x \in X_n, p_0(x) \leqslant 1; x' \in X_{n+1}, p'(x') \leqslant 1; \alpha,$$

$$\alpha' \in \mathbb{R}_+, \alpha + \alpha' = 1\}$$

has a gauge $= p_0$ on X_n since $A \cap X_n = \{x \in X_n: p_0(x) \leqslant 1\}$.

Iterating the result, we get a sequence $(p_k)_{k \in \mathbb{N}}$, $p_k \in csn(X_{n+k}, \Pi_{n+k})$, such that $p_k = p_{k+1}|_{X_{n+k}}$; setting $q(x) = p_k(x)$ for all $k \in \mathbb{N}$ such that $x \in X_{n+k}$ (which does not depend on k), we obtain a seminorm $q \in csn(X, \Pi)$.

b) Since $y + X_n$ is closed in (X_{n+1}, Π_{n+1}) and does not contain the origin, the seminorm p' in the proof of a) can be chosen so that $p' \geqslant M$ (a given number) on $y + X_n$; then $y \in \beta A$ or $y = \alpha \beta x + \alpha' \beta x'$ (with the properties of x, x', α, α' listed in the above definition of A) implies $p'(\alpha' \beta x') = p'(y - \alpha \beta x) \geqslant M$, $\beta \geqslant M$, the gauge of A at y is larger than M. $\square$

Proposition 1.5.3. *Let* (X, Π) *be the strict inductive limit of a sequence of complete spaces* (X_n, Π_n).

a) *For each n,* X_n *is closed in* (X_{n+1}, Π_{n+1}) *and* (X, Π).
b) *Any bounded set in* (X, Π) *is contained in some* X_n *and then bounded in* (X_n, Π_n).
c) (X, Π) *is q.c.*

Proof. a) Since X_n is complete for the topology induced by the Hausdorff topology (X_{n+1}, Π_{n+1}), X_n is closed in (X_{n+1}, Π_{n+1}); the closure of X_n in X_{n+1} for the topology (X, Π) is again X_n by Proposition 1.5.2.a. Then, if $x \in X_{n+k}$, $k \in \mathbb{N}^*$, lies in the closure of X_n for the topology (X, Π), a fortiori in the closure of X_{n+k-1} for the same topology, this implies $x \in X_{n+k-1}$, and so on down to $x \in X_n$.
b) Let B be bounded in (X, Π) but contained in no X_n: there is an increasing sequence of integers n_k ($n_0 = $ the smallest n) and $y_k \in B \cap (X_{n_{k+1}} \backslash X_{n_k})$ for all $k \in \mathbb{N}$. Given $p_{n_0} \in \mathrm{csn}(X_{n_0}, \Pi_{n_0})$, we may now use Proposition 1.5.2.b to obtain a sequence $(p_n)_{n \geqslant n_0}$, $p_n \in \mathrm{csn}(X_n, \Pi_n)$, such that $p_n = p_{n+1}|_{X_n}$ and $p_{n_{k+1}}(y_k) \geqslant k$ for all $k \in \mathbb{N}^*$. Setting again $q(x) = p_n(x)$ for all $n \geqslant n_0$ such that $x \in X_n$, we obtain a seminorm $q \in \Pi$, which therefore should be bounded on B but is not.
So there is an X_n containing B, and then B is bounded for the topology (X_n, Π_n) which is induced by (X, Π).
c) Let $\mathscr{F}$ be a Cauchy filter in (X, Π) for which there is a bounded set B such that $A \cap B \neq \emptyset$ for all $A \in \mathscr{F}$: this B is contained in some X_n and the $A \cap X_n$, $A \in \mathscr{F}$, make up a Cauchy filter in (X_n, Π_n). $\square$

Examples 1.5.4. (A) $X = c_{00}(\mathbb{N}^*)$, the space of complex sequences $x = (\xi_1, \xi_2, \ldots)$ with only a finite number of nonzero terms, is a strict inductive limit, satisfying the assumptions of all three above propositions, if X_n is the subspace of those sequences x for which $\xi_k = 0$ for all $k > n$, identified with $\mathbb{C}^n$. The inductive limit topology is not metrizable (since the space is q.c. but not Baire), and defined by all possible seminorms on X.

(B) The space $\mathscr{A}(K, \mathbb{C})$ of germs of analytic functions on a compact set K in $\mathbb{C}^m$ is the inductive limit of the Banach spaces defined as follows. Let $U_n = \left\{ \zeta \in \mathbb{C}^m : \mathrm{dist.}(\zeta, K) < \dfrac{1}{n} \right\}$: X_n is the space, with the sup norm, of continuous functions on $\overline{U_n}$ which are analytic on U_n; since $\overline{U_{n+1}}$ is a compact subset of U_n, any ball in (X_n, Π_n) is relatively compact in (X_{n+1}, Π_{n+1}).

(C) The space of Radon measures on a locally compact space U which is countable at infinity (resp.: of distributions on an open set U in R^m) is the adjoint space to the strict inductive limit of a sequence of Banach (resp.: Fréchet) spaces X_n associated as follows to an exhaustion of U by compact sets K_n: X_n is the space of functions $x \in \mathscr{C}(U)$ (resp.: $\mathscr{C}^\infty(U)$) vanishing on $U \backslash K_n$, with the norm $x \mapsto \sup |x|$ (resp.: the seminorms $x \mapsto \sup |D^k x|$, $k \in \mathbb{N}^m$). The barrelledness of this inductive limit implies the important fact that the pointwise limit of a sequence of measures or distributions on U is again a measure or distribution on U.

(D) Let X be a Fréchet space whose topology is defined by an increasing sequence of seminorms p_n; then the adjoint space X' is the union of the increasing sequence of subspaces $X'_n = \{x' \in X': |x'| \leqslant \alpha p_n \text{ for some } \alpha > 0\}$, which are Banach spaces with the norms $\|x'\|_n = \inf\{\alpha > 0: |x'| \leqslant \alpha p_n\}$. Each injection $X'_n \to X'_{n+1}$ is continuous, and the inductive limit topology finer than the strong topology on X': in fact, if B is bounded in X, then

$$\sup_B |x'| \leqslant \left(\sup_B p_n \right) \|x'\|_n \text{ for all } x' \in X'_n.$$

If X is also a Schwartz space ([Ho], Chap. 3, §15, especially Prop. 5), to each integer n corresponds an integer $m > n$ such that the ball $\{x' \in X': |x'| \leqslant p_n\}$ of X'_n is relatively compact in X'_m.

In the next chapters, since the interest will no longer be in topological matters, a simpler notation will be used: X instead of (X, Π) for a l.c. space, $\mathrm{csn}(X)$ for the set of all continuous seminorms on X.

Chapter 2
Gâteaux-analyticity

Summary

Def. 2.1.1.   Analyticity, holomorphy, scalar analyticity of a map $U \to Z$ (U a finite dimensional open set, Z a l.c. space).

Th. 2.1.2.   The m-uple Cauchy formula; the inequalities of Cauchy.

Th. 2.1.3.   The above three notions are equivalent if Z is s.c.

Th. 2.1.5.   Bounded subsets of $\mathscr{A}(U, Z)$.

Prop. 2.1.7.   The Vitali theorem [HiPh].

Prop. 2.2.1.   Polynomials in one complex variable with coefficients in Z.

Prop. 2.2.4.   The Bernstein-Walsh inequality [Sic].

Coroll. 2.2.5.   The Leja polynomial lemma [Sic].

Def. 2.2.6.   Polynomial maps.

Def. 2.2.8.   Homogeneous polynomial maps.

Th. 2.2.9.   The relation between n-homogeneous polynomial maps and n-linear maps.

Prop. 2.2.11.   A translation lemma.

Def. 2.3.1. $\mathscr{G}$-analytic maps.

Th. 2.3.5.   The homogeneous polynomial expansion of a $\mathscr{G}$-analytic map; the generalized inequalities of Cauchy.

Prop. 2.3.6.   A converse statement [Sic].

Prop. 2.3.8.   The Hartogs property for $\mathscr{G}$-analytic maps.

Prop. 2.3.10. $\mathscr{G}$-analytic maps do not enjoy the composition property.

Prop. 2.4.1(2)   Boundedness and continuity of polynomial ($\mathscr{G}$-analytic) maps.

Th. 2.4.4.   Zorn's theorem [Zo].

Prop. 2.4.7,8,9. Discontinuous $\mathscr{G}$-analytic functions.

Chapter 2
Gâteaux-analyticity

2.1 Vector valued functions of several complex variables

In this section, the data are an open set U in $\mathbb{C}^m$ and a vector space Z with a given locally convex topology. When needed, the generic point in U will be written as $x = \sum_{j=1}^{m} \xi_j e_j$, where the ξ_j are the complex coordinates of x in the canonical basis $(e_1, \ldots, e_m)$ of $\mathbb{C}^m$; for

$$x = \sum_{j=1}^{m} \xi_j e_j, \quad a = \sum_{j=1}^{m} \alpha_j e_j, \quad k = (k_1, \ldots, k_m) \in \mathbb{N}^m,$$

we write $(x - a)^k$ instead of $(\xi_1 - \alpha_1)^{k_1} \ldots (\xi_m - \alpha_m)^{k_m}$, and set $k! = k_1! \ldots k_m!$.

Definition 2.1.1. *A map $f: U \to Z$ is*

a) *analytic if, for each $a \in U$, there is a family $(c_k)_{k \in \mathbb{N}^m}$ of vectors in Z such that the power series $(c_k(x - a)^k)_{k \in \mathbb{N}^m}$ is summable to $f(x)$ for x sufficiently near a;*

b) *holomorphic if, at any point $a \in U$, each first order partial derivative*

$$\lim_{0 \neq \xi_j \to 0} \frac{1}{\xi_j} [f(a + \xi_j e_j) - f(a)] \text{ exists;}$$

c) *scalarly analytic or holomorphic if each complex valued function $z' \circ f$, $z' \in Z'$, is analytic or holomorphic on U: these are equivalent properties by the classical Hartogs theorem ([He]$_1$, Th. 2 in II.2.1)*

The analytic (resp. holomorphic, scalarly analytic) maps $U \to Z$ are the elements of a vector space over $\mathbb{C}$; the first vector space thus defined will be denoted by $\mathscr{A}(U, Z)$.

(A) Obviously a) or b) implies c); for more information, we first investigate the properties of a power series $(c_k(x - a)^k)_{k \in \mathbb{N}^m}$. If this series is summable to $f(x)$ for $|\xi_j - \alpha_j| \leqslant \rho_j, j = 1, \ldots, m$, then (Prop. 1.2.1. a) for each $q \in \mathrm{csn}(Z)$ there is a number $M(q)$ such that $\rho^k q(c_k) = q(c_k \rho^k) \leqslant M(q)$ for all $k \in \mathbb{N}^m$, and $|\xi_j - \alpha_j| \leqslant \theta \rho_j, j = 1, \ldots, m, 0 < \theta < 1$, implies

$$q[f(x) - f(a)] \le M(q)\left[\left(\frac{1}{1-\theta}\right)^m - 1\right],$$

hence f continuous at a, while $0 < |\xi_j| \le \theta \rho_j$, with the notation $k = \sum_{j=1}^{m} k_j e'_j$, $e'_j \in \mathbb{N}^m$, implies

$$q\left(\frac{1}{\xi_j}[f(a + \xi_j e_j) - f(a)] - c_{e'_j}\right) \le \frac{M(q)}{\rho_j} \frac{\theta}{1-\theta},$$

which proves the existence of a first order partial derivative at a, with respect to each ξ_j, namely the coefficients of the $\xi_j - \alpha_j$ in the power series. So f analytic, denoted by $f \in \mathscr{A}(U, Z)$, implies f continuous and holomorphic. For converse results, we need the assumption that Z is sequentially complete (See Remark 2.1.8 below).

(B) *From now on let Z be s.c.*
Then the boundedness of the set $\{c_k \rho^k : k \in \mathbb{N}^m\}$ for a given multi-radius ρ has the following consequences on account of Proposition 1.2.1.c:

a) *Abel's lemma*
The power series is summable for $|\xi_j - \alpha_j| < \rho_j, j = 1, \ldots, m$, and uniformly summable for $|\xi_j - \alpha_j| \le r_j$ if $r_j < \rho_j$ for all $j \in \{1, \ldots, m\}$; therefore its sum $f(x)$ is continuous for $|\xi_j - \alpha_j| < \rho_j, j = 1, \ldots, m$.

b) Let $|\alpha'_j - \alpha_j| < \rho_j$ for all $j \in \{1, \ldots, m\}$. The family $\left(\frac{(k+l)!}{k! \, l!} c_{k+l}(a' - a)^k \times (x - a')^l\right)_{k, l \in \mathbb{N}^m}$ is summable for $|\alpha'_j - \alpha_j| + |\xi_j - \alpha'_j| < \rho_j, j = 1, \ldots, m$; its sum is $f(x)$ since $\sum_{k+l=n} \frac{(k+l)!}{k! \, l!}(a' - a)^k(x - a')^l = (x - a)^n$ for all $n \in \mathbb{N}^m$.

On the other hand, we get a summable subfamily by fixing l and x, e.g. $\xi_j = \alpha'_j + \frac{1}{2}(\rho_j - |\alpha'_j - \alpha_j|)$ for all $j \in \{1, \ldots, m\}$; so we may set for each $l \in \mathbb{N}^m$

(1) $$l! \, c'_l = \sum_{k \in \mathbb{N}^m} c_{k+l} \frac{(k+l)!}{k!}(a' - a)^k = \sum_{k \ge l} c_k \frac{k!}{(k-l)!}(a' - a)^{k-l}$$

and we obtain a power series expansion of f around a':

(2) $$f(x) = \sum_{l \in \mathbb{N}^m} c'_l(x - a')^l \quad \text{for } |\xi_j - \alpha'_j| < \rho_j - |\alpha'_j - \alpha_j|.$$

c) Putting (2) and the result of **(A)** together, we know that f has, at each point a' in the open polydisc $\prod_{j=1}^{m} \Delta(\alpha_j, \rho_j)$, first order partial derivatives given by (1) (with $|l| = l_1 + \cdots + l_m = 1$), that is, by one termwise differentiation of the initial power series; the argument leading to (1) shows that all series obtained by termwise differentiations of the initial one are also summable on the same open polydisc (and perhaps elsewhere).

Then by induction, throughout the open polydisc $\prod_{j=1}^{m} \Delta(\alpha_j, \rho_j)$, f has partial derivatives $D^l f = \dfrac{\partial^{|l|} f}{\partial \xi_1^{l_1} \ldots \partial \xi_m^{l_m}}$ of any multi-order l, obtained by termwise differentiations of the initial power series; their values at a' are the $l! c_l'$ by (1), so that (2) is the *Taylor expansion* of f around a':

$$ f(x) = \sum_{l \in \mathbb{N}^m} \frac{D^l f(a')}{l!} (x - a')^l, \qquad x \in \prod_{j=1}^{m} \Delta(\alpha_j', \rho_j - |\alpha_j' - \alpha_j|). $$

We conclude that an analytic map $f: U \to Z$ has partial derivatives of any multi-order, which again are analytic maps $U \to Z$. Moreover, the coefficients c_k in Definition 2.1.1.a are *uniquely determined* by $c_k = \dfrac{1}{k!} D^k f(a)$ for all $k \in \mathbb{N}^m$ and the *principle of analytic continuation* follows: if U is connected and f vanishes on some open nonempty subset V of U [resp.: $f(x) \in Z_0$ for all $x \in V$, where Z_0 is a closed vector subspace of Z], then $f(x) = 0$ (resp.: $\in Z_0$) for all $x \in U$, since $\{x \in U: D^k f(x) = 0 \text{ for all } k \in \mathbb{N}^m\}$ or $\{x \in U: D^k f(x) \in Z_0 \text{ for all } k \in \mathbb{N}^m\}$ is closed and open in U.

(C) We now use Proposition 1.2.3.

Theorem 2.1.2. a) *Let U contain the closed polydisc $\prod_{j=1}^{m} \bar{\Delta}(\alpha_j, \rho_j)$. On the open polydisc $\prod_{j=1}^{m} \Delta(\alpha_j, \rho_j)$ with the same centre a and multi-radius ρ, the partial derivatives of any $f \in \mathscr{A}(U, Z)$ are given by the Cauchy m-uple integral formula*

$$ D^k f(x) = \frac{k!}{(2\pi)^m \rho^k} \int (m) \int_{-\pi}^{\pi} e^{-ik \cdot \theta} \frac{f(\alpha_1 + \rho_1 e^{i\theta_1}, \ldots, \alpha_m + \rho_m e^{i\theta_m})}{\prod_{j=1}^{m} \left(1 - \dfrac{\xi_j - \alpha_j}{\rho_j} e^{-i\theta_j}\right)^{k_j+1}} d\theta_1 \ldots d\theta_m $$

with the notations $\int (m) \int$ for the m-uple integral, $k \cdot \theta = \sum_{j=1}^{m} k_j \theta_j$.

b) *Given $f \in \mathscr{A}(U, Z)$ and $a \in U$, the Taylor expansion of f around a:*

$\sum_{k \in \mathbb{N}^m} c_k(x - a)^k$, $c_k = \dfrac{1}{k!} D^k f(a)$, *is valid (i.e. summable to f) on any polydisc with centre a contained in U; whenever $U \supset \prod_{j=1}^{m} \bar{\Delta}(\alpha_j, \rho_j)$ and for each $q \in \text{csn}(Z)$, the family $(q(c_k \rho^k))_{k \in \mathbb{N}^m}$ is summable and all its terms majorized by*

$$\sup_{|\xi_j - \alpha_j| = \rho_j} q \circ f(x) \text{ (inequalities of Cauchy); in particular}$$

(3) $\displaystyle q(c_0) = q \circ f(a) \leqslant \sup_{|\xi_j - \alpha_j| = \rho_j} q \circ f(x).$

c) *Each linear map* $\mathscr{A}(U, Z) \ni f \mapsto D^k f \in \mathscr{A}(U, Z)$ *is continuous for the topology of compact convergence, defined on* $\mathscr{A}(U, Z)$ *by the seminorms* $f \mapsto \sup_K (q \circ f)$, $q \in \mathrm{csn}(Z)$, K *compact* $\subset U$.

Proof. a) Since f is continuous, the right hand member $F(x)$ makes sense and $\langle F(x), z' \rangle = D^k(z' \circ f)(x) = \langle D^k f(x), z' \rangle$ for all $z' \in Z'$ by the m-uple integral formula for $z' \circ f$, which is obtained by iteration of the classical Cauchy formula for one complex variable.

b) For each $j \in \{1, \ldots, m\}$, choose $\rho_j' > \rho_j$ so that U also contains $\prod_{j=1}^{m} \bar{\Delta}(\alpha_j, \rho_j')$. From the m-uple integral formula with $x = a$ follow $q(c_k \rho^k) \leqslant$
$\displaystyle \sup_{|\xi_j - \alpha_j| = \rho_j} q \circ f(x)$ and $\displaystyle \left(\frac{\rho}{\rho'} \right)^k \sup_{|\xi_j - \alpha_j| = \rho_j'} q \circ f(x)$, hence the last two statements in b).

As to the validity of the Taylor expansion: for a given

$$x \in \prod_{j=1}^{m} \bar{\Delta}(\alpha_j, \rho_j), \quad \sum_{|k| \leqslant n} \left(\frac{x - a}{\rho'} \right)^k e^{-ik \cdot \theta} \to \frac{1}{\displaystyle\prod_{j=1}^{m} \left(1 - \frac{\xi_j - \alpha_j}{\rho_j'} e^{-i\theta_j} \right)}$$

uniformly as $n \to +\infty$, therefore $\sum_{|k| \leqslant n} c_k(x - a)^k \to f(x)$, $f(x)$ also given by the m-uple integral formula, with $|k| = 0$ and ρ' instead of ρ.

c) Given a compact set $K \subset U$: for sufficiently small radii ρ_j, U also contains the compact set $K' = \{x \in \mathbb{C}^m : \exists a \in K \text{ with } |\xi_j - \alpha_j| \leqslant \rho_j, j = 1, \ldots, m\}$, and then $\displaystyle \sup_K (q \circ D^k f) \leqslant \frac{k!}{\rho^k} \sup_{K'} (q \circ f)$ for all $f \in \mathscr{A}(U, Z)$, by the inequalities of Cauchy. □

Theorem 2.1.3. *For maps into a s.c.l.c. space: analyticity, holomorphy and scalar analyticity are equivalent properties.*

Proof. Let Z be s.c., f scalarly analytic $U \to Z$, and $K = \prod_{j=1}^{m} \bar{\Delta}(\alpha_j, \rho_j)$ any closed polydisc contained in U. We first prove that $f|_K$ is continuous by considering, for each $j \in \{1, \ldots, m\}$, the set

$$B_j = \left\{ \frac{f(x) - f(y)}{\xi_j - \eta_j} : x \in K, y \in K, x - y = (\xi_j - \eta_j)e_j \neq 0 \right\},$$

which is bounded (Prop. 1.4.2.) since, for any such x, y and

$$z' \in Z', \text{ we have } \left| \left\langle \frac{f(x) - f(y)}{\xi_j - \eta_j}, z' \right\rangle \right| \leqslant \sup_K \left| \frac{\partial}{\partial \xi_j} (z' \circ f) \right|;$$

for all $q \in \mathrm{csn}(Z)$, $z \in B_j$ implies $q(z) \leqslant M_j(q)$, hence

$$q[f(x) - f(y)] \leqslant \sum_{j=1}^m M_j(q) |\xi_j - \eta_j| \quad \text{for all } x, y \in K \text{ and } f|_K$$
$$\text{continuous.}$$

Now the integral formulas

$$F(x) = \frac{1}{(2\pi)^m} \int_{-\pi}^{\pi} (m) \int \frac{f(\alpha_1 + \rho_1 e^{i\theta_1}, \ldots, \alpha_m + \rho_m e^{i\theta_m})}{\prod_{j=1}^m \left(1 - \frac{\xi_j - \alpha_j}{\rho_j} e^{-i\theta_j} \right)} d\theta_1 \ldots d\theta_m, \ x \in \overset{\circ}{K},$$

$$c_k = \frac{1}{(2\pi)^m \rho^k} \int_{-\pi}^{\pi} (m) \int e^{-ik \cdot \theta} f(\alpha_1 + \rho_1 e^{i\theta_1}, \ldots, \alpha_m + \rho_m e^{i\theta_m}) d\theta_1 \ldots d\theta_m$$

make sense; for $x \in \overset{\circ}{K}$, the power series $(c_k (x - a)^k)_{k \in \mathbb{N}^m}$ is summable to $F(x)$ by the argument used in the proof of Theorem 2.1.2.b.; but $\langle F(x), z' \rangle = \langle f(x), z' \rangle$ for all $z' \in Z'$ since the Cauchy m-uple integral formula holds for each complex valued analytic function $z' \circ f$. Thus $f \in \mathscr{A}(U, Z)$. □

Corollary 2.1.4. *Let g be a map $U \to Z$ with the property: for each $z' \in Z'$, each compact set $K \subset U$ and $\varepsilon > 0$, there is an $f \in \mathscr{A}(U, Z)$ such that $\sup_K |z' \circ (g - f)| \leqslant \varepsilon$. If Z is s.c., then $g \in \mathscr{A}(U, Z)$.*

(D) The equivalence between analyticity and scalar analyticity can be used in many other problems, for example:

a) to show that $f \in \mathscr{A}(U, Z)$ implies $hf \in \mathscr{A}(U, Z)$ for all $h \in \mathscr{A}(U, \mathbb{C})$ and $f \circ g \in \mathscr{A}(V, Z)$ for any analytic map $g: V \to U$, where V is open in $\mathbb{C}^n$; in particular, the definition of $\mathscr{A}(U, Z)$ does not depend on the basis chosen for the finite dimensional space in which U is an open set;
b) to prove the Liouville theorem for a map $\mathbb{C} \to Z$;
c) to find cases when an analytic map $U \to Z$ remains analytic for a finer topology on Z (it obviously remains analytic for a coarser one): here are two such cases.

   If f is analytic $U \to Z$ for the weakened topology on Z and Z is s.c. for the initial one, then f is also analytic for the initial topology on Z since the adjoint space to Z is the same (Prop. 1.1.1.c).

   If f is analytic $U \to Z'$ for the weak* topology on Z' and Z is barrelled (which implies Z' strongly quasi-complete by Th. 1.4.4.b), then f is also analytic for the strong topology on Z', although the adjoint space to Z' is not always the same. In fact, since $U \ni x \mapsto \langle z, f(x) \rangle$ is analytic for all $z \in Z$, the sets B_j in the proof of Theorem 2.1.3. are weak* bounded, hence also strongly bounded (Th. 1.4.4.b), f is strongly continuous and the power series $(c_k(x - a)^k)$ in the proof of Theorem 2.1.3. strongly summable to $f(x)$.

(E) *A set $\mathcal{B}$ in $\mathcal{A}(U, Z)$ is said to be* bounded *if it is bounded for the topology of compact convergence (defined in Th. 2.1.2.c), i.e.: for each $q \in \mathrm{csn}(Z)$ and each compact set $K \subset U$, there is a number $M(q, K) > 0$ such that $\sup_K (q \circ f) \leqslant M(q, K)$ for all $f \in \mathcal{B}$.*

By Theorem 2.1.2.c, $\mathcal{B}$ bounded implies $D^k \mathcal{B} = \{D^k f : f \in \mathcal{B}\}$ bounded for all $k \in \mathbb{N}^m$.

Theorem 2.1.5. *Let $\mathcal{B} \subset \mathcal{A}(U, Z)$ be bounded and Z s.c.*

a) *$\mathcal{B}$ is equicontinuous; therefore, if a sequence $(f_n) \subset \mathcal{B}$ converges pointwise to f, it actually converges uniformly on compact sets, and $f \in \mathcal{A}(U, Z)$ by Coroll 2.1.4, $(D^k f_n)$ converges to $D^k f$, uniformly on compact sets, by Theorem 2.1.2.c.*

b) *Let Z be not only s.c. but s.r. too (see Rem. 1.4.3.). Given a sequence $(f_n) \subset \mathcal{B}$, there is a $g \in \mathcal{A}(U, Z)$ with the property: for each $z' \in Z'$ and a subsequence (f_{n_k}) depending on z', $z' \circ (g - f_{n_k}) \to 0$ uniformly on compact sets. More generally, given a filter $\mathcal{F}$ on $\mathcal{B}$, there is a $g \in \mathcal{A}(U, Z)$ with the property: for each $z' \in Z'$, each compact set $K \subset U$ and $\varepsilon > 0$, the set*

$$\left\{ f \in \mathcal{B} : \sup_K |z' \circ (g - f)| \leqslant \varepsilon \right\} \text{ meets all sets } \in \mathcal{F}.$$

If only one g has the above property, then $g = \lim f_n$ or $g = \lim_{\mathcal{F}} f$ for the weakened topology, uniformly on compact sets.

c) *Let μ be a real or complex measure on a compact metric space T. If $f(., t) \in \mathcal{B}$ for all $t \in T$ and $f(x, .)$ is continuous for all $x \in U$, then the integral formula $F(x) = \int f(x, t) \, d\mu(t)$ defines a $F \in \mathcal{A}(U, Z)$, $D_x^k f(x, .)$ is continuous and $D^k F(x) = \int D_x^k f(x, t) \, d\mu(t)$ for all $x \in U$, $k \in \mathbb{N}^m$.*

Proof. a) Let $K = \prod_{j=1}^{m} \bar{\Delta}(\alpha_j, \rho_j) \subset U$; for $f \in \mathcal{B}$, $b \in K$, we have

$$f(b) - f(a) = \int_0^1 \sum_{j=1}^{m} (\beta_j - \alpha_j) \frac{\partial f}{\partial \xi_j} [(1 - t)a + tb] \, dt$$

since for all $z' \in Z'$ this is true with $z' \circ f$ instead of f; hence follows

$$q[f(b) - f(a)] \leqslant \sum_{j=1}^{m} |\beta_j - \alpha_j| \sup_K \left(q \circ \frac{\partial f}{\partial \xi_j} \right) \text{ for all } q \in \mathrm{csn}(Z), \text{ and the}$$

boundedness of $D^k \mathcal{B}$ for $|k| = 1$ implies the equicontinuity of $\mathcal{B}$.

b) Let Φ be any ultrafilter on $\mathbb{N}$ finer than the Fréchet filter; for all $x \in U$, since the set $\{f_n(x): n \in \mathbb{N}\}$ is relatively compact for the weakened topology, there is a $g(x) \in Z$ such that each weakened neighbourhood of $g(x)$ contains $\{f_n(x): n \in \Omega\}$ for some $\Omega \in \Phi$. Since $\mathscr{B}$ is equicontinuous: given $z' \in Z'$, a compact set $K \subset U$ and $\varepsilon > 0$, there is an $\Omega \in \Phi$ such that $\sup_K |z' \circ (g - f_n)| \leqslant \varepsilon$ for all $n \in \Omega$, and $g \in \mathscr{A}(U, Z)$ by Corollary 2.1.4.

The ultrafilter Φ can be chosen as follows. Given $x \in X$, if $z \in Z$ is such that each weakened neighbourhood V of z meets $\{f_n(x): n \geqslant n_0\}$ for all n_0, then the sets $A(n_0, V) = \{n \geqslant n_0: f_n(x) \in V\}$ are the basis of a filter finer than the Fréchet; if Φ is chosen still finer, then $g(x) = z$. So the uniqueness of g implies the uniqueness of z, i.e. $z = \lim f_n(x)$ for the weakened topology, and this holds for all $x \in X$.

The argument may be repeated with an ultrafilter Φ on $\mathscr{B}$ finer than the given $\mathscr{F}$: for all $x \in U$, $g(x)$ is such that each weakened neighbourhood of $g(x)$ contains $\{f(x): f \in \Omega\}$ for some $\Omega \in \Phi$; given z', K and ε, there is an $\Omega \in \Phi$ such that $\sup_K |z' \circ (g - f)| \leqslant \varepsilon$ for all $f \in \Omega$; the uniqueness of g implies the uniqueness of z such that each weakened neighbourhood V of z meets $\{f(x): f \in A\}$ for all $A \in \Phi$, i.e. $z = \lim_{\mathscr{F}} f(x)$.

c) F is the pointwise limit of the sequence

$$F_n(x) = \sum_{j=1}^{J(n)} \mu(T_n^j) f(x, t_n^j)$$

used in the proof of Proposition 1.2.3., which is in the bounded set

$$\mathscr{B}' = \{f \in \mathscr{A}(U, Z): \sup_K (q \circ f) \leqslant \|\mu\| \cdot M(q, K) \text{ for all } q \text{ and } K\}$$

with $M(q, K)$ as in the beginning of (E). Then $F \in \mathscr{A}(U, Z)$, $D^k F$ is the limit of the sequence

$$D^k F_n(x) = \sum_{j=1}^{J(n)} \mu(T_n^j) D_x^k f(x, t_n^j)$$

and $f(x, .)$ continuous, $f(., t) \in \mathscr{B}$ entail $D_x^k(x, .)$ continuous for all $x \in U$.

$\square$

Proposition 2.1.6. *Conversely, let $\mathscr{B} \subset \mathscr{A}(U, Z)$ be equicontinuous.*

a) *If each connected component of U contains at least one point x such that $\{f(x): f \in \mathscr{B}\}$ is bounded, then $\mathscr{B}$ is bounded.*

b) *$D^k \mathscr{B}$ is bounded, hence equicontinuous, for all $k \in \mathbb{N}^m$ with $|k| \geqslant 1$.*

Proof. a) Let $q \in \text{csn}(Z)$: each point $x_0 \in U$ has a neighbourhood V in U such that $x \in V$ implies $q[f(x) - f(x_0)] \leqslant 1$ for all $f \in \mathscr{B}$; therefore $\sup\limits_{f \in \mathscr{B}} q \circ f(x_0)$ finite (resp. : infinite) entails $\sup\limits_{f \in \mathscr{B}} q \circ f(x)$ finite (resp. : infinite) for all $x \in V$. So the assumption entails : first $\{f(x) : x \in \mathscr{B}\}$ bounded for all $x \in U$: secondly $\sup\limits_{f \in \mathscr{B}} q \circ f(x)$ bounded for $x \in K$ compact $\subset U$.

b) Assume U connected, choose $x_0 \in U$ and replace each $f \in \mathscr{B}$ by $f - f(x_0)$, which does not change $D^k \mathscr{B}$. □

The first part of the next Proposition, which extends a theorem of Vitali, will be proved as a direct application of Theorem 2.1.5.b; a quite different proof, which appears in [HiPh] (Th. 3.14.1) for the case of a Banach space Z and uses only Cauchy sequences, is given for the second part, but can also be used for the first part, since s.r. implies q.c. for the weakened topology as well as the initial one (Th. 1.1.2.e).

Proposition 2.1.7. *Let U be a connected open set in $\mathbb{C}$, (f_n) a bounded sequence in $\mathscr{A}(U, Z)$, $(\zeta_m) \subset U \setminus \{\alpha\}$ a sequence converging to $\alpha \in U$. a) If Z is s.r. and the sequence (f_n) converges for the weakened topology at each point ζ_m, it also converges for the same topology at all points $\in U$, uniformly on compact sets.*
b) *If Z is only s.c., the same statement holds with "initial" instead of "weakened".*

Proof. a) The map g in Theorem 2.1.5.b is uniquely determined by the $z' \circ g(\zeta_m)$, $z' \in Z'$, and these are determined by the assumption.
b) It is enough to show the uniform convergence on any closed disc $\bar{\Delta}(\alpha, r) \subset U$. Let $U \supset \bar{\Delta}(\alpha, \rho)$ and, for each $q \in \text{csn}(Z)$, $\sup\limits_{\bar{\Delta}(\alpha, \rho)} (q \circ f_n) \leqslant M(q)$ for all n; if $\sum\limits_{k=0}^{\infty} c_n^k (\zeta - \alpha)^k$ is the Taylor expansion of f_n around α, the inequalities of Cauchy give $q(c_n^k \rho^k) \leqslant M(q)$ for all k, n, q and therefore

$$q[f_n(\zeta) - c_n^0] \leqslant M(q) \sum_{k=1}^{\infty} \left(\frac{|\zeta - \alpha|}{\rho} \right)^k \leqslant 2 \frac{M(q)}{\rho} |\zeta - \alpha|$$

$$\text{if} \quad |\zeta - \alpha| \leqslant \frac{\rho}{2}.$$

Choosing m so that $|\zeta_m - \alpha| \leqslant \frac{\rho}{2}$, we have

$$q(c_n^0 - c_{n'}^0) \leqslant 4 \frac{M(q)}{\rho} |\zeta_m - \alpha| + q[f_n(\zeta_m) - f_{n'}(\zeta_m)]$$

which proves that (c_n^0) is a Cauchy sequence in Z since $|\zeta_m - \alpha|$ is arbitrarily small: $c_n^0 \to c^0, q(c^0) \leqslant M(q)$.

Now $g_n(\zeta) = \dfrac{f_n(\zeta) - c_n^0}{\zeta - \alpha}$ defines another bounded sequence (g_n) in $\mathscr{A}(U, Z)$, which again converges at each point ζ_m; so by the same argument, $c_n^1 \to c^1, \ q(c^1) \leqslant \dfrac{M(q)}{\rho}$, and so on. Since $q(c^k) \leqslant \dfrac{M(q)}{\rho^k}$ for all $k \in \mathbb{N}$, the

series $\displaystyle\sum_{k=0}^{\infty} c^k (\zeta - \alpha)^k$ is summable to $f(\zeta)$ for $|\zeta - \alpha| < \rho, f \in \mathscr{A}[\Delta(\alpha, \rho), Z]$
and

$$
q[f_n(\zeta) - f(\zeta)] \leqslant \sum_{j=0}^{k} r^j \, q(c_n^j - c^j) + 2M(q) \sum_{j=k+1}^{\infty} \left(\frac{r}{\rho}\right)^j
$$

for $|\zeta - \alpha| \leqslant r < \rho$: $f = \lim f_n$ uniformly on $\bar{\Delta}(\alpha, r)$ for all $r < \rho$. $\square$

Remark 2.1.8. In the Banach space $L^1(\mathbb{R}_+, e^{-t}\, dt)$ (of classes of functions of the variable $t \in \mathbb{R}_+$ which are integrable for the measure $e^{-t}\, dt$), consider the (non closed) subspace Z made up of finite sums $\displaystyle\sum_{j=1}^{m} (\alpha_j + \beta_j t)e^{-tx_j}$, $\alpha_j, \beta_j, x_j \in \mathbb{C}, \mathscr{R}e\, x_j > -1$, and set $U = \{x \in \mathbb{C}: \mathscr{R}e\, x > -1\}$. If, for $x \in U$, we denote by $f(x)$ the function $t \mapsto e^{-tx}$, we define a map $f: U \to Z$ which is continuous and holomorphic: $f'(x)$ is the function $t \mapsto -te^{-tx}$.

If f were analytic, the function $t \mapsto t^2$ would lie in Z, hence a relation $t^2 = \displaystyle\sum_{j=1}^{m} (\alpha_j + \beta_j t)e^{-tx_j}$ for all $t \in \mathbb{C}$, which is impossible: this is left to the reader.

2.2 Polynomials and polynomial maps

In this section, we consider polynomials with coefficients in a vector space Z over $\mathbb{C}$, which needs not be endowed with a topology.

Proposition 2.2.1. *Let f be a map $\mathbb{C}^m \to Z$ with the property: for any x, $y \in \mathbb{C}^m$, $f(x + \zeta y)$ is a polynomial in the complex variable ζ. Then $f(x)$ is a polynomial in the coordinates ξ_j of x (the converse statement being obvious), and the degree (denoted by $d^{\circ}f$) of this polynomial is the upper bound of the degree of $\zeta \mapsto f(x + \zeta y)$ as x and y run through $\mathbb{C}^m$.*

Proof by induction on m. Let $m \geqslant 2$, let f satisfy the above assumption, and set $x = x' + \xi_m e_m$, where $(e_1, \ldots, e_m)$ is the canonical basis of $\mathbb{C}^m$. If the proposition is true for a map $\mathbb{C}^{m-1} \to Z$, then $x' \mapsto f(x' + \xi_m e_m)$ is a polynomial P_{ξ_m} in $\xi_1, \ldots, \xi_{m-1}$, whose degree and coefficients depend on ξ_m. First assume that $X_0 = \{\xi_m \in \mathbb{C} : P_{\xi_m} \text{ constant}\}$ is infinite: then, given $a' \in \mathbb{C}e_1 + \cdots + \mathbb{C}e_{m-1}$, we have $f(x' + \xi_m e_m) = f(a' + \xi_m e_m)$ for each x' and for all $\xi_m \in X_0$, hence for all $\xi_m \in \mathbb{C}$ since both members are polynomials in ξ_m.

Now let p be the smallest positive integer such that $X_p = \{\xi_m \in \mathbb{C} : P_{\xi_m}$ has degree $p\}$ is infinite: p exists because $\mathbb{C}$ is uncountable. One can find $N = C_{m+p-1}^{m-1}$ points $x'_j \in \mathbb{C}e_1 + \cdots + \mathbb{C}e_{m-1}$, $j = 1, \ldots, N$, and N polynomials $A_j(x')$, with degrees $\leqslant p$ and complex coefficients, such that

$$P(x') = \sum_{j=1}^{N} P(x'_j) A_j(x')$$ for any polynomial P with degree $\leqslant p$ and coefficients in Z: for the sake of completeness, a proof of this algebraic fact is given in Remark 2.2.2. (A) below. In particular:

$$f(x' + \xi_m e_m) = \sum_{j=1}^{N} f(x'_j + \xi_m e_m) A_j(x')$$

for each x' and for all $\xi_m \in X_p$, hence for all $\xi_m \in \mathbb{C}$ since both members are polynomials in ξ_m.

Now let n be the degree of the polynomial $f(x)$: the degree of $\zeta \mapsto f(x + \zeta y)$ is $\leqslant n$, with equality if and only if the sum of the terms of degree n in $f(x)$ does not vanish at y. $\square$

Remark 2.2.2. (A) The generic polynomial in m variables, with degree $\leqslant p$, can be written as $P(x) = \sum_{k=1}^{N} c_k M_k(x)$, where $N = C_{m+p}^m$ and the M_k are all monomials with degrees $\leqslant p$ in the m coordinates of $x \in \mathbb{C}^m$.

If the coefficients c_k are complex, the maps $c = (c_1, \ldots, c_N) \mapsto P(x)$, $x \in \mathbb{C}^m$, are linear maps $\mathbb{C}^N \to \mathbb{C}$ which generate the N-dimensional vector space of all linear maps $\mathbb{C}^N \to \mathbb{C}$; then N points x_j can be found in $\mathbb{C}^m$ so that the N linear maps $c \mapsto P(x_j)$ are independent, i.e. $\det M_k(x_j) \neq 0$. Denoting by $(\alpha_{j,k})_{j,k=1,\ldots,N}$ the inverse of the matrix $(M_j(x_k))$, we have $c_k = \sum_{j=1}^{N} \alpha_{j,k} P(x_j)$, these relations still hold if the c_k lie in any vector space over $\mathbb{C}$, and they imply

$$P(x) = \sum_{k=1}^{N} c_k M_k(x) = \sum_{j=1}^{N} P(x_j) A_j(x), \qquad A_j(x) = \sum_{k=1}^{N} \alpha_{j,k} M_k(x).$$

(B) If the space Z is l.c. and s.c., the following simpler proof of Proposition

2.2.1 proceeds from the results of Section 2.1. The assumption implies that f is holomorphic, hence analytic; for all $x \in \mathbb{C}^m$, $\dfrac{\partial^{p_1} f}{\partial \xi_1^{p_1}}(x) = 0$ for any integer p_1 greater than the degree of the polynomial $\xi_1 \mapsto f(x + \xi_1 e_1)$, so that $\mathbb{C}^m$ is the union of the closed sets $\left\{ x \in \mathbb{C}^m : \dfrac{\partial^{p_1} f}{\partial \xi_1^{p_1}}(x) = 0 \right\}$, and one of these must have a non empty interior, hence an integer p_1 such that $\dfrac{\partial^{p_1} f}{\partial \xi_1^{p_1}}$ vanishes identically. We obtain the other p_j in the same way, and then $f(x) = \sum\limits_{\substack{k \in \mathbb{N}^m \\ k_j < p_j}} \dfrac{D^k f(0)}{k!} x^k$.

Proposition 2.2.3. *Let* $f(\zeta) = \sum\limits_{k=0}^{n} c_k h_k(\zeta)$, *where each* $c_k \in Z$ *and the* h_k *are complex valued holomorphic functions on a connected open set* U *in* $\mathbb{C}$: *for any seminorm* q *on* Z, $q \circ f$ *and* $\ln q \circ f$ *are continuous functions on* U, *with values in* $[0, +\infty[$ *and* $[-\infty, +\infty[$ *respectively,* $q \circ f$ *subharmonic,* $\ln q \circ f$ *subharmonic or* $\equiv -\infty$.

For subharmonic functions, more generally for potential theory on the complex plane, we refer to [Brel]$_3$.

Proof. $q \circ f$ is continuous on U because

$$|q \circ f(\zeta) - q \circ f(\zeta_0)| \leqslant q[f(\zeta) - f(\zeta_0)] \leqslant \sum_{k=0}^{n} q(c_k) |h_k(\zeta) - h_k(\zeta_0)|.$$

For any linear map $z^* : Z \to \mathbb{C} : z^* \circ f(\zeta) = \sum\limits_{k=0}^{n} z^*(c_k) h_k(\zeta)$ is a holomorphic function on U, therefore $|z^* \circ f|$ subharmonic, $\ln|z^* \circ f|$ subharmonic or $\equiv -\infty$. By the proof of Proposition 1.1.1.a, q is a supremum of moduli of linear maps $Z \to \mathbb{C}$; then $q \circ f$, a continuous supremum of subharmonic functions, is subharmonic and $\ln q \circ f$ subharmonic or $\equiv -\infty$. $\square$

Proposition 2.2.4. *Let* f *be a polynomial* $\mathbb{C} \to Z$ *with degree* n, q *a seminorm on* Z, K *a nonpolar compact set in* $\mathbb{C}$, U *the unbounded connected component of* $\mathbb{C} \setminus K$, G *the Green function of* U *with pole at infinity. Then (Bernstein-Walsh inequality):*

$$q \circ f(\zeta) \leqslant \sup_{K} (q \circ f) e^{nG(\zeta)} \quad \text{for all } \zeta \in U,$$

$$q \circ f(\zeta) \leqslant \sup_{K} (q \circ f) \quad \text{for all } \zeta \in \complement U.$$

G can be defined as the smallest positive harmonic function h on U such that $h(\zeta) - \ln|\zeta|$ has a finite limit as $\zeta \to \infty$; $\gamma = \lim_{\zeta \to \infty} [G(\zeta) - \ln|\zeta|]$ is known as the Robin constant of the compact set $\complement U$.

Proof. Since $q \circ f$ is subharmonic (Prop. 2.2.3), the second inequality is an immediate consequence of the maximum principle: K contains the boundary of the open set $\complement U \setminus K$. In order to prove the first one, we may assume $0 \in K$ and consider $V = \left\{ \xi = \dfrac{1}{\zeta} : \zeta \in U \text{ or } \zeta = \infty \right\}$, an open set in $\mathbb{C}$: $\xi \mapsto G\left(\dfrac{1}{\xi}\right) + \ln|\xi|$ is a harmonic function on V, therefore, on each open disc $\delta \subset V$, the real part of a holomorphic function h_δ. With the notation $f(\zeta) = \sum_{k=0}^{n} c_k \zeta^k$, $c_n \neq 0$, Proposition 2.2.3 also asserts that

$$q\left[\sum_{k=0}^{n} c_k \xi^{n-k} e^{-nh_\delta(\xi)} \right] = q\left[\sum_{k=0}^{n} c_k \left(\frac{1}{\xi}\right)^k \right] \cdot \exp\left[-nG\left(\frac{1}{\xi}\right) \right]$$

is a subharmonic function of $\xi \in \delta$. By the local nature of subharmonicity, the right hand member is a subharmonic function of $\xi \in V$, which by the maximum principle is smaller than the maximum of $q\left[\sum_{k=0}^{n} c_k \left(\frac{1}{\xi}\right)^k \right]$ for ξ on the boundary of V (∞ included if V is unbounded), i.e. $\sup_K (q \circ f)$.

Corollary 2.2.5. *Let q, K, U, G be as in Proposition 2.2.4; assume moreover that no connected component of K consists in a single point, so that the function L equal to e^G on U, to 1 on $\complement U$, is continuous. Then, given $\lambda > 1$ and a family $\mathscr{B}$ of polynomials $\mathbb{C} \to Z$ such that $\sup_{f \in \mathscr{B}} q \circ f(\zeta) < +\infty$ for all $\zeta \in K$, there is a number m such that $q \circ f \leqslant m(\lambda L)^{d \circ f}$ everywhere for all $f \in \mathscr{B}$ (the Leja polynomial lemma), in particular $q \circ f \leqslant m\lambda^{d \circ f}$ on K for all $f \in \mathscr{B}$.*

Proof. Since $\ln L$ is continuous, $\ln L$ is subharmonic on $\mathbb{C}$, there is a positive Radon measure μ, carried by the boundary of $\complement U$, hence by K, and a harmonic function h on $\mathbb{C}$ such that

$$\ln L(\xi) = h(\xi) + \int \ln|\xi - \eta| \, d\mu(\eta) \quad \text{for all } \xi \in \mathbb{C};$$

a comparison with $\ln|\xi|$ as $\xi \to \infty$ shows that $\mu(K) = 1$ and actually h is the constant γ: therefore $\int \ln|\xi - \eta| \, d\mu(\eta) = -\gamma$ for all $\xi \in \complement U$. Without loss of generality, we may assume that $\operatorname{diam}(\complement U) \leqslant 1$.

For all $m \in \mathbb{N}^*$, $K_m = \{\xi \in K : q \circ f(\xi) \leqslant m$ for all $f \in \mathscr{B}\}$ is a compact subset of K, and the sequence (K_m) increases to K; we consider only the indeces m for which $\mu_m = \mu|_{K_m}$ is not 0. Since $\xi \mapsto \int \ln|\xi - \eta| \, d\mu(\eta)$ is continuous, the upper semicontinuous functions $\xi \mapsto \int \ln|\xi - \eta| \, d\mu_m(\eta)$ and $\xi \mapsto \int \ln|\xi - \eta| \, d(\mu - \mu_m)(\eta)$ are continuous too; the first one decreases on $\complement U$ to $\int \ln|\xi - \eta| \, d\mu(\eta)$, uniformly on $\complement U$ by the classical Dini lemma: let $\varepsilon_m - \gamma$ be its lower bound on $\complement U$, ε_m decreasing to 0. Then

$$\ln L_m(\xi) = \frac{1}{\mu(K_m)} \left[\gamma - \varepsilon_m + \int \ln|\xi - \eta| \, d\mu_m(\eta) \right]$$

defines a positive continuous function $\ln L_m$, harmonic on U_m (the unbounded connected component of $\mathbb{C} \backslash K_m$), whose difference with $\ln|\xi|$ has a finite limit as $\xi \to \infty$: by Proposition 2.2.4 we have

$$q \circ f \leqslant m . L_m^{d \circ f} \quad \text{for all } f \in \mathscr{B} \text{ everywhere.}$$

But $\ln \dfrac{L_m}{L}$ tends to 0 uniformly on $\complement U$, hence everywhere since it is harmonic on U and has a finite limit at ∞: given $\lambda > 1$, we can choose m so that $L_m \leqslant \lambda L$ everywhere. $\square$

Definition 2.2.6. *Let X be an infinite dimensional vector space. A map $\varphi \colon X \to Z$ is a polynomial map if, for one or any integer $m \in \mathbb{N}^*$, φ has the property: for any $a, a_1, \ldots, a_m \in X$, $\varphi (a + \zeta_1 a_1 + \cdots + \zeta_m a_m)$ is a polynomial in the complex variables ζ_j, and the degree (depending on a, $a_1, \ldots, a_m$) of this polynomial has a finite upper bound n, which by definition is the degree $d° \varphi$ of the polynomial map φ.*

By Proposition 2.2.1: this property does not depend on m, nor does the integer n under the assumption of bounded degrees, which is not superfluous. Consider, for example, the space c_{00} ($\mathbb{N}^*$) of sequences $x = (\xi_1, \xi_2, \ldots)$ of complex numbers ξ_n with $\xi_n = 0$ for sufficiently large n: $\varphi(x) = \sum_{n=1}^{\infty} \xi_n^n$ is a map $c_{00} \to \mathbb{C}$ such that $\zeta \to \varphi(x + \zeta a)$ is a polynomial with unbounded degree.

If ψ is a polynomial map $Y \to Z$, then $\psi(b + \zeta b_1 + \cdots + \zeta^m b_m)$ is also a polynomial in the complex variable ζ. Therefore the composed map of polynomial maps $X \to Y$ and $Y \to Z$ is a polynomial map $X \to Z$.

Proposition 2.2.7. a) *If φ is a polynomial map of degree n, $\varphi(x)$ is a linear combination, with rational coefficients depending only on n, of $\varphi(x + a)$, $\varphi(x + 2a), \ldots, \varphi(x + (n + 1) a)$.*
b) *If φ is a polynomial map of degree 1, $\varphi - \varphi(0)$ is linear.*

Proof. a) Let $\varphi(x + \zeta a) = \sum\limits_{k=0}^{n} c_k \zeta^k$: given $\varphi(x + a)$, $\varphi(x + 2a)$, ... , $\varphi(x +$

$(n + 1)a)$, the c_k are the solution of a linear system with coefficients $\in \mathbb{N}^*$, whose Vandermonde determinant is not 0.

b) Let $\varphi(0) = 0$ and $\varphi(\zeta_1 a_1 + \zeta_2 a_2) = c_0 + \zeta_1 c_1 + \zeta_2 c_2$: first take $\zeta_1 = \zeta_2 = 0$ and get $c_0 = 0$; then take $\zeta_1 = 1$, $\zeta_2 = 0$ or $\zeta_1 = 0$, $\zeta_2 = 1$ and get $c_1 = \varphi(a_1)$ or $c_2 = \varphi(a_2)$. $\square$

Definition 2.2.8. *A polynomial map $\varphi: X \to Z$ is n-homogeneous, $n \in \mathbb{N}^*$, if $\varphi(\zeta x) = \zeta^n \varphi(x)$ for all $x \in X$, $\zeta \in \mathbb{C}$.*

Theorem 2.2.9. a) *Any n-homogeneous polynomial map has degree n unless it vanishes identically.*

b) *For $n \in \mathbb{N}^*$, any n-linear map $\lambda: X^n \to Z$ generates an n-homogeneous polynomial map $\varphi: X \to Z$ (which may vanish identically even if λ does not) by the relation*

(1) $\varphi(x) = \lambda(x, \ldots, x)$ *(x written n times).*

c) *Conversely, given an n-homogeneous polynomial map $\varphi: X \to Z$, there is a unique symmetric n-linear map $\lambda: X^n \to Z$ associated to φ by the relation* (1).

Proof. a) Any n-homogeneous polynomial $\mathbb{C}^2 \to Z$ has degree n or vanishes identically; since this is true for the polynomial $(\xi, \zeta) \mapsto \varphi(\xi x + \zeta a)$, the polynomial $\zeta \mapsto \varphi(x + \zeta a)$ has a degree $\leqslant n$. But the polynomial $\zeta \mapsto \varphi(\zeta a)$ has degree n unless $\varphi(a) = 0$.

c) Let λ be a symmetric n-linear map $X^n \to Z$ and $\hat{\lambda}(x)$ the right hand member of (1). For $a_1, \ldots, a_n \in X$, $\hat{\lambda}(\zeta_1 a_1 + \cdots + \zeta_n a_n)$ is a n-homogeneous polynomial in the complex variables $\zeta_1, \ldots, \zeta_n$, in which the coefficient of ζ^k, $k \in \mathbb{N}^n$, $|k| = k_1 + \cdots + k_n = n$, is

$$\frac{n!}{k!} \lambda(a_1 \text{ written } k_1 \text{ times}, \ldots, a_n \text{ written } k_n \text{ times}).$$

Now replace $\zeta = (\zeta_1, \ldots, \zeta_n) \in \mathbb{C}^n$ by $\varepsilon = (\varepsilon_1, \ldots, \varepsilon_n) \in \{-1, +1\}^n$ and remark that $\varepsilon_1 \ldots \varepsilon_n = (-1)^{(n-|\varepsilon|)/2}$; then, for all $k \in \mathbb{N}^n$, $|k| = n$,

$$\sum_\varepsilon (-1)^{(n-|\varepsilon|)/2} \varepsilon^k = \sum_\varepsilon \varepsilon_1^{1+k_1} \ldots \varepsilon_n^{1+k_n}$$

is 0 if some k_j is even, 2^n if all k_j are odd, which happens only if all k_j are 1. Therefore a symmetric n-linear λ satisfying (1) must be

(2) $\lambda(a_1, \ldots, a_n) = \dfrac{1}{2^n n!} \sum_\varepsilon (-1)^{(n-|\varepsilon|)/2} \varphi(\varepsilon_1 a_1 + \cdots + \varepsilon_n a_n).$

This formula defines a symmetric map $X^n \to Z$; but is it n-linear and does it satisfy (1)?

In order to prove that it depends linearly on a_1, set $\varepsilon' = (\varepsilon_2, \ldots, \varepsilon_n) \in \{-1, +1\}^{n-1}$ and

$$\psi(x) = \sum_{\varepsilon'} \varepsilon_2 \ldots \varepsilon_n \varphi(x + \varepsilon_2 a_2 + \cdots + \varepsilon_n a_n) = \Delta_{a_2} \ldots \Delta_{a_n} \varphi(x)$$

with the notation $\Delta_h \varphi(x) = \varphi(x + h) - \varphi(x - h)$. For any polynomial map φ of degree $\leqslant n$, $\Delta_h \varphi$ is a polynomial map of degree $\leqslant n - 1$, since $(\zeta, \varepsilon) \mapsto \varphi(x + \zeta a + \varepsilon h)$ of degree $\leqslant n$ implies $\zeta \mapsto \varphi(x + \zeta a + h) - \varphi(x + \zeta a - h)$ of degree $\leqslant n - 1$. Then ψ is a polynomial map of degree 0 or 1, $x \mapsto \psi(x) - \psi(-x)$ is linear by Proposition 2.2.7.b, and the sum in the right hand member of (2) is $\psi(a_1) - \psi(-a_1)$.

Finally take $a_1 = \cdots = a_n = a$: since φ is n-homogeneous, the right hand member of (2) becomes

$$\frac{1}{2^n n!} \sum_\varepsilon (-1)^{(n-|\varepsilon|)/2} (\varepsilon_1 + \cdots + \varepsilon_n)^n \varphi(a) = \varphi(a)$$

by the above computation, repeated for the n-linear map $\lambda \colon \mathbb{C}^n \to \mathbb{C}$ defined by $\lambda(\zeta_1, \ldots, \zeta_n) = \zeta_1 \ldots \zeta_n$. $\square$

Proposition 2.2.10. *Any polynomial map $\varphi \colon X \to Z$ with degree n can be written uniquely as*

(3) $\varphi(x) = \varphi(0) + \displaystyle\sum_{k=1}^{n} \varphi_k(x),$

where each φ_k is a k-homogeneous polynomial map $X \to Z$. Consequently: the degree of the polynomial $\zeta \mapsto \varphi(x + \zeta a)$ is actually n if and only if $\varphi_n(a) \neq 0$, and then the term of highest degree is $\zeta^n \varphi_n(a)$, independent of x.

Proof. If (3) holds and each φ_k is k-homogeneous, then

(4) $\varphi(\xi x) = \varphi(0) + \displaystyle\sum_{k=1}^{n} \xi^k \varphi_k(x) \colon$

the $\varphi_k(x)$ must be the coefficients of the polynomial $\varphi(\xi x)$ in the complex variable ξ. Conversely, the φ_k defined by (4) are k-homogeneous polynomial maps $X \to Z$ because the $\varphi_k(x + \zeta a)$ are the coefficients of the polynomial $\xi \mapsto \varphi(\xi x + \xi \zeta a)$, especially if $x = 0$. The last statement follows from Theorem 2.2.9.c for φ_n. $\square$

Proposition 2.2.11. *Let φ be a polynomial map $X \to Z$ with degree n, p and q seminorms on X and Z respectively.*

a) *If $p(x) \leqslant 1$ implies $q \circ \varphi(x) \leqslant M$, then everywhere $\ln(q \circ \varphi) \leqslant \ln M + n . \ln^+ p$, and more precisely $q \circ \varphi \leqslant Mp^n$ if φ is n-homogeneous.*
b) *Let φ be n-homogeneous and V a balanced set in X: $q \circ \varphi \leqslant M$ on $a + V$ implies $q \circ \varphi \leqslant M$ on V (a translation lemma).*

Proof. a) Let us prove the first inequality for $x = \zeta a$, where ζ is a complex variable and $p(a) = 1$: set $\varphi(\zeta a) = \sum\limits_{k=0}^{n} c_k \zeta^k$. By Proposition 2.2.3., $\mathbb{C} \ni \zeta \mapsto \ln q \circ \varphi(\zeta a)$ is either $\equiv -\infty$ or a subharmonic function satisfying

$$\limsup_{\zeta \to \infty} \frac{\ln q \circ \varphi(\zeta a)}{\ln|\zeta|} \leqslant n$$

since $q \circ \varphi(\zeta a) \leqslant \left[\sum\limits_{k=0}^{n} q(c_k)\right] |\zeta|^n$ for $|\zeta| \geqslant 1$. In the latter case:

$$L(r) = \sup_{|\zeta|=r} \ln q \circ \varphi(\zeta a)$$

is a convex function of $\ln r$ such that $L(1) \leqslant \ln M$ and $\limsup\limits_{r \to \infty} \dfrac{L(r)}{\ln r} \leqslant n$, which implies $L(r) \leqslant \ln M + n . \ln r$ for all $r \geqslant 1$.
b) Since $\zeta \mapsto q \circ \varphi(x + \zeta a)$ is subharmonic, for $x \in V$ we have

$$q \circ \varphi(x) \leqslant \sup_{\theta \in \mathbb{R}} q \circ \varphi(x + e^{i\theta}a) = \sup_{\theta \in \mathbb{R}} q \circ \varphi(a + e^{-i\theta}x) \leqslant M. \quad \square$$

2.3 Gâteaux-analyticity

In this section, a s.c.l.c. topology is given on the vector space Z, but none on the vector space X, for the class $\mathscr{G}(\Omega, Z)$ can be defined on any *finitely open* set Ω in X.

Definition 2.3.1. a) Ω *is finitely open in X if, for any finite dimensional affine subspace E of X, endowed with the euclidean topology, $E \cap \Omega$ is open in E.*
b) *Given such an Ω, a map $f: \Omega \to Z$ is $\mathscr{G}$-analytic, or $f \in \mathscr{G}(\Omega, Z)$, if $f_{E \cap \Omega}$ is analytic for any finite dimensional affine subspace E of X or, equivalently, for any complex line E in X.*

Since any m-dimensional affine subspace of X is embedded in an $(m + 1)$-dimensional linear subspace, replacing the word "affine" by "linear" does not alter Definition a).

$f|_{E \cap \Omega}$ analytic is understood in the sense of Definition 2.1.1; by Theorem 2.1.3 and its consequences, it is equivalent to $z' \circ f|_{E \cap \Omega}$ analytic for all $z' \in Z'$ and does not depend on the basis chosen for E; so the equivalence stated in the above definition is a consequence of the Hartogs theorem ([He]$_1$, Th. 2 in II.2.1). $f \in \mathscr{G}(\Omega, Z)$ if and only if $z' \circ f \in \mathscr{G}(\Omega, \mathbb{C})$ for all $z' \in Z'$; $\mathscr{G}(\Omega, \mathbb{C})$ is a ring and $\mathscr{G}(\Omega, Z)$ a $\mathscr{G}(\Omega, \mathbb{C})$-module; $\mathscr{G}(\Omega, Z)$ contains a limit of maps $\in \mathscr{G}(\Omega, Z)$ which is uniform, either on any compact set $\subset \Omega$ and located in a finite dimensional affine subspace of X (Coroll. 2.1.4), or on any closed disc $\{a + \zeta b : |\zeta| \leqslant \rho\} \subset \Omega$.

Gâteaux-analyticity has the following *sheaf property*. Let $\Omega = \bigcup_{i \in I} \Omega_i$, where each Ω_i is finitely open in X; then a map $f : \Omega \to Z$ is $\mathscr{G}$-analytic if and only if each $f|_{\Omega_i}$ is $\mathscr{G}$-analytic. In fact, analytic maps on finite dimensional open sets have the same property by Definition 2.1.1.

Any polynomial map $\in \mathscr{G}(X, Z)$, even if the assumption of bounded degree made in Definition 2.2.6 were dropped. Any open set, for any vector space topology on X, is finitely open. Conversely:

Proposition 2.3.2. *If X has a countable dimension, i.e. countable Hamel bases, then the finitely open sets are exactly the open sets for the finest locally convex topology on X, defined by all possible seminorms on X.*

Proof. Let $(e_n)_{n \in \mathbb{N}^*}$ be a countable Hamel basis of X, $E_n = \mathbb{C}e_1 + \cdots + \mathbb{C}e_n$ be endowed with the euclidean topology, and Ω a finitely open set containing the origin. Since any compact subset of the open set $E_n \cap \Omega$ is also a compact subset of the open set $E_{n+1} \cap \Omega$, we can by induction find $\alpha_n > 0$ for all $n \in \mathbb{N}^*$ such that $|\zeta_1| \leqslant \alpha_1, \ldots, |\zeta_n| \leqslant \alpha_n$ imply $\zeta_1 e_1 + \cdots + \zeta_n e_n \in \Omega$. Each $x \in X$ can be written uniquely as $x = \sum_{n \in \mathbb{N}^*} \langle x, x_n^* \rangle e_n$, where the x_n^* are linear maps $X \to \mathbb{C}$, and $\langle x, x_n^* \rangle \neq 0$ only for a finite number of indeces n (depending on x); then $p(x) = \sup_{n \in \mathbb{N}^*} \frac{1}{\alpha_n} |\langle x, x_n^* \rangle|$ is a seminorm on X and $p(x) \leqslant 1$ implies $x \in \Omega$: this means that, for the finest locally convex topology on X, Ω is a neighbourhood of the origin, and also of any point $\in \Omega$.   □

Remark 2.3.3. Proposition 2.3.2 is due to Kakutani-Klee [KaKl], who also proved that it no longer holds if X has an uncountable dimension: in this case the finitely open sets are the open sets for a Hausdorff topology on X which is not a vector space topology because the addition $(x, y) \mapsto x + y$ is no longer continuous.

Now let X be any infinite dimensional space: given a Hamel basis $(e_i)_{i \in I}$, each $x \in X$ is uniquely written as $x = \sum_{i \in I} \langle x, x_i^* \rangle e_i$, where the x_i^* are linear maps $X \to \mathbb{C}$, and $\langle x, x_i^* \rangle \neq 0$ only for a finite number of indeces i (depending on x); then the k-linear maps $\lambda: X^k \to \mathbb{C}$ have the form

$$\lambda(x_1, \ldots, x_k) = \sum_{i_1, \ldots, i_k} \lambda(e_{i_1}, \ldots, e_{i_k}) \langle x_1, x_{i_1}^* \rangle \ldots \langle x_k, x_{i_k}^* \rangle$$

with arbitrary complex coefficients $\lambda(e_{i_1}, \ldots, e_{i_k})$.

If the topology on X is again defined by all possible seminorms on X: all linear maps $X \to \mathbb{C}$ (especially the x_i^*) are continuous, whereas λ is continuous if and only if one can find seminorms $p_1, \ldots, p_k$ on X such that

$$|\lambda(x_1, \ldots, x_k)| \leqslant p_1(x_1) \ldots p_k(x_k) \quad \text{for all } x_1, \ldots, x_k \in X,$$

which, for fixed $i_2, \ldots, i_k$, implies

(*) $|\lambda(e_{i_1}, \ldots, e_{i_k})| \leqslant c(i_2, \ldots, i_k) p_1(e_{i_1})$ for all $i_1 \in I$.

Assume moreover that I has the continuum power: given a countable infinite subset D of I, since the set of maps $D \to \mathbb{N}$ has the same power, it can be indexed by I, i.e. denoted by $\{\sigma_i: i \in I\}$; let $\tilde{\sigma}_i = \sigma_i$ on D, $\tilde{\sigma}_i = 0$ on $I \setminus D$. Consider the k-linear symmetric map $\lambda: X^k \to \mathbb{C}$ defined by

$$\lambda(e_{i_1}, \ldots, e_{i_k}) = \sup\{\tilde{\sigma}_{i_m}(i_n): m, n = 1, \ldots, k\}:$$

if the inequality (*) did hold for this λ, it would imply $\tilde{\sigma}_{i_1}(i) \leqslant c_1(i) p_1(e_{i_1})$ for all $i_1 \in I$, in other words any map $D \to \mathbb{N}$ would be majorized modulo a constant factor by the map $c_1|_D$, which obviously is impossible.

This argument, which is taken from [Ba Ma Na] (lemma 19), shows that, if X has Hamel bases with the continuum power, then for all $k \geqslant 2$ there are k-linear symmetric maps $X^k \to \mathbb{C}$ (hence also k-homogeneous polynomial maps $X \to \mathbb{C}$ by formula (2) in the proof of Th. 2.2.9) which are not continuous for any l.c. topology on X.

But any $\mathscr{G}$-analytic map $\Omega \to Z$, especially any polynomial map $X \to Z$, is continuous for the topology on X whose open sets are the finitely open sets, since the inverse image of any open set in Z is finitely open in X. Other general properties of this topology are the continuity of the translations $x \mapsto x + t$, $t \in X$, and moreover:

Proposition 2.3.4. a) *The multiplication $\mathbb{C} \times X \ni (\alpha, x) \to \alpha x$ is continuous for the topology on X whose open sets are the finitely open sets.*
b) *If the finitely open set Ω contains the origin, then its largest balanced subset*

$$\omega = \{x \in X : |\zeta| \leqslant 1 \text{ implies } \zeta x \in \Omega\}$$

is finitely open too.

Proof. a) Let Ω be a finitely open set containing the origin. Given $\alpha_0 \in \mathbb{C}$ and $x_0 \in X$, choose $\rho > 0$ such that $|\zeta| \leqslant \rho$ implies $\zeta x_0 \in \Omega$; then, for any $x_1, \ldots, x_m \in X$,

$$\{(\zeta, \zeta_1, \ldots, \zeta_m) \in \mathbb{C}^{m+1} : \zeta x_0 + (\zeta + \alpha_0)(\zeta_1 x_1 + \cdots + \zeta_m x_m) \in \Omega\}$$

is an open set in $\mathbb{C}^{m+1}$ containing $\bar{\Delta}(0, \rho) \times (0, \ldots, 0)$ and therefore the set of $(\zeta_1, \ldots, \zeta_m) \in \mathbb{C}^m$ such that $|\zeta| \leqslant \rho$ implies $\zeta x_0 + (\zeta + \alpha_0)(\zeta_1 x_1 + \cdots + \zeta_m x_m) \in \Omega$ is open in $\mathbb{C}^m$. This, with the notation

$$\omega_0 = \{x \in X : |\alpha - \alpha_0| \leqslant \rho \text{ implies } \alpha x - \alpha_0 x_0 \in \Omega\},$$

means that $E \cap \omega_0$ is open in E for any finite dimensional subspace $E = \{x_0 + \zeta_1 x_1 + \cdots + \zeta_m x_m : (\zeta_1, \ldots, \zeta_m) \in \mathbb{C}^m\}$ going through x_0, hence also for any finite dimensional affine subspace E. So we have found $\rho > 0$ and ω_0 finitely open containing x_0 such that $|\alpha - \alpha_0| \leqslant \rho$, $x \in \omega_0$ imply $\alpha x - \alpha_0 x_0 \in \Omega$.

b) If $x_0 \in \omega$, the above argument can be repeated with $\rho = 1$, $\alpha_0 = 0$. $\square$

Theorem 2.3.5. *Let Ω be finitely open in X, $f \in \mathscr{G}(\Omega, Z)$, $a \in \Omega$.*

a) *There is a sequence of k-homogeneous polynomial maps $(\hat{D}_a^k f) : X \to Z$ ($k \in \mathbb{N}^*$) such that the expansion*

(1) $$f(a + x) = f(a) + \sum_{k \in \mathbb{N}^*} \frac{1}{k!} (\hat{D}_a^k f)(x)$$

(the notation $\sum_{k \in \mathbb{N}^*}$ implying summability of the expansion) holds for all x in $\omega(a)$, the largest balanced subset of $\Omega - a$; the $(\hat{D}_a^k f)$ are uniquely determined by

(2) $$\frac{1}{k!} (\hat{D}_a^k f)(x) = \frac{1}{2\pi} \int_{-\pi}^{\pi} e^{-ik\theta} f(a + e^{i\theta} x) d\theta, \quad x \in \omega(a),$$

from which follow, for each $q \in \mathrm{csn}(Z)$, $k \in \mathbb{N}^*$, the generalized inequalities of Cauchy:

(3) $$\frac{1}{k!} q \circ (\hat{D}_a^k f)(x) \leqslant \sup_{\theta \in \mathbb{R}} q \circ f(a + e^{i\theta} x) \quad \text{for all } x \in \omega(a).$$

b) *$\mathscr{G}(\Omega, Z)$ also contains each generalized derivative of $f : \Omega \ni a \to (\hat{D}_a^k f)(x)$, $x \in X$; setting $(\hat{D}_a^k f)(x_0) = D^k f(a)$ for a given x_0, one has $D^j(D^k f) = D^{j+k} f$ for all $j, k \in \mathbb{N}^*$.*

Proof. a) $x \in \omega(a)$ means that the open set $U(a, x) = \{\zeta \in \mathbb{C}: a + \zeta x \in \Omega\}$ contains the closed disc $\bar{\Delta}(0, 1)$; then the analytic map $U(a, x) \ni \zeta \mapsto f(a + \zeta x)$ has a Taylor expansion around 0 which, by Theorem 2.1.2.b, is summable to $f(a + \zeta x)$ for $|\zeta| \leqslant 1$; writing it as

$$(4) \qquad f(a + \zeta x) = f(a) + \sum_{k \in \mathbb{N}^*} \frac{\zeta^k}{k!} (\hat{D}_a^k f)(x), \qquad |\zeta| \leqslant 1,$$

we have (2) as a special case of the integral formula in Theorem 2.1.2.a. Conversely, formula (1) for all $x \in \omega(a)$, with k-homogeneous $(\hat{D}_a^k f)$, implies (4), which proves the uniqueness stated in the theorem.

In order to show that the $(\hat{D}_a^k f)$ defined by (4) are k-homogeneous polynomial maps, we also consider the Taylor expansion around $(0, \dots, 0)$ of $f(a + \zeta_1 x_1 + \cdots + \zeta_k x_k)$ and the coefficient $(c_k)_a (x_1, \dots, x_k)$ of the product $\zeta_1 \dots \zeta_k$ in this expansion: we claim that, for all $a \in \Omega$, for all $k \in \mathbb{N}^*$, $(c_k)_a$ is a k-linear (obviously symmetric) map and

$$(5) \qquad (\hat{D}_a^k f)(x) = (c_k)_a(x, \dots, x) \qquad (x \text{ written } k \text{ times}).$$

First let $k = 1$. The latter statement is obvious, and $(c_1)_a(\alpha x) = \alpha(c_1)_a(x)$ for all $\alpha \in \mathbb{C}$; as to the former statement, the homogeneous part of degree 1 in the Taylor expansion of $f(a + \zeta_1 x_1 + \zeta_2 x_2)$ is $\zeta_1 (c_1)_a(x_1) + \zeta_2 (c_1)_a(x_2)$, and therefore the coefficient of ζ in the expansion of $f(a + \zeta(\alpha_1 x_1 + \alpha_2 x_2))$ is $\alpha_1 (c_1)_a(x_1) + \alpha_2 (c_1)_a(x_2)$. For $k \geqslant 2$, we proceed by an induction on k.

If the inequalities $|\zeta| \leqslant \rho, |\zeta_1| \leqslant \rho_1, \dots, |\zeta_{k-1}| \leqslant \rho_{k-1}$ imply $a + \zeta x + \zeta_1 x_1 + \cdots + \zeta_{k-1} x_{k-1} \in \Omega$, they also imply the validity of the Taylor expansion around $(0, \dots, 0)$ of $f(a + \zeta x + \zeta_1 x_1 + \cdots + \zeta_{k-1} x_{k-1})$; for $|\zeta| \leqslant \rho$, the part of this expansion in which the degrees of $\zeta_1, \dots, \zeta_{k-1}$ are all equal to 1 is summable to $\zeta_1 \dots \zeta_{k-1} (c_{k-1})_{a+\zeta x} (x_1, \dots, x_{k-1})$, which proves: first, that $\mathscr{G}(\Omega, Z)$ contains the map $a \mapsto (c_{k-1})_a(x_1, \dots, x_{k-1})$; secondly, that $(c_k)_a(x, x_1, \dots, x_{k-1})$ is the coefficient of ζ in the Taylor expansion around 0 of $(c_{k-1})_{a+\zeta x}(x_1, \dots, x_{k-1})$. Then $x_1 \mapsto (c_{k-1})_{a+\zeta x}(x_1, \dots, x_{k-1})$ linear implies $x_1 \mapsto (c_k)_a(x, x_1, \dots, x_{k-1})$ linear, hence $(c_k)_a$ actually k-linear since $(c_k)_a$ is symmetric.

If $|\zeta| \leqslant \rho$ implies $a + \zeta x \in \Omega$, each family $\left(\dfrac{\rho^k}{k!} q \circ (\hat{D}_a^k f)(x) \right)_{k \in \mathbb{N}^*}$, $q \in$ csn(Z), is summable by Theorem 2.1.2.b and therefore, by Theorem 1.2.1.c, the family $\left(\dfrac{\zeta^k \alpha^l}{k! \, l!} (\hat{D}_a^{k+l} f)(x) \right)_{(k, l) \in \mathbb{N}^2 \setminus (0, 0)}$ is summable to $f(a + \alpha x + \zeta x) - f(a)$ for $|\alpha| + |\zeta| \leqslant \rho$. By considering the subfamily in which the degree of

ζ is $k - 1$ $(k \geqslant 2)$, we get $(\hat{D}_{a+\alpha x}^{k-1} f)(x) = \sum_{l \in \mathbb{N}} \frac{\alpha^l}{l!} (\hat{D}_a^{k+l-1} f)(x)$, and $(\hat{D}_a^k f)(x)$ is the coefficient of α in the Taylor expansion around 0 of $(\hat{D}_{a+\alpha x}^{k-1} f)(x)$.

Putting all results together, we see that $(c_{k-1})_a$ $(k - 1)$-linear for all $a \in \Omega$ implies $(c_k)_a$ k-linear for all $a \in \Omega$, and (5) for all $a \in \Omega$ for the degree $k - 1$ implies (5) for all $a \in \Omega$ for the degree k.

b) Since $\mathscr{G}(\Omega, Z)$ contains each map $a \mapsto (c_k)_a (x_1, \ldots, x_k)$, by (5) $\mathscr{G}(\Omega, Z)$ also contains $a \mapsto (\hat{D}_a^k f)(x)$. The fact that $(\hat{D}_a^k f)(x_0)$ is the coefficient of ζ in the expansion of $(\hat{D}_{a+\zeta x_0}^{k-1} f)(x_0)$ is now written $D^k f(a) = \hat{D}_a^1 (D^{k-1} f)(x_0)$ or $D^k f = D^1 (D^{k-1} f)$; the general formula follows by an induction on j. $\square$

Proposition 2.3.6. *Conversely: let Ω be finitely open and balanced, φ_k a k-homogeneous polynomial map $X \to Z$ for each $k \in \mathbb{N}^*$; if $\{\varphi_k(x): k \in \mathbb{N}^*\}$ is bounded for all $x \in \Omega$, then the series $\sum\limits_{k \in \mathbb{N}^*} \varphi_k$ is summable on Ω to a map $f \in \mathscr{G}(\Omega, Z)$, $\varphi_k = (\hat{D}_0^k f)/k!$ for all $k \in \mathbb{N}^*$.*

Note that the set of $x \in X$ such that $\{\varphi_k(x): k \in \mathbb{N}^*\}$ is bounded and the set of $x \in X$ where the series $\sum\limits_{k \in \mathbb{N}^*} \varphi_k(x)$ is summable are both balanced. (See Ex. 1.2.2)

Proof. We will show that, for $(a, b) \in X^2$, the series $\sum\limits_{k \in \mathbb{N}^*} \varphi_k(a + \zeta b)$ is uniformly summable for $|\zeta| \leqslant 1$ whenever $\Omega \supset \{a + \zeta b: |\zeta| \leqslant 1\}$. In fact, since $\{(\alpha, \beta) \in \mathbb{C}^2: \alpha a + \beta b \in \Omega\}$ is an open set containing $\{1\} \times \bar{\Delta}(0, 1)$, there is a $\lambda > 1$ such that $\lambda(a + \zeta b) \in \Omega$ for all $\zeta \in \bar{\Delta}(0, 1)$; by the assumption, for each $q \in \text{csn}(Z)$ we have $\sup\limits_{k \in \mathbb{N}^*} q \circ \varphi_k[\lambda(a + \zeta b)] < +\infty$ for all $\zeta \in \bar{\Delta}(0, 1)$; then, by Corollary 2.2.5, there is a number m such that

$$q \circ \varphi_k[\lambda(a + \zeta b)] \leqslant m\lambda^{k/2} \quad \text{or} \quad q \circ \varphi_k(a + \zeta b) \leqslant \frac{m}{\lambda^{k/2}}$$

for all $k \in \mathbb{N}^*$, $\zeta \in \bar{\Delta}(0, 1)$. $\square$

Proposition 2.3.7. *Let Ω be a finitely open set in X with the following connectedness property: given $a \in \Omega$, $b \in \Omega$, there are a finite number of points $a_0 = a$, a_1, $\ldots$, $a_p = b$ such that Ω contains each line segment $\{(1 - t)a_{q-1} + ta_q: 0 \leqslant t \leqslant 1\}$, $q = 1, \ldots, p$.*

a) *Let $f \in \mathscr{G}(\Omega, Z)$ and $a \in \Omega$ be such that $(\hat{D}_a^k f)(x) = 0$ for all $k \in \mathbb{N}^*$, $x \in X$: then f is constant on Ω.*

b) *Let $f \in \mathscr{G}(\Omega, \mathbb{C})$ and $a \in \Omega$ be such that, for all $x \in X$, $|f(a + \zeta x)| \leqslant |f(a)|$ for sufficiently small $|\zeta|$: then f is constant on Ω (the local maximum modulus principle).*

Proof. a) It is enough to check that $(\hat{D}_{a_1}^k f)(x) = 0$ for all $x \in X$ if Ω contains the line segment $\{a + t(a_1 - a): 0 \leqslant t \leqslant 1\}$. In fact the open set

$$U(a; a_1 - a, x) = \{(\alpha, \zeta) \in \mathbb{C}^2: a + \alpha(a_1 - a) + \zeta x \in \Omega\}$$

contains $[0, 1] \times \{0\}$; by formula (1) in Theorem 2.3.5, $f[a + \alpha(a_1 - a) + \zeta x] = f(a)$ for (α, ζ) in any balanced subset of $U(a; a_1 - a, x)$, hence also for (α, ζ) in the connected component of $U(a; a_1 - a, x)$ which contains $[0, 1] \times \{0\}$; then $f(a_1 + \zeta x) = f(a)$ for sufficiently small $|\zeta|$, and $(\hat{D}_{a_1}^k f)(x) = 0$ for all $k \in \mathbb{N}^*$ by formula (4) in the proof of the same theorem.
b) $|f(a + \zeta x)| \leqslant |f(a)|$ for sufficiently small $|\zeta|$ implies $(\hat{D}_a^k f)(x) = 0$ for all $k \in \mathbb{N}^*$ by the same formula (4) and the classical maximum modulus principle. $\square$

Proposition 2.3.8. *The class of $\mathscr{G}$-analytic maps has the Hartogs property: let Ω be finitely open in a cartesian product $X_1 \times X_2$, which implies*

$$\Omega_1(x_2) = \{x_1 \in X_1: (x_1, x_2) \in \Omega\}, \Omega_2(x_1) = \{x_2 \in X_2: (x_1, x_2) \in \Omega\}$$

finitely open in X_1 for all $x_2 \in X_2$, in X_2 for all $x_1 \in X_1$. Then $\mathscr{G}(\Omega, Z)$ contains any map $f: \Omega \to Z$ which is separately $\mathscr{G}$-analytic, i.e.

$$f(., x_2) \in \mathscr{G}(\Omega_2(x_1), Z) \text{ for all } x_1 \in X_2$$

and $f(x_1, .) \in \mathscr{G}(\Omega_2(x_1), Z) \text{ for all } x_1 \in X_1.$

Proof. By the Hartogs theorem (for the $z' \circ f$, $z' \in Z'$), $(\zeta_1, \zeta_2) \mapsto f(a_1 + \zeta_1 x_1, a_2 + \zeta_2 x_2)$ is analytic on the open set

$$\{(\zeta_1, \zeta_2) \in \mathbb{C}^2: (a_1 + \zeta_1 x_1 + \zeta_2 o_1, a_2 + \zeta_1 o_2 + \zeta_2 x_2) \in \Omega\}$$

(where o_1, o_2 are the origins in X_1, X_2), hence $\zeta \mapsto f(a_1 + \zeta x_1, a_2 + \zeta x_2)$ analytic on the open set $\{\zeta \in \mathbb{C}: (a_1 + \zeta x_1, a_2 + \zeta x_2) \in \Omega\}$. $\square$

The following refinement of Proposition 2.3.8, essentially due to Nguyen [Ng], will be useful later on (see 3.2).

Proposition 2.3.9. *Let again Ω be finitely open in $X_1 \times X_2$, and moreover $\Omega = \bigcup_{x_1 \in \Omega_1} \{x_1\} \times \Omega_2(x_1)$, where each $\Omega_2(x_1)$ has the connectedness property in Proposition 2.3.7 and contains a nonempty finitely open set ω_2 independent of $x_1 \in \Omega_1$. Then $\mathscr{G}(\Omega, Z)$ contains any map $f: \Omega \to Z$ such that $f(., x_2) \in \mathscr{G}(\Omega_1, Z)$ for all $x_2 \in \omega_2$ and $f(x_1, .) \in \mathscr{G}(\Omega_2(x_1), Z)$ for all $x_1 \in \Omega_1$.*

Proof. By Proposition 2.3.8, f is $\mathscr{G}$-analytic on $\Omega_1 \times \omega_2$, and now we have to check that, for all $(a, b) \in \Omega$, $(x, y) \in X_1 \times X_2$, $z' \in Z': (\xi, \eta) \mapsto z' \circ f(a + \xi x, b + \eta y)$ is analytic on some neighbourhood of the origin in $\mathbb{C}^2$.

This is true if $b \in \omega_2$; if not, by the connectedness property of $\Omega_2(a)$ we can find: first, a finite number of points $z_1, \ldots, z_{p-1}, z_p \in \omega_2 - b$ such that $\Omega_2(a)$ contains the line segments joining b and $b + z_1$, $b + z_1$ and $b + z_2$, $\ldots$, $b + z_{p-1}$ and $b + z_p \in \omega_2$; secondly, $\rho > 0$ such that Ω contains the cartesian product of $\{a + \xi x : |\xi| \leqslant \rho\}$ and the same line segments. Let U be the connected component containing $(0, 0, \ldots, 0, 0)$ and $(0, 0, \ldots, 0, 1)$ of the open set $\{(\eta, \zeta_1, \ldots, \zeta_p) \in \mathbb{C}^{p+1} : (a + \xi x, b + \eta y + \zeta_1 z_1 + \cdots + \zeta_p z_p) \in \Omega$ for all $\xi \in \bar{\Delta}(0, \rho)\}$: if we prove that $g(\xi; \eta, \zeta_1, \ldots, \zeta_p) = z' \circ f(a + \xi x, b + \eta y + \zeta_1 z_1 + \cdots + \zeta_p z_p)$ is analytic on $\Delta(0, \rho) \times U$, from that will ensue that $z' \circ f(a + \xi x, b + \eta y) = g(\xi; \eta, 0, \ldots, 0)$ is analytic for $\xi \in \Delta(0, \rho)$, $(\eta, 0, \ldots, 0) \in U$, a neighbourhood of the origin in $\mathbb{C}^2$.

Assume that g is not analytic on $\Delta(0, \rho) \times U$; since g is analytic on the cartesian product of $\Delta(0, \rho)$ and some open neighbourhood of $(0, 0, \ldots, 0, 1)$ in U, let: V be the largest open subset of U such that g is analytic on $\Delta(0, \rho) \times V$; $\prod$ an open polydisc in $\mathbb{C}^{p+1}$, contained in U, centred in V but not contained in V; finally $\bar{\prod}$ a closed polydisc in $\mathbb{C}^{p+1}$, with the same centre, contained in V. Then the Taylor expansion of $g(\xi; .)$ around this centre is summable for all $\xi \in \bar{\Delta}(0, \rho)$ on $\prod$ and its general term bounded for $\xi \in \bar{\Delta}(0, \rho - \varepsilon)$, $(\eta, \zeta_1, \ldots, \zeta_p) \in \bar{\prod}$ for all $\varepsilon > 0$; by another theorem of Hartogs ([He]$_1$, Th. 1 in II.2.1), the expansion is normally summable on the compact subsets of $\Delta(0, \rho - \varepsilon) \times \prod$, hence g analytic on $\Delta(0, \rho) \times \prod$, a contradiction. $\square$

Proposition 2.3.10. *The class of $\mathcal{G}$-analytic maps does not enjoy the composition property, but only the following partial one: if φ is a polynomial map $X \to Y$ and Ω finitely open in Y, then $\varphi^{-1}(\Omega)$ is finitely open in X and $f \in \mathcal{G}(\Omega, Z)$ implies $f \circ \varphi \in \mathcal{G}(\varphi^{-1}(\Omega), Z)$.*

Proof. For any $a, a_1, \ldots, a_m \in X$, $\varphi(a + \zeta_1 a_1 + \cdots + \zeta_m a_m)$ is a polynomial in the complex variables ζ_j, and therefore runs over a finite dimensional affine subspace of Y; since Ω is finitely open in Y, $\{(\zeta_1, \ldots, \zeta_m) : a + \zeta_1 a_1 + \cdots + \zeta_m a_m \in \varphi^{-1}(\Omega)\}$ is an open subset of $\mathbb{C}^m$.

Given $a \in X$, $x \in X$, we have $\varphi(a + \zeta x) = b + \zeta b_1 + \cdots + \zeta^m b_m$; since $(\zeta_1, \ldots, \zeta_m) \mapsto f(b + \zeta_1 b_1 + \cdots + \zeta_m b_m)$ is analytic for $b + \zeta_1 b_1 + \cdots + \zeta_m b_m \in \Omega$, $\zeta \mapsto f \circ \varphi(a + \zeta x)$ is analytic for $a + \zeta x \in \varphi^{-1}(\Omega)$, by the composition property in 2.1(D)a.

On the contrary, if φ is a polynomial map $Y \to Z$, both Y and Z endowed with s.c.l.c. topologies: $f \in \mathcal{G}(\Omega, Y)$ does not imply $\varphi \circ f \in \mathcal{G}(\Omega, Z)$, unless φ is continuous, as we shall see later on (Th. 3.1.10); in other words, if U is an open set in $\mathbb{C}^m$, $f \in \mathcal{A}(U, Y)$ does not imply $\varphi \circ f \in \mathcal{A}(U, Z)$, unless φ is continuous.

In fact, let $U = \Delta(0, 1)$; choose a bounded sequence $(b_k)_{k \in \mathbb{N}}$ in Y such that $b_0, b_1, \ldots, b_k$ are linearly independent for all $k \in \mathbb{N}$ (a choice which is not always possible: see the following Remark). Then, by 2.1(B),

$$f_n(\xi) = \sum_{k \in \mathbb{N}} b_{k+n} \xi^k, \qquad |\xi| < 1, n \in \mathbb{N},$$

defines $f_n \in \mathscr{A}(U, Y)$. Assume that φ is linear, each $\varphi \circ f_n \in \mathscr{A}(U, Z)$, and set $\varphi \circ f_0(\xi) = \sum_{k \in \mathbb{N}} c_k \xi^k$: since $f_0(\xi) = b_0 + \xi b_1 + \cdots + \xi^{n-1} b_{n-1} + \xi^n f_n(\xi)$, we have $\varphi \circ f_0(\xi) = \varphi(b_0) + \xi \varphi(b_1) + \cdots + \xi^{n-1} \varphi(b_{n-1}) + \xi^n \varphi \circ f_n(\xi)$ for all $n \in \mathbb{N}$, therefore $\varphi(b_k) = c_k$ for all $k \in \mathbb{N}$, which by the inequalities of Cauchy implies the boundedness of the sequence $\varphi(b_k/2^k)$. But, by the assumption of linear independence, the sequence $(b_k/2^k)$ may be completed into a Hamel basis of Y, and the images, under a linear map $\varphi: Y \to Z$, of the elements of this Hamel basis may be arbitrarily chosen in Z, so that the sequence $\varphi(b_k/2^k)$ may be unbounded. □

Remark 2.3.11. A bounded sequence (b_k) in Y such that $b_0, b_1, \ldots, b_k$ are linearly independent for all $k \in \mathbb{N}$ certainly exists if Y has an infinite dimension and is metrizable, for one can choose, first the sequence (y_k) with the property of linear independence, and then the $\alpha_k \in \mathbb{C}^*$ such that $b_k = \alpha_k y_k$ tends to the origin.

It also exists if Y is an infinite product of l.c. spaces Y_i, $i \in I$, with the seminorms $p_i \circ pr_i$, where $p_i \in \operatorname{csn}(Y_i)$ and pr_i is the projection $Y \to Y_i$: in fact, one can choose a sequence (Y_{i_k}) of distinct factors in the product, and $b_k \in Y$ with $pr_{i_k} b_k \neq 0$, $pr_i b_k = 0$ for all $i \neq i_k$.

On the contrary, a bounded sequence (b_k) with the property of linear independence cannot exist if the topology of Y is the finest l.c. one, defined by all possible seminorms: if it did exist, it could be completed into a Hamel basis by other linearly independent vectors b_i', $i \in I$, each $y \in Y$ could be uniquely written as

$$y = \sum_{k \in \mathbb{N}} \beta_k(y) b_k + \sum_{i \in I} \beta_i'(y) b_i'$$

with only a finite number of nonvanishing $\beta_k(y)$ and $\beta_i'(y)$; the seminorm $p = \sum_{k \in \mathbb{N}} k |\beta_k|$ would be unbounded on the sequence (b_k).

2.4 Boundedness and continuity of Gâteaux-analytic maps

In this section, l.c. topologies are given on both spaces X and Z. In view of the expansion (1) in Theorem 2.3.5, we begin with polynomial maps $X \to Z$;

their k-homogeneous parts were defined by formula (3) in Proposition 2.2.10.

Proposition 2.4.1. *Let φ be a polynomial map $X \to Z$.*

a) *For each $a \in X$, the following properties are equivalent:*
(i) *for all $q \in \mathrm{csn}(Z)$, $q \circ \varphi$ is bounded on some neighbourhood of a;*
(ii) *for all $q \in \mathrm{csn}(Z)$, $q \circ \varphi$ is continuous at the point a;*
(iii) *φ is continuous at the point a.*
b) *for all $q \in \mathrm{csn}(Z)$, $q \circ \varphi$ continuous at one point implies $q \circ \varphi$ continuous everywhere; consequently, φ continuous at one point implies φ continuous everywhere.*
c) *φ is (everywhere) continuous if and only if its k-homogeneous parts are (everywhere) continuous.*

Proof. a) In order to show that (i) implies (iii), write formula (3) in Proposition 2.2.10 for the polynomial map $x \mapsto \varphi(a + x)$: $\varphi(a + x) = \varphi(a) + \sum_{k=1}^{n} \varphi_k(x)$, where each φ_k is a k-homogeneous polynomial map $X \to Z$, namely $\varphi_k = \dfrac{1}{k!}(\hat{D}_a^k \varphi)$; then, by the generalized inequalities of Cauchy:

$$q \circ \varphi_k(x) \leqslant \sup_{\theta \in \mathbb{R}} q \circ \varphi(a + e^{i\theta}x) \quad \text{for all } k \in \{1, \ldots, n\}.$$

Thus $q \circ \varphi \leqslant M$ on $a + V$, where V is a balanced neighbourhood of the origin in X, implies $q \circ \varphi_k \leqslant M$ on V for all $k \in \{1, \ldots, n\}$ and $q[\varphi(a + x) - \varphi(a)] \leqslant M(\theta + \theta^2 + \cdots + \theta^n)$ for all $x \in \theta V, 0 < \theta < 1$.
b) Let $q \circ \varphi$ be continuous at the point a, i.e. $q \circ \varphi \leqslant M$ on $a + V$ for some balanced neighbourhood V of the origin in X. By the proof of a) and with the same notation:

$$q \circ \varphi_k \leqslant M \text{ on } V \text{ for all } k \in \{1, \ldots, n\} \qquad \text{and therefore}$$

$$q \circ \varphi \leqslant M(1 + r + r^2 + \cdots + r^n) \qquad \text{on } a + rV.$$

But $r > 0$ can be chosen so that $a + rV$ is a neighbourhood of an arbitrarily given point b.
c) If $a = 0$, the φ_k in the proof of b) are the k-homogeneous parts of φ. $\square$

Proposition 2.4.2. *Let Ω be open in X, Z s.c., $f \in \mathscr{G}(\Omega, Z)$, $a \in \Omega$.*

a) *The following properties are equivalent:*
(i) *f is continuous at the point a;*
(ii) *for all $q \in \mathrm{csn}(Z)$, $q \circ f$ is continuous at the point a;*

(iii) *for all $q \in \mathrm{csn}(Z)$, $q \circ f$ is bounded on some neighbourhood of a;*

(iv) *for all $q \in \mathrm{csn}(Z)$, there is a $p \in \mathrm{csn}(X)$ such that $\dfrac{1}{k!} q \circ (\hat{D}_a^k f) \leqslant p^k$ for*

all $k \in \mathbb{N}^*$ [which by Prop. 2.4.1, implies the continuity of each $(\hat{D}_a^k f)$].

(v) *for all $q \in \mathrm{csn}(Z)$, the functions $\dfrac{1}{k!} q \circ (\hat{D}_a^k f)$, $k \in \mathbb{N}^*$, are bounded*

together on some neighbourhood of the origin in X.
b) *If X is a Baire space: $(\hat{D}_a^k f)$ continuous for all $k \in \mathbb{N}^*$ is also equivalent to the above properties.*

Proof. a) (iii) means the existence of a $p_0 \in \mathrm{csn}(X)$ such that $p_0(x) \leqslant 1$ implies $x \in \omega(a)$ (which is now open) and $q \circ f(a + x) \leqslant M$, therefore $\dfrac{1}{k!} q \circ (\hat{D}_a^k f)(x) \leqslant M$ for all $k \in \mathbb{N}^*$ by the generalized inequalities of Cauchy.
Then, for any $x \in X$, if $M \geqslant 1$: $\dfrac{1}{k!} q \circ (\hat{D}_a^k f)(x) \leqslant M[p_0(x)]^k \leqslant [M p_0(x)]^k$,
i.e. (iv) with $p = M p_0$. So it only remains to show that (v) implies (i).

Let $V \subset \Omega - a$ be a balanced neighbourhood of the origin in X: by formula (1) in Theorem 2.3.5, $\dfrac{1}{k!} q \circ (\hat{D}_a^k f) \leqslant M$ for all $k \in \mathbb{N}^*$ on V implies

$q[f(a + x) - f(a)] \leqslant M \dfrac{\theta}{1 - \theta}$ for all $x \in \theta V$, $0 < \theta < 1$.

b) $g(x) = \sup\limits_{k \in \mathbb{N}^*} \dfrac{1}{k!} q \circ (\hat{D}_a^k f)(x)$ is finite for all $x \in \omega(a)$ since the expansion (1) in Theorem 2.3.5 is summable; $(\hat{D}_a^k f)$ continuous for all $k \in \mathbb{N}^*$ entails g lower semi-continuous. As an open set in X, $\omega(a)$ is Baire and there is a set $b + V \subset \omega(a)$, where V is a balanced neighbourhood of the origin in X, such that $g \leqslant M$ on $b + V$ or $q \circ (\hat{D}_a^k f)(b + x) \leqslant M k!$ for all $k \in \mathbb{N}^*$, $x \in V$, hence also

$\qquad q \circ (\hat{D}_a^k f)(x) \leqslant M k! \quad$ for all $k \in \mathbb{N}^*$, $x \in V$,

by the translation lemma 2.2.11.b.   □

Remarks 2.4.3. (A) f bounded on some neighbourhood of the point a is equivalent to the properties in a) if and only if the topology of Z can be defined by a norm, i.e. the origin has bounded neighbourhoods: if not, the identity map $Z \to Z$ is continuous but not locally bounded.
(B) Statement b) may no longer hold without the Baire assumption, and once more counter-examples are easily obtained in the space $X = c_{00}(\mathbb{N})$.
With the notation $x = (\xi_0, \xi_1, \xi_2, \ldots)$, $\varphi(x) = \sum\limits_{n=1}^{\infty} (n\xi_n)^n$ is a map $X \to \mathbb{C}$

such that $\zeta \mapsto \varphi(a + \zeta x)$ is a polynomial with unbounded degree, and $(\hat{D}_0^k \varphi)(x) = k! \, (k\xi_k)^k$: for the topology on X defined by $\|x\| = \sup_{n \in \mathbb{N}} |\xi_n|$, each $(\hat{D}_0^k \varphi)$ is continuous, but φ is not continuous since $\varphi(x) = 1$ and $\|x\| = \dfrac{1}{n}$ are compatible.

Statement b) is a first step in the proof of the next theorem, which is due to Max Zorn [Zo] in the case of Banach spaces, and again may no longer hold without the Baire assumption. A counter-example in $X = c_{00}(\mathbb{N})$, with the notation as above, is $f(x) = \sum_{n=1}^{\infty} \xi_0^n \xi_n$: each term in this series is a continuous function $X \to \mathbb{C}$, the sum $f \in \mathscr{G}(X, \mathbb{C})$; f is continuous on the open unit ball of X since the series converges normally for $\|x\| \leqslant r < 1$, but discontinuous at the point $(1, 0, 0, \ldots)$ since $\dfrac{1}{n}\left(1 + \dfrac{1}{\sqrt{n}}\right)^n$ tends to $+\infty$ with n.

Theorem 2.4.4. *Let Ω be open in the Baire space X, $f \in \mathscr{G}(\Omega, Z)$.*

a) *If f is the pointwise limit of a sequence of continuous maps $g_n \colon \Omega \to Z$, then f is continuous too.*
b) *If Ω is connected and f continuous at one point $\in \Omega$, then f is continuous at all points $\in \Omega$.*

Proof. a) In what follows, $x \in \Omega$, $q \in \mathrm{csn}(Z)$ and $k \in \mathbb{N}^*$ are given; we claim the existence of points $y_0 \in \omega(x)$ (depending on q) such that $q \circ (\hat{D}_x^k f)$ is bounded on some neighbourhood of y_0: by Proposition 2.4.1., this will imply $q \circ (\hat{D}_x^k f)$ continuous everywhere, and this in turn for all $q \in \mathrm{csn}(Z)$ will imply $(\hat{D}_x^k f)$ continuous everywhere; finally, by Proposition 2.4.2.b, the last result for all $k \in \mathbb{N}^*$ will imply f continuous at x.

Since $q \circ f$ is the pointwise limit of the sequence of real functions $q \circ g_n$, by Proposition 1.3.1, Ω contains a meagre set M_0 such that $q \circ f$ is continuous at all points $\in \Omega \backslash M_0$. Now the relation

$$\frac{1}{2\pi} \int_{-\pi}^{\pi} q \circ f(x + e^{i\theta} y) \, d\theta = \lim_{J \to \infty} \frac{1}{J} \sum_{j=1}^{J} q \circ f(x + e^{2\pi i j / J} y)$$

holds for all $y \in \omega(x)$, and $q \circ f$ is continuous at each point $\in [x + \omega(x)] \backslash M_0$; $\bigcup_{r \in \mathbb{Q}} [\omega(x) \cap e^{2\pi i r} (M_0 - x)]$ is a meager subset M_1 of the open set $\omega(x)$, and $y \in \omega(x) \backslash M_1$ implies $q \circ f$ continuous at each point $x + e^{2\pi i r} y$, $r \in \mathbb{Q}$. A fortiori each function $y \mapsto \dfrac{1}{J} \sum_{j=1}^{J} q \circ f(x + e^{2\pi i j / J} y)$ has a continuous restriction to $\omega(x) \backslash M_1$, which is again a Baire space (see 1.3.A).

Then there is another meagre subset $M \supset M_1$ of $\omega(x)$ such that $y \mapsto$ $\frac{1}{2\pi} \int\limits_{-\pi}^{\pi} q \circ f(x + e^{i\theta}y)\, d\theta$ restricted to $\omega(x) \backslash M_1$ is continuous at each point $y_0 \in \omega(x) \backslash M$, which we keep fixed from now on. On account of formula (2) in Theorem 2.3.5, the origin in X has a balanced neighbourhood V such that $y_0 + V + V \subset \omega(x)$ and

$$q \circ (\hat{D}_x^k f)(y) \leqslant \frac{k!}{2\pi} \int\limits_{-\pi}^{\pi} q \circ f(x + e^{i\theta}y)\, d\theta \leqslant A$$

$$\text{for all } y \in (y_0 + V + V) \backslash M_1,$$

and the proof will be over if we obtain $q \circ (\hat{D}_x^k f)(y) \leqslant B$ for all $y \in y_0 + V$.

In fact, given $y \in y_0 + V$, we can find $u \in V$ such that $M_1 \cap \left\{ y + \frac{u}{k+1}, \right.$ $\left. y + \frac{2u}{k+1}, \ldots, y + u \right\} = \emptyset$ and, by Proposition 2.2.7.a, $(\hat{D}_x^k f)(y)$ is a linear combination, with coefficients depending only on k, of the $(\hat{D}_x^k f) \left(y + \frac{ju}{k+1} \right), j = 1, \ldots, k+1$.

b) It is enough to prove that f continuous at $a \in \Omega$ entails f continuous at all points in the open set $a + \omega(a)$. In fact, this will imply that $\{x \in \Omega: f \text{ continuous at } x\}$ is an open subset of Ω, but also a closed one: if $y \in \Omega$ lies in its closure, $x \in y + \frac{1}{2}\omega(y)$ implies $y + (1 - \zeta)(x - y) \in \Omega$ for all $\zeta \in \bar{\Delta}(0, 1)$ or $y \in x + \omega(x)$; since Ω is connected, the proof will be over.

By Proposition 2.4.2.a, the assumption f continuous at a implies the continuity everywhere of all k-homogeneous polynomial maps $(\hat{D}_a^k f)$, hence also of each function g_n defined by

$$g_n(x) = f(a) + \sum_{k=1}^{n} \frac{1}{k!} (\hat{D}_a^k f)(x - a);$$

on $a + \omega(a)$, by Theorem 2.3.5.a, f is the pointwise limit of the sequence (g_n). $\square$

Corollary 2.4.5. *Let Ω be open and balanced in the Baire space X, φ_k a continuous k-homogeneous polynomial map $X \to Z$ for each $k \in \mathbb{N}^*$; if $\{\varphi_k(x): k \in \mathbb{N}^*\}$ is bounded for all $x \in \Omega$, then the series $\sum\limits_{k \in \mathbb{N}^*} \varphi_k$ is summable on Ω to a continuous map $f \in \mathcal{G}(\Omega, Z)$, and $\varphi_k = (\hat{D}_0^k f)/k!$ for all $k \in \mathbb{N}^*$.*

A consequence of Proposition 2.3.6 and Theorem 2.4.4.a.

Proposition 2.4.6. *Let X be metrizable and $f \in \mathcal{G}(\Omega, Z)$: f is continuous at $a \in \Omega$ if and only if $z' \circ f$ is continuous at a for all $z' \in Z'$.*

Proof. Let the latter property hold. For any sequence $(x_n)_{n \in \mathbb{N}} \subset \Omega$ with $\lim x_n = a$, the set $\{z' \circ f(x_n): n \in \mathbb{N}\}$ is bounded for all $z' \in Z'$, hence also (Prop. 1.4.2) the set $\{q \circ f(x_n): n \in \mathbb{N}\}$ for all $q \in \mathrm{csn}(Z)$; then $q \circ f$ is bounded on some neighbourhood of a and f continuous at a by Proposition 2.4.2a. $\square$

Since $z' \circ f \in \mathcal{G}(\Omega, \mathbb{C})$ for all $z' \in Z'$, this result leads us to consider the special case of $\mathcal{G}$-analytic complex valued functions.

Proposition 2.4.7. *Let Ω be open and connected in X, $f \in \mathcal{G}(\Omega, \mathbb{C})$ and non constant.*

a) *$f(\Omega)$ is connected and f is an open mapping.*
b) *If φ is a continuous and locally injective map $f(\Omega) \to \mathbb{C}$, e.g. a non constant analytic function $f(\Omega) \to \mathbb{C}$: f and $\varphi \circ f$ are continuous at the same points $\in \Omega$.*

Proof. a) Since Ω is connected: given $a \in \Omega$, $b \in \Omega$, there are a finite number of points $a_0 = a, a_1, \ldots, a_p = b$ such that Ω contains each line segment $\{(1 - t)a_{q-1} + ta_q: 0 \leqslant t \leqslant 1\}$, $q = 1, \ldots, p$. From this follow that $f(\Omega)$ is connected and also, by Proposition 2.3.7.a, that f cannot be constant on any open nonempty set $\omega \subset \Omega$.

Now assume that, for some $a \in \omega$, $f(\omega)$ is not a neighbourhood of $f(a)$: then, for all $x \in X$, $f(a + \zeta x) = f(a)$ for sufficiently small $|\zeta|$, and f is constant on Ω by Proposition 2.3.7.b.
b) Let $\varphi \circ f$ be continuous at $a \in \Omega$. Since φ is continuous and locally injective, for any sufficiently small closed disc $\bar{\Delta} \subset f(\Omega)$, with centre $f(a)$, there is an $\alpha > 0$ such that $|\varphi - \varphi \circ f(a)| \geqslant \alpha$ on the circumference of $\bar{\Delta}$; the point a has an open connected neighbourhood $\omega \subset \Omega$ where $|\varphi \circ f - \varphi \circ f(a)| < \alpha$, which implies $f(\omega) \subset \Delta$ since $f(\omega)$ is connected. $\square$

The following Proposition and Theorem show that $\mathcal{G}$-analytic scalar functions which are not continuous behave as badly ([Ta]$_2$, Th. 3.5.E) as linear functionals which are not continuous. Again we begin with polynomial maps, which by Proposition 2.4.1.b are continuous either everywhere or nowhere.

Proposition 2.4.8. *Let φ be a polynomial map $X \to \mathbb{C}$: either φ is continuous or $\varphi^{-1}(\alpha)$ is dense for all $\alpha \in \mathbb{C}$; therefore $\varphi^{-1}(\alpha)$ closed for one $\alpha \in \mathbb{C}$ implies φ continuous.*

Proof. By an affine transformation on φ, we may assume $\varphi(a) = 1$, $\varphi \neq 0$ on $a + V$, where V is a balanced neighbourhood of the origin in X; let n

be the degree of φ and $x \in V$. Then $\zeta \mapsto \varphi(a + \zeta x)$ is a polynomial with degree $\leqslant n$, which does not vanish for $|\zeta| \leqslant 1$, $\zeta \to \zeta^n \varphi \left(a + \dfrac{x}{\zeta} \right)$ a polynomial with degree n and leading coefficient 1, which does not vanish for $|\zeta| \geqslant 1$; since all its zeroes have moduli < 1, the coefficient of ζ^k is smaller than C_n^k for all $k \in \{0, \ldots, n-1\}$, and $|\varphi(a+x)| < \sum\limits_{k=0}^{n} C_n^k = 2^n$, i.e. $|\varphi| < 2^n$ on $a + V$. $\square$

Theorem 2.4.9. *Let Ω be open and connected in X, $f \in \mathcal{G}(\Omega, \mathbb{C})$.*

a) *A sufficient (and obviously necessary) condition for the continuity of f at a point $a \in \Omega$ is the existence of a neighbourhood of a in Ω where f omits at least 2 values.*
b) *A sufficient (and obviously necessary) condition for the continuity of f on all Ω is the existence of at least 2 values α, β such that $f^{-1}(\alpha) \cup f^{-1}(\beta)$ is closed in Ω.*

Proof. a) Let V be a balanced neighbourhood of the origin in X. If 2 values are omitted by f on $a + V$, they are omitted on the unit disc by the holomorphic functions $\zeta \mapsto f(a + \zeta x)$, $x \in V$, which therefore form a normal family in the sense of Montel. Since they assume the same value $f(a)$ for $\zeta = 0$, they are bounded together for $\zeta = \dfrac{1}{2}$, which means f bounded on $a + \dfrac{V}{2}$.

The omission of one value is not a sufficient condition since, by Proposition 2.4.7.b, f and e^f are simultaneously continuous. The same counterexample serves for statement b).
b) We may assume that $g = (f - \alpha)(f - \beta)$ is not identically 0; $g^{-1}(0) = f^{-1}(\alpha) \cup f^{-1}(\beta)$ is closed in Ω and, by part a), f is continuous on $\Omega \backslash g^{-1}(0)$. Given $a \in g^{-1}(0)$, we shall prove that f is bounded on some neighbourhood of a.

Choose $b \in X$ so that $(\hat{D}_a^k g)(b) \neq 0$ for some $k \in \mathbb{N}^*$, then choose $r > 0$ so that

$$K = \{a + \zeta b : |\zeta| \leqslant r\} \subset \Omega \text{ and } L = \{a + \zeta b : |\zeta| = r\} \subset \Omega \backslash g^{-1}(0).$$

Since K and L are compact sets, the origin in X has a neighbourhood V such that $K + V \subset \Omega$ and f is bounded, $|f| \leqslant M$, on $L + V$; then, for $x \in a + V$, $f(x + \zeta b)$ is a holomorphic function of ζ on an open set containing the closed disc $\bar{\Delta}(0, r)$, with a modulus $\leqslant M$ on the circumference of this disc, hence also at its centre: $|f(x)| \leqslant M$ for all $x \in a + V$. $\square$

Exercises

2.1.2. Let the entire function $g \in \mathscr{A}(\mathbb{C}, \mathbb{C})$ have only a finite number of zeros. Find all maps $f \in \mathscr{A}(\mathbb{C}, Z)$ such that $\limsup\limits_{\mathbb{C} \ni x \to \infty} \dfrac{q \circ f(x)}{|g(x)|} < \infty$ for all $q \in \mathrm{csn}(Z)$.

2.3.1. Let F be a closed subset of the open set Ω in X, Z s.c.; assume that every finite dimensional affine subspace of X is contained in another one, say E, such that $E \cap F$ is compact and $E \cap (\Omega \backslash F)$ connected. Show that every map $\in \mathscr{G}(\Omega \backslash F, Z)$ has a unique extension $\in \mathscr{G}(\Omega, Z)$.

2.3.3. [Lel]$_2$ Let θ be any Hausdorff topology, on a vector space X, making the translations $x \mapsto x + a$ and the map $\mathbb{C} \times X \ni (\alpha, x) \mapsto \alpha x$ continuous. Show that:
a) if $e_1, \ldots, e_m$ are linearly independent, θ makes the 1-1 map

$$\mathbb{C}^m \ni (\zeta_1, \ldots, \zeta_m) \mapsto a + \zeta_1 e_1 + \cdots + \zeta_m e_m$$

a homeomorphism and the map

$$X \times \mathbb{C}^m \ni (x, \zeta_1, \ldots, \zeta_m) \mapsto x + \zeta_1 e_1 + \cdots + \zeta_m e_m$$

continuous;
b) the topology on X whose open sets are the finitely open sets is the finest topology θ.

2.3.7. Let Ω be a finitely open set in X with the connectedness property in Proposition 2.3.7, $a \in \Omega$: for all $f \in \mathscr{G}(\Omega, Z)$, the sets $f(\Omega)$ and $\{f(a); (\hat{D}_a^k f)(x), k \in \mathbb{N}^*, x \in X\}$ generate the same closed linear subspace of X.

2.4.2. [Col] a) Let the space X be metrizable and the space Z contain a countable family of bounded sets such that every bounded subset of Z is contained in one of them: show that any continuous map $X \to Z$ is locally bounded.
b) Let X be the Banach space $\mathscr{C}([0, 1])$, $Z = \mathbb{C}^{\mathbb{N}^*}$ (the Fréchet space of complex sequences, with the product topology); for $x \in X$, let

$$I(x) = \int_0^1 x(t)\, dt, \quad \varphi(x) = \sum_{n=0}^{\infty} \left[x\left(\frac{1}{2n+2}\right) - x\left(\frac{1}{2n+1}\right) \right]^{2n},$$

$$\xi_k = \varphi[I(x^k) \cdot x] \quad \text{for all } k \in \mathbb{N}^*, \quad f(x) = (\xi_1, \xi_2, \ldots).$$

Show that $f \in \mathscr{G}(\Omega, Z)$, is continuous, bounded on some neighbourhood of the constant 0, unbounded on any neighbourhood of the constant 2.

Chapter 3
Analyticity, or Fréchet-analyticity

Summary

3.1.   Analyticity in the sense of Fréchet: the class $\mathscr{A}(\Omega, Z)$, where Ω is open in the space X and Z s.c.l.c.

Th. 3.1.5
   and 3.1.7. Locally bounded subsets of $\mathscr{A}(\Omega, Z)$.

Th. 3.1.8.   A Banach-Steinhaus theorem for continuous polynomial maps [Lel]$_5$.

Th. 3.1.10.   The composition property for analytic maps.

Prop. 3.2.1. The Hartogs property for a separately analytic map $f \colon \Omega_1 \times \Omega_2 \to Z$, Ω_1 in a Baire space, Ω_2 in a metrizable one.

Prop. 3.2.3. The special case where $f(., x_2)$ is a continuous polynomial map.

Th. 3.2.4.   If $f(x_1, .)$ is analytic for all $x_1 \in \Omega_1$, then $f(., x_2)$ $\mathscr{G}$-analytic (resp.: analytic) for x_2 in a nonmeagre subset of Ω_2 implies the same for all $x_2 \in \Omega_2$ [Ng].

Def. 3.2.6.   Analyticity in the sense of Fantappiè.

Th. 3.2.7.   Relations between Fantappiè-and Fréchet-analyticity [Te].

Prop. 3.3.2. Necessary and sufficient conditions for the existence of an entire map $X \to Z$ with a given homogeneous polynomial expansion around the origin.

Def. and
   Th. 3.3.5 (p, q)-radius of boundedness of an entire map $X \to Z$.

Prop. 3.3.7. The Liouville theorem.

Th. 3.3.9.   An extension theorem for maps of exponential type [Lel]$_5$.

Th. 3.3.10.   General types of growth [Lel]$_5$.

Def. 3.4.1.   Bounding sets [Di]$_2$.

Th. 3.4.3.   If X is the union of an increasing sequence of open sets V_n, every bounding subset of X is contained in the closed convex hull of some V_n [Di]$_5$.

Th. 3.4.6.   Bounding sets in $l^\infty(A)$ [Di]$_1$.

Chapter 3
Analyticity, or Fréchet-analyticity

3.1 Equivalent definitions

In this section, X and Z are l.c. spaces, Z is s.c., Ω an open set in X.

Definition 3.1.1. *A map $f \in \mathscr{G}(\Omega, Z)$ is analytic, or $f \in \mathscr{A}(\Omega, Z)$, if f has the equivalent properties: f continuous; $q \circ f$ continuous for all $q \in \mathrm{csn}(Z)$; $q \circ f$ locally bounded for all $q \in \mathrm{csn}(Z)$.*

The properties are equivalent by Proposition 2.4.2.a; their first consequences are: $\mathscr{A}(\Omega, \mathbb{C})$ is a ring and $\mathscr{A}(\Omega, Z)$ an $\mathscr{A}(\Omega, \mathbb{C})$-module; $\mathscr{A}$ and $\mathscr{G}$ have the same sheaf property; a uniform limit of analytic maps is again analytic.

Proposition 3.1.2. *The properties*

(i) $f \in \mathscr{A}(\Omega, Z)$,
(ii) $z' \circ f \in \mathscr{A}(\Omega, \mathbb{C})$ *for all* $z' \in Z'$,
(iii) $f \in \mathscr{G}(\Omega, Z)$ *and* $f(K)$ *bounded for every compact set* $K \subset \Omega$,
are equivalent in the following 2 cases:

a) X *metrizable;*
b) X *inductive limit of a sequence of Hausdorff l.c. spaces* X_n *in which the origin has a neighbourhood relatively compact in* X_{n+1}.

Proof. Without any assumption on X:

(i) implies (ii) since $f \in \mathscr{G}(\Omega, Z)$ is equivalent to $z' \circ f \in \mathscr{G}(\Omega, \mathbb{C})$ for all $z' \in Z'$;
(ii) implies $z' \circ f(K)$ bounded for all $z' \in Z'$ or $f(K)$ bounded by Proposition 1.4.2. Now we assume (iii) and prove (i) in cases a) and b).

a) See Proposition 2.4.6.
b) For each n, $f|_{\Omega \cap X_n} \in \mathscr{G}(\Omega \cap X_n, Z)$ and, by Proposition 1.5.1.b, it is enough to check $f|_{\Omega \cap X_n} \in \mathscr{A}(\Omega \cap X_n, Z)$. Let $a \in \Omega \cap X_n$; if a neighbourhood V_n of the origin in X_n is relatively compact in X_{n+1}, there is a $\lambda_n > 0$ such that $a + \lambda_n V_n$ is included in a compact subset of $\Omega \cap X_{n+1}$, also of Ω. $\quad \square$

The equivalence between (i) and (ii) fails without a suitable assumption on X (or Z, for it is obvious if Z is endowed with a weakened topology). For instance, if Z is an infinite dimensional Banach space with the norm topology, and X the same space with the weakened one, the identity map $X \to Z$ is $\mathscr{G}$-analytic but not continuous, which, by the way, also shows that the weakened topology is not metrizable.

Proposition 3.1.3. *A map* $f: \Omega \to Z$ *is analytic if and only if, for each* $a \in \Omega$, *there is a continuous linear map* $\lambda_a: X \to Z$ *which is the first order derivative of* f *at* a *in the following sense. With each* $q \in \mathrm{csn}(Z)$ *can be associated* $p \in \mathrm{csn}(X)$ *(depending only on* a, q*) such that, for all* $\varepsilon > 0$: $q[f(a + x) - f(a) - \lambda_a(x)] \leqslant \varepsilon p(x)$ *for sufficiently small* $p(x)$. *This* λ_a *is unique, namely* $\lambda_a = \hat{D}_a^1 f$.

Proof. First of all: if λ is a linear map $X \to Z$ and the inequality $q \circ \lambda(x) \leqslant \varepsilon p(x)$ holds for sufficiently small $p(x)$, it holds for all $x \in X$; if moreover p depends only on $q \in \mathrm{csn}(Z)$, $q \circ \lambda(x) = 0$ for all $x \in X$ and also for all $q \in \mathrm{csn}(Z)$, proving that λ_a, if it does exist, is unique.

Let $f \in \mathscr{G}(\Omega, Z)$ be continuous at the point $a \in \Omega$. By Proposition 2.4.2.a: $(\hat{D}_a^1 f)$ is a continuous linear map $X \to Z$ and, given $q \in \mathrm{csn}(Z)$, there is a $p \in \mathrm{csn}(X)$ such that $p(x) \leqslant 1$ implies $a + x \in \Omega$ and $\dfrac{1}{k!} q \circ (\hat{D}_a^k f)(x) \leqslant M$ for all $k \in \mathbb{N}^*$; then, by formula (1) in Theorem 2.3.5, $p(x) \leqslant \alpha < 1$ implies

$$q[f(a + x) - f(a) - (\hat{D}_a^1 f)(x)] \leqslant M \sum_{k=2}^{\infty} \alpha^k = \frac{M\alpha^2}{1 - \alpha}$$

and $p(x) = \alpha \leqslant \dfrac{\varepsilon}{M + \varepsilon}$ implies $q[f(a + x) - f(a) - (\hat{D}_a^1 f)(x)] \leqslant \varepsilon p(x)$.

Conversely, let λ_a exist for all $a \in \Omega$. Given $a \in \Omega$, since p depends only on q:

$$q\left[\frac{f(a + \zeta x) - f(a)}{\zeta} - \lambda_a(x)\right] \to 0 \quad \text{as } 0 \neq \zeta \to 0$$

for all $x \in X$ and $q \in \mathrm{csn}(Z)$; this means $f|_{E \cap \Omega}$ holomorphic for any complex line E in X, or $f \in \mathscr{G}(\Omega, Z)$. Moreover $q[f(a + x) - f(a)] \to 0$ for all $q \in \mathrm{csn}(Z)$ as x goes to the origin in X, f is continuous at the point a. $\square$

Corollary 3.1.4. *Let* X *be a normed space and* Z *a Banach space. Then a map* $f: \Omega \to Z$ *is analytic if and only if it is differentiable at each point* $\in \Omega$.

Theorem 3.1.5. a) *For all* $f \in \mathscr{A}(\Omega, Z)$, $a \in \Omega$, *the expansion*

$$f(a + x) = f(a) + \sum_{k \in \mathbb{N}^*} \frac{1}{k!} (\hat{D}_a^k f)(x), \qquad x \in \omega(a),$$

is uniformly summable for x in any compact subset of $\omega(a)$.

b) *If $f \in \mathscr{A}(\Omega, Z)$, then $\mathscr{A}(\Omega, Z)$ also contains each generalized derivative $D^k f$, $k \in \mathbb{N}^*$, defined for a given $x_0 \in X$ by $D^k f(a) = (\hat{D}_a^k f)(x_0)$, and each linear map $\mathscr{A}(\Omega, Z) \ni f \mapsto D^k f \in \mathscr{A}(\Omega, Z)$ is continuous for the topology of compact convergence, defined by the seminorms $f \mapsto \sup_K (q \circ f)$, $q \in \mathrm{csn}(Z)$, K compact $\subset \Omega$.*

c) *Let $\mathscr{B} \subset \mathscr{A}(\Omega, Z)$ be such that $\sup_{f \in \mathscr{B}} (q \circ f)$ is locally bounded for all $q \in \mathrm{csn}(Z)$; then, for all $k \in \mathbb{N}^*$, $\sup_{f \in \mathscr{B}} (q \circ D^k f)$ is locally bounded for all $q \in \mathrm{csn}(Z)$, $\mathscr{B}$ and $D^k \mathscr{B}$ are equicontinuous; therefore, if a sequence $(f_n) \subset \mathscr{B}$ converges pointwise to f, it actually converges uniformly on compact sets, and $f \in \mathscr{A}(\Omega, Z)$.*

Proof. a) Given $q \in \mathrm{csn}(Z)$ and $b \in \omega(a)$, we can find $\lambda > 1$ and a balanced neighbourhood V of the origin in X such that $\lambda(b + V) \subset \omega(a)$ and $q \circ f \leqslant M$ on $a + \lambda e^{i\theta}(b + V) = a + \lambda(e^{i\theta} b + V)$ for all $\theta \in \mathbb{R}$; then, by the generalized inequalities of Cauchy, $\frac{1}{k!} q \circ (\hat{D}_a^k f) \leqslant M$ on $\lambda(b + V)$ or $\frac{1}{k!} q \circ (\hat{D}_a^k f) \leqslant \frac{M}{\lambda^k}$ on $b + V$.

b) $D^k f \in \mathscr{G}(\Omega, Z)$ by Theorem 2.3.5.b and, by the same inequalities, $q \circ f \leqslant M$ on $a + V \subset \Omega$, where V is balanced and convex, implies $\frac{1}{k!} q \circ (\hat{D}_b^k f)(x_0) = \frac{1}{k!} (q \circ D^k f)(b) \leqslant M$ if $b \in a + \frac{V}{2}$, $x_0 \in \frac{V}{2}$: $q \circ D^k f$ is locally bounded for all $q \in \mathrm{csn}(Z)$.

Given a compact set $K \subset \Omega$, if x_0 is taken sufficiently near the origin in X, $K_0 = \{a + \zeta x_0 : a \in K, |\zeta| \leqslant 1\}$ is another compact subset of Ω, and

$$\frac{1}{k!} \sup_K (q \circ D^k f) \leqslant \sup_{K_0} (q \circ f).$$

c) With the same M and for all $f \in \mathscr{B}$: $q \circ f \leqslant M$ on $a + V \subset \Omega$ implies $\frac{1}{k!} q \circ (\hat{D}_a^k f) \leqslant M$ on V for all $k \in \mathbb{N}^*$ and

$$q[f(a + x) - f(a)] \leqslant M \frac{\theta}{1 - \theta}$$

$$\text{for all } f \in \mathscr{B}, \qquad x \in \theta V, \qquad 0 < \theta < 1.$$

The equicontinuity of the sequence (f_n) implies: on the one hand, its uniform convergence on compact sets, hence $f \in \mathscr{G}(\Omega, Z)$; on the other hand, f continuous. $\square$

Proposition 3.1.6. *Conversely, let* $\mathscr{B} \subset \mathscr{A}(\Omega, Z)$ *be equicontinuous.*

a) *If each connected component of* Ω *contains at least one point* x *such that* $\{f(x): f \in \mathscr{B}\}$ *is bounded, then* $\sup_{f \in \mathscr{B}} q \circ f$ *is locally bounded for all* $q \in \mathrm{csn}(Z)$.

b) *For all* $k \in \mathbb{N}^*$, $\sup_{f \in \mathscr{B}} q \circ D^k f$ *is locally bounded for all* $q \in \mathrm{csn}(Z)$, *hence* $D^k \mathscr{B}$ *equicontinuous.*

Proof. See Proposition 2.1.6.

Theorem 3.1.7. *Let* $\mathscr{B} \subset \mathscr{A}(\Omega, Z)$ *be such that* $\sup_{f \in \mathscr{B}} (q \circ f)$ *is locally bounded for all* $q \in \mathrm{csn}(Z)$.

a) *Let either* X *be metrizable and* Z *s.r. (see Remark 1.4.3) or* Z *be a reflexive Banach space. Given a sequence* $(f_n) \subset \mathscr{B}$, *there is a* $g \in \mathscr{A}(\Omega, Z)$ *with the property: for each* $z' \in Z'$, *each compact set* $K \subset \Omega$ *and a subsequence* (f_{n_k}) *depending on* z' *and* K, $z' \circ (g - f_{n_k}) \to 0$ *uniformly on* K. *More generally, given a filter* $\mathscr{F}$ *on* $\mathscr{B}$, *there is a* $g \in \mathscr{A}(\Omega, Z)$ *with the property: for each* $z' \in Z'$, *each compact set* $K \subset \Omega$ *and* $\varepsilon > 0$, *the set* $\{f \in \mathscr{B}: \sup_K |z' \circ (g - f)| \leq \varepsilon\}$ *meets all sets* $\in \mathscr{F}$.

   If only one g *has the above property, then* $g = \lim f_n$ *or* $g = \lim_{\mathscr{F}} f$ *for the weakened topology, uniformly on compact sets.*

b) *Let* μ *be a real or complex measure on a compact metric space* T. *If* $f(.,t) \in \mathscr{B}$ *for all* $t \in T$ *and* $f(x, .)$ *is continuous for all* $x \in \Omega$: *then the integral formula* $F(x) = \int f(x, t) \, d\mu(t)$ *defines a* $F \in \mathscr{A}(\Omega, Z)$; $D^k f(., t)$ *denoting the generalized derivatives of* $f(., t)$, $D^k f(a, .)$ *is continuous and* $D^k F(a) = \int D^k f(a, t) \, d\mu(t)$ *for all* $a \in \Omega$, $k \in \mathbb{N}^*$.

   Note that one may take $\mathscr{B} = \{f(., t): t \in T\}$ *if the data are* f *continuous on* $\Omega \times T$, $f(., t) \in \mathscr{A}(\Omega, Z)$ *for all* $t \in T$.

Proof. a) If we repeat the proof of Theorem 2.1.5.b, one statement only remains unproved: by Corollary 2.1.4 we have $g \in \mathscr{G}(\Omega, Z)$ and not $g \in \mathscr{A}(\Omega, Z)$. But $\sup_K |z' \circ g - z' \circ f| \leq \varepsilon$ for some continuous $z' \circ f$ proves that $g(K)$ is bounded for each compact set $K \subset \Omega$, hence $g \in \mathscr{A}(\Omega, Z)$ if X is metrizable by Proposition 3.1.2. If Z is a Banach space: each point $\in \Omega$ has a neighbourhood in Ω where $\|f\| \leq M$, $|z' \circ f| \leq M \|z'\|$ for all $f \in \mathscr{B}$, $z' \in Z'$, therefore $|z' \circ g| \leq M \|z'\|$ for all $z' \in Z'$, $\|g\| \leq M$.

b) F is the pointwise limit of the sequence (F_n) used in the proof of Proposition 1.2.3, which satisfies $q \circ F_n \leqslant \|\mu\| \sup_{f \in \mathscr{B}} (q \circ f)$ for all $q \in \operatorname{csn}(Z)$; then $F \in \mathscr{A}(\Omega, Z)$ by Proposition 3.1.5.c and the rest follows from Theorem 2.1.5.c applied to

$$F(a + \zeta x) = \int f(a + \zeta x, t) \, d\mu(t), \qquad a + \zeta x \in \Omega,$$

since $D^k F(a)$ [resp.: $D^k f(a, t)$] is the kth derivative for $\zeta = 0$ of $\zeta \mapsto F(a + \zeta x_0)$ [resp.: $f(a + \zeta x_0, t)$]. $\square$

Now we prove a *Banach-Steinhaus theorem* for continuous polynomial maps, which will be improved later on (Th. 5.4.2) in the special case of a Banach space X.

Theorem 3.1.8. *Let $\mathscr{B}$ be a family of continuous polynomial maps $X \to Z$ and $q \in \operatorname{csn}(Z)$ have the property*
(N) *the set $N = \{x \in X : \sup_{\varphi \in \mathscr{B}} [q \circ \varphi(x)]^{1/d \circ \varphi} < \infty\}$ is not meagre; e.g. has a nonempty interior, if X is a Baire space.*

a) *If the polynomial maps $\in \mathscr{B}$ are homogeneous, there is a $p \in \operatorname{csn}(X)$ such that $q \circ \varphi \leqslant p^{d \circ \varphi}$ for all $\varphi \in \mathscr{B}$.*
b) *In the general case, there are a number M and a $p \in \operatorname{csn}(X)$ such that $q \circ \varphi \leqslant [\sup(p, M)]^{d \circ \varphi}$ for all $\varphi \in \mathscr{B}$: if N is not meagre, then $N = X$.*

Proof. $F^{(m)} = \{x \in X : q \circ \varphi(x) \leqslant m^{d \circ \varphi}$ for all $\varphi \in \mathscr{B}\}$ is closed in X for all $m \in \mathbb{N}^*$ and some $F^{(m)}$ has a nonempty interior since $\bigcup_{m \in \mathbb{N}^*} F^{(m)}$ is the nonmeagre set N: there are $a \in X$, $m \in \mathbb{N}^*$, $p_0 \in \operatorname{csn}(X)$ such that

(1) $p_0(x) \leqslant 1$ implies $q \circ \varphi(a + e^{i\theta} x) \leqslant m^{d \circ \varphi}$ for all $\varphi \in \mathscr{B}, \theta \in \mathbb{R}$.

a) In this case, by Proposition 2.2.11.b: $p_0(x) \leqslant 1$ also implies $q \circ \varphi(x) \leqslant m^{d \circ \varphi}$, hence $q \circ \varphi \leqslant (m p_0)^{d \circ \varphi}$ everywhere, for all $\varphi \in \mathscr{B}$, and a) is proved with $p = m p_0$
b) Let a, m, p_0 be as above and $\varphi(a + y) = \varphi(a) + \sum_{k=1}^{d \circ \varphi} \varphi_k(y)$, where each φ_k is a continuous k-homogeneous polynomial map: by the generalized inequalities of Cauchy, from (1) above follows that $p_0(y) \leqslant 1$ implies $q \circ \varphi_k(y) \leqslant m^{d \circ \varphi}$ for all $\varphi \in \mathscr{B}, k \in \mathbb{N}^*$; then $p_0(y) \leqslant \lambda, \lambda > 1$, implies

$$q \circ \varphi(a + y) < \frac{\lambda}{\lambda - 1} (\lambda m)^{d \circ \varphi} \quad \text{for all } \varphi \in \mathscr{B}.$$

If we choose $\lambda \geqslant 1 + p_0(a)$, $p_0(x) \leqslant 1$ implies

$$q \circ \varphi(x) < \frac{\lambda}{\lambda - 1}(\lambda m)^{d \circ \varphi} \leqslant \left(\frac{\lambda^2 m}{\lambda - 1}\right)^{d \circ \varphi} \quad \text{for all } \varphi \in \mathscr{B};$$

by Proposition 2.2.11,b) is proved with $M = \dfrac{\lambda^2 m}{\lambda - 1}$, $p = M p_0$. $\square$

Corollary 3.1.9. a) *If the continuous polynomial maps $\in \mathscr{B}$ are homogeneous and each seminorm $\in \mathrm{csn}(Z)$ has property* (**N**) *[resp.: and Z is a Banach space with a norm satisfying* (**N**)*], then $\mathscr{B}$ is equicontinuous at the origin (resp.: on some neighbourhood of the origin) in X.*
b) *If the continuous polynomial maps $\in \mathscr{B}$ have bounded degrees and each seminorm $\in \mathrm{csn}(Z)$ satisfies* (**N**)*, then* $\sup_{\varphi \in \mathscr{B}} (q \circ \varphi)$ *is locally bounded for all $q \in \mathrm{csn}(Z)$, hence $\mathscr{B}$ equicontinuous.*

Proof of a). If Z is a Banach space: the above proof gives a $p \in \mathrm{csn}(X)$ such that $\|\varphi(x)\| \leqslant [p(x)]^{d \circ \varphi}$ for all $\varphi \in \mathscr{B}$, $x \in X$, and this entails $\mathscr{B}$ equi-continuous for $p(x) < 1$ (Th. 3.1.5.c). $\square$

Theorem 3.1.10. *The class of analytic maps has the composition property. More precisely, if Ω is open in X and Γ open in another s.c.l.c. space Y: $f \in \mathscr{G}(\Omega, Y)$, $f(\Omega) \subset \Gamma$ and $g \in \mathscr{A}(\Gamma, Z)$ imply $g \circ f \in \mathscr{G}(\Omega, Z)$ and $\hat{D}_a^1(g \circ f) = (\hat{D}_{f(a)}^1 g) \circ (\hat{D}_a^1 f)$ for all $a \in \Omega$.*

Proof. Let us prove the following equivalent (by 2.1(D) statement. If U is an open set in $\mathbb{C}$: $f \in \mathscr{A}(U, Y)$, $f(U) \subset \Gamma$ and $g \in \mathscr{A}(\Gamma, \mathbb{C})$ imply $g \circ f \in \mathscr{A}(U, \mathbb{C})$ and $(g \circ f)'(\alpha) = (\hat{D}_{f(\alpha)}^1 g)(f'(\alpha))$ for all $\alpha \in U$.
Let $U \supset \bar{\Delta}(\alpha, \rho)$: for $\xi \in \bar{\Delta}(\alpha, \rho)$, the expansion $f(\xi) = \sum\limits_{k=0}^{\infty} c_k(\xi - \alpha)^k$ is uniformly summable, $\varphi_n(\xi) = \sum\limits_{k=0}^{n} c_k(\xi - \alpha)^k \in \Gamma$ for sufficiently large n, and $g \circ \varphi_n \to g \circ f$ uniformly on $\bar{\Delta}(\alpha, \rho)$ because g is continuous.
On the other hand: $(\zeta_1, \ldots, \zeta_n) \mapsto g(c_0 + \zeta_1 c_1 + \cdots + \zeta_n c_n)$ is analytic for $c_0 + \zeta_1 c_1 + \cdots + \zeta_n c_n \in \Gamma$, with 1st order partial derivatives $(\hat{D}_{c_0}^1 g)(c_j)$, $j = 1, \ldots, n$, at the origin; therefore $g \circ \varphi_n \in \mathscr{A}[\Delta(\alpha, \rho), \mathbb{C}]$ for sufficiently large n, with $(g \circ \varphi_n)'(\alpha) = (\hat{D}_{c_0}^1 g)(c_1)$.
The necessity of the assumption "g continuous" has already been shown in the proof of Proposition 2.3.10. $\square$

Example 3.1.11. *Let A be an uncountable infinite set and $X = c_0(A)$ the Banach space, with the sup norm, of all maps $x: A \to \mathbb{C}$ such that $\{\alpha \in A: |x(\alpha)| > \varepsilon\}$ is finite for all $\varepsilon > 0$. For every analytic function f on a balanced open set Ω, there is a countable subset D of A such that, for x and y in Ω, $x(\alpha) = y(\alpha)$ for all $\alpha \in D$ implies $f(x) = f(y)$.*

Proof. By Theorem 2.3.5.a, it is enough to consider a continuous n-homogeneous polynomial map φ. Let $|\varphi| \leqslant M$ on the unit ball and λ be the symmetric n-linear map $X^n \to \mathbb{C}$ associated to φ by the relation $\varphi(x) = \lambda(x,\ldots,x)$ (Th. 2.2.9.c); also let $x_\alpha = \mathbf{1}_{\{\alpha\}}$ for all $\alpha \in A$.

First let $\alpha_1, \ldots, \alpha_m$ be distinct elements of A: in the n-homogeneous polynomial $\varphi(\zeta_1 x_{\alpha_1} + \cdots + \zeta_m x_{\alpha_m})$, the coefficient of ζ^k, $k \in \mathbb{N}^m$, $|k| = n$, is

$$\frac{n!}{k!} \lambda \quad (x_{\alpha_1} \text{ written } k_1 \text{ times}, \ldots, x_{\alpha_m} \text{ written } k_m \text{ times}),$$

from which follows by the Gutzmer formula

$$\sum_{\substack{k \in \mathbb{N}^m \\ |k|=n}} \left(\frac{n!}{k!}\right)^2 |\lambda(x_{\alpha_1} \text{ written } k_1 \text{ times}, \ldots, x_{\alpha_m} \text{ written } k_m \text{ times})|^2 \leqslant M^2$$

since $\|e^{i\theta_1} x_{\alpha_1} + \cdots + e^{i\theta_m} x_{\alpha_m}\| = 1$ for all $(\theta_1,\ldots,\theta_m) \in \mathbb{R}^m$.

With another notation, for the family of real numbers $|\lambda(x_{\alpha_1},\ldots,x_{\alpha_n})|^2$ indexed by A^n, all finite sums are $\leqslant M^2$; therefore the family is summable and the set of its nonzero terms at most countable: there is a countable subset D of A such that $\lambda(x_{\alpha_1},\ldots,x_{\alpha_n}) \neq 0$ only if $\alpha_1, \ldots, \alpha_n \in D$. Then

$$\lambda(x_1,\ldots,x_n) = \lambda(\mathbf{1}_D x_1,\ldots,\mathbf{1}_D x_n)$$

if $x_1, \ldots, x_n$ are finite linear combinations of the x_α, and also in the general case since λ is continuous by formula (2) in 2.2.9. $\square$

3.2 Separate analyticity

In this section, X_1, X_2, Z are l.c. spaces, Z s.c., Ω an open set in $X = X_1 \times X_2$; $\Omega_1(x_2)$ and $\Omega_2(x_1)$ have the same meaning as in Proposition 2.3.8 proving that $\mathscr{G}$-analyticity enjoys the Hartogs property. This is no longer true, in general, for analyticity.

In fact, let $X_1 = X_2 = c_{00}$ with the notation $x_1 = (\xi_1, \xi_2,\ldots)$, $x_2 = (\eta_1, \eta_2,\ldots)$, $\|x_1\| = \sup_{n \in \mathbb{N}^*} |\xi_n|$, $\|x_2\| = \sup_{n \in \mathbb{N}^*} |\eta_n|$: then $f(x_1, x_2) = \sum_{n=1}^{\infty} \xi_n \eta_n$ is a continuous linear function of x_1 or x_2 since $|f(x_1,x_2)| \leqslant \|x_1\| \sum_{n=1}^{\infty} |\eta_n|$ or $\|x_2\| \sum_{n=1}^{\infty} |\xi_n|$; but $f(x_1,x_2) = 1$ when, in each sequence x_1 or x_2, the first n numbers are $1/\sqrt{n}$ and the others 0. This f is discontinuous everywhere, in accordance with Proposition 2.4.1.b since f is a polynomial map $X_1 \times X_2 \to \mathbb{C}$ with degree 2.

Since separate analyticity implies $\mathcal{G}$-analyticity (Prop. 2.3.8), the question is: under what topological assumption does it imply continuity? In view of the above counter-example, continuity at some points $\in \Omega$ (obtained below under the assumption that X_1 or X_2 is a Baire space) may be considered a satisfactory result; by Theorem 2.4.4.b, continuity on all Ω will follow if Ω is connected and $X = X_1 \times X_2$ a Baire space (which implies that X_1, X_2 are Baire spaces too, since M_1 meagre in X_1 implies $M_1 \times X_2$ meagre in $X_1 \times X_2$).

As we did for $\mathcal{G}$-analyticity in Proposition 2.3.9, we shall also consider situations in which $f(., x_2)$ is assumed analytic for some x_2 only.

Proposition 3.2.1. *Let X_1 (or X_2) be a Baire space, X_2 (or X_1) metrizable, Z a Banach; let $\Omega = \bigcup_{x_1 \in \Omega_1} \{x_1\} \times \Omega_2(x_1)$, where each $\Omega_2(x_1)$ is connected and contains a nonempty open set ω_2 independent of $x_1 \in \Omega_1$. If a map $f : \Omega \to Z$ satisfies $f(., x_2) \in \mathcal{A}(\Omega_1, Z)$ for all $x_2 \in \omega_2$ and $f(x_1, .) \in \mathcal{A}(\Omega_2(x_1), Z)$ for all $x_1 \in \Omega_1$, then $f \in \mathcal{G}(\Omega, Z)$ and f is bounded, hence continuous, on some nonempty open subset of $\Omega_1 \times \omega_2$.*

Proof. $f \in \mathcal{G}(\Omega, Z)$ by Proposition 2.3.9. Replacing $\Omega_2(x_1)$ by ω_2 makes the assumptions symmetric; so let X_1 be Baire and X_2 metrizable, with distance d_2. Given $a_2 \in \omega_2$, for each $m \in \mathbb{N}^*$ and $n \in \mathbb{N}^*$ such that $d(a_2, \complement\omega_2) > \dfrac{1}{n}$:

$$F_1^{(m,n)} = \{x_1 \in \Omega_1 : d_2(a_2, x_2) \leqslant \frac{1}{n} \text{ implies } \|f(x_1, x_2)\| \leqslant m\}$$

is a closed set in Ω_1; since Ω_1 is the union of the $F_1^{(m,n)}$, some $F_1^{(m,n)}$ has a nonempty interior. $\square$

The simplest statement seems to be

Corollary 3.2.2. *If X_1 and X_2 are Fréchet spaces, Z l.c.: $\mathcal{A}(\Omega, Z)$ contains any map $f : \Omega \to Z$ which is separately analytic, i.e. $f(., x_2) \in \mathcal{A}(\Omega_1(x_2), Z)$ for all $x_2 \in X_2$ and $f(x_1, .) \in \mathcal{A}(\Omega_2(x_1), Z)$ for all $x_1 \in X_1$. See also Theorem 3.2.7.*

Proof. Since $X = X_1 \times X_2$ is also a Fréchet space, it is enough (Prop. 3.1.2.a) to obtain $z' \circ f \in \mathcal{A}(\Omega, \mathbb{C})$ for all $z' \in Z'$, which (Th. 2.4.4.b) follows from $z' \circ f$ bounded on some nonempty open subset of each connected component of Ω: this is included in Proposition 3.2.1. $\square$

Weaker assumptions are sufficient if $f(., x_2)$ is a polynomial map for all $x_2 \in \Omega_2$:

Proposition 3.2.3. *Let X_1 be a Baire space, $X_1 \times X_2$ metrizable, Z l.c., Ω_2 an open set in X_2: $\mathscr{A}(X_1 \times \Omega_2, Z)$ contains any map $f\colon X_1 \times \Omega_2 \to Z$ such that $f(., x_2)$ is a continuous polynomial map for all $x_2 \in \Omega_2$, with a bounded degree, and $f(x_1, .) \in \mathscr{A}(\Omega_2, Z)$ for all $x_1 \in X_1$.*

Proof. Let K_2 be any compact set in Ω_2. The family $\mathscr{B} = \{f(., x_2)\colon x_2 \in K_2\}$ of continuous polynomial maps $X_1 \to Z$ satisfies the assumptions of Corollary 3.1.9.b: in fact, each $q \in \mathrm{csn}(Z)$ satisfies (N) with $N = X_1$ since, for any given x_1, $q \circ f(x_1, x_2)$ is a continuous function of $x_2 \in \Omega_2$, hence bounded for $x_2 \in K_2$. Then $q \circ f$ is bounded on $K_1 \times K_2$ for any pair of compact sets $K_1 \subset X_1$, $K_2 \subset \Omega_2$, and we conclude by Proposition 3.1.2.a. $\square$

Let Ω_1, Ω_2 be open sets in X_1, X_2, Ω_2 connected, Z a Banach space and $f\colon \Omega_1 \times \Omega_2 \to Z$. By Proposition 3.2.1, if X_1, X_2 are Fréchet spaces (for instance) and $f(x_1, .) \in \mathscr{A}(\Omega_2, Z)$ for all $x_1 \in \Omega_1$, then $f(., x_2) \in \mathscr{A}(\Omega_1, Z)$ for x_2 in an open subset of Ω_2 eventually implies $f(., x_2) \in \mathscr{A}(\Omega_1, Z)$ for all $x_2 \in \Omega_2$. We shall now replace an open subset of Ω_2 by a nonmeagre subset of Ω_2 and obtain the same result under an additional assumption: $f(x_1, .)$ bounded for all $x_1 \in \Omega_1$.

Theorem 3.2.4. *Let Ω_1, Ω_2 be open sets in X_1, X_2, Ω_2 connected, N_2 a nonmeagre subset of Ω_2, Z a reflexive Banach space and f a map $\Omega_1 \times \Omega_2 \to Z$ such that $f(x_1, .) \in \mathscr{A}(\Omega_2, Z)$ and $\|f(x_1, .)\| \leqslant M(x_1)$ for all $x_1 \in \Omega_1$.*

a) *If $\mathbb{C} \times X_2$ is a Baire space: $f(., x_2) \in \mathscr{G}(\Omega_1, Z)$ for all $x_2 \in N_2$ implies $f(., x_2) \in \mathscr{G}(\Omega_1, Z)$ for all $x_2 \in \Omega_2$.*
b) *If X_1 and $\mathbb{C} \times X_2$ are Baire spaces: $f(., x_2) \in \mathscr{A}(\Omega_1, Z)$ for all $x_2 \in N_2$ implies $f(., x_2) \in \mathscr{A}(\Omega_1, Z)$ for all $x_2 \in \Omega_2$.*

Proof. a) Since the intersections of Ω_1 with complex lines in X_1 are concerned, we may replace Ω_1 by a connected open set U in $\mathbb{C}$, write ζ instead of x_1 and delete the index 2 for simplicity. Let V be the largest open subset of U such that f is analytic on $V \times \Omega$: we shall prove $V \neq \emptyset$, then $V = U$, from which a) will follow.
1) Our first step towards $V \neq \emptyset$ is the existence of a nonempty open subset ω of U such that f is bounded on $\omega \times \Omega$. Setting

(*) $F_1^{(m)} = \{\zeta \in U\colon \|f(\zeta, .)\| \leqslant m\}$,

so that $\mathrm{U} = \bigcup\limits_{m \in \mathbb{N}^*} F_1^{(m)}$ by the assumption, we claim that each $F_1^{(m)}$ is closed in U; let $\alpha \in U$ be the limit of a sequence $(\zeta_n) \subset F_1^{(m)}$. Since Z is reflexive, by Theorem 3.1.7.a there is a $g \in \mathscr{A}(\Omega, Z)$ with the property: for each $z' \in Z'$, $x \in \Omega$, $z' \circ g(x)$ is the limit of a subsequence $z' \circ f(\zeta_{n_k}, x)$. This implies: on the one hand, $|z' \circ g(x)| \leqslant m\|z'\|$ for all $z' \in Z'$, $x \in \Omega$ or $\|g\| \leqslant m$; on the other hand, $z' \circ g(x) = z' \circ f(\alpha, x)$ for all $z' \in Z'$, $x \in N$ (since $f(., x)$ is continuous for $x \in N$) or $g = f(\alpha, .)$ on N.

Since both g and $f(\alpha, .)$ are analytic on Ω and N nonmeagre, $g = f(\alpha, .)$ holds on all Ω, which proves $\alpha \in F_1^{(m)}$ (our claim) and also determines g. By Theorem 3.1.7.a again, the uniqueness of g proves that $f(\alpha, x) = \lim f(\zeta_n, x)$ for the weakened topology, i.e. $z' \circ f(., x)$ continuous on $F_1^{(m)}$ for all $z' \in Z'$, $x \in \Omega$. Now some $F_1^{(m)}$ contains a nonempty open set ω, such that $\|f\| \leqslant m$ on $\omega \times \Omega$, and moreover $z' \circ f(., x)$ is continuous on ω for all $z' \in Z', x \in \Omega$.
2) Next we show that $f(., x) \in \mathscr{A}(\omega, Z)$ for all $x \in \Omega$ or $z' \circ f(., x) \in \mathscr{A}(\omega, \mathbb{C})$ for all $z' \in Z'$, $x \in \Omega$. Let $\bar{\Delta}(\alpha, \rho) \subset \omega$; since $\mathbb{R} \ni \theta \mapsto z' \circ f(\alpha + \rho e^{i\theta}, x)$ is continuous, its modulus $\leqslant m\|z'\|$, by Theorem 3.1.7.b

$$h(\zeta, x) = \frac{1}{2\pi} \int_{-\pi}^{\pi} \frac{z' \circ f(\alpha + \rho e^{i\theta}, x)}{1 - \dfrac{\zeta - \alpha}{\rho} e^{-i\theta}} d\theta$$

is an analytic function of (ζ, x) for $|\zeta - \alpha| < \rho$, $x \in \Omega$; but $h(\zeta, x) = z' \circ f(\zeta, x)$ for $|\zeta - \alpha| < \rho$, $x \in N$, consequently for $|\zeta - \alpha| < \rho$, $x \in \Omega$, since both members of the last relation are analytic functions of $x \in \Omega$.

Putting the results of 1) and 2) together, we have f bounded and $\mathscr{G}$-analytic (Prop. 2.3.8) on $\omega \times \Omega$, i.e. $f \in \mathscr{A}(\omega \times \Omega, Z)$: $V \neq \emptyset$ is proved.
3) Now assume $V \subsetneq U$, choose $\alpha \in V$ and $\rho > 0$ so that the open disc $\Delta(\alpha, \rho)$ is contained in U but not in V; let $\Delta(\alpha, r)$ be the largest disc with centre α contained in V. The coefficient of $(\zeta - \alpha)^k$ in the Taylor expansion of $\zeta \mapsto f(\zeta, x)$ around α, namely

$$c_k(x) = \frac{1}{2\pi(r/2)^k} \int_{-\pi}^{\pi} e^{-ik\theta} f\left(\alpha + \frac{r}{2} e^{i\theta}, x\right) d\theta,$$

is an analytic function of $x \in \Omega$ by Theorem 3.1.7.b (Note); moreover, for $x \in N$, $f(., x)$ is analytic on U, hence $\limsup_{k \to \infty} \|c_k(x)\|^{1/k} \leqslant \dfrac{1}{\rho}$ by the inequalities of Cauchy;

$$F_2^{(n)} = \left\{x \in \Omega: \|c_k(x)\|^{1/k} \leqslant \frac{2}{r + \rho} \text{ for all } k \geqslant n\right\}$$

is a closed subset of Ω and $N \subset \bigcup_{n \in \mathbb{N}^*} F_2^{(n)}$.

Now consider the open set $G(N) = \{x \in \Omega: N \cap W \text{ meagre for some neighbourhood } W \text{ of } x \text{ in } \Omega\}$. The union of the closed sets $F_2^{(n)} \cap \complement G(N)$, which are closed in Ω, contains $N \cap \complement G(N)$, which is nonmeagre by Proposition 1.3.2; so this union cannot be meagre, i.e. some $F_2^{(n)} \cap \complement G(N)$ contains a nonempty open set Ω', which can be contracted so that c_k is bounded on Ω' for each $k < n$.

The series $\sum_{k=0}^{\infty} (\zeta - \alpha)^k c_k(x)$ is uniformly summable for $|\zeta - \alpha| \leqslant \dfrac{r + \rho}{2} - \varepsilon$, $\varepsilon > 0$, $x \in \Omega'$, to a function analytic on $\Delta\left(\alpha, \dfrac{r + \rho}{2}\right) \times \Omega'$ and bounded on $\Delta\left(\alpha, \dfrac{r + \rho}{2} - \varepsilon\right) \times \Omega'$, which coincides with $f(\zeta, x)$ on $\Delta\left(\alpha, \dfrac{r + \rho}{2}\right) \times (N \cap \Omega')$, hence on $\Delta\left(\alpha, \dfrac{r + \rho}{2}\right) \times \Omega'$ since $N \cap \Omega'$ is not meagre.

Finally f is analytic on $\Delta\left(\alpha, \dfrac{r + \rho}{2}\right) \times \Omega'$ and bounded on $\Delta\left(\alpha, \dfrac{r + \rho}{2} - \varepsilon\right) \times \Omega'$, hence $\mathscr{G}$-analytic on $\Delta\left(\alpha, \dfrac{r + \rho}{2}\right) \times \Omega$ by Proposition 2.3.9, and actually analytic since $\mathbb{C} \times X_2$ is a Baire space: $V \supset \Delta\left(\alpha, \dfrac{r + \rho}{2}\right)$, a contradiction which proves $V = U$ and completes the proof of a).

b) Thanks to the additional assumption that X_1 is Baire, we only have to check that each connected component of Ω_1 contains a nonempty open set ω_1 such that f is bounded on $\omega_1 \times \Omega_2$, and this is achieved by the argument already used in the first part of the proof of a): define $F_1^{(m)}$ by (*) with x_1 instead of ζ; for $a_1 \in \Omega_1 \cap F_1^{(m)}$, if X_1 is not assumed metrizable, consider the sets $A(V_1) = \{f(x_1, .) : x_1 \in F_1^{(m)} \cap V_1\}$, where V_1 is any neighbourhood of a_1 in Ω_1, and the filter on $\mathscr{B} = \{f_2 \in \mathscr{A}(\Omega_2, Z) : \|f_2\| \leqslant m\}$ which has $\{A(V_1)\}$ as a basis. $\square$

Corollary 3.2.5. *Let Ω_1, Ω_2 be open sets in X_1, X_2, Ω_2 connected, N_2 a nonmeagre subset of another open set $\Omega'_2 \subset \Omega_2$, Z l.c. and f a map $\Omega_1 \times \Omega_2 \to Z$ such that $f(x_1, .) \in \mathscr{G}(\Omega_2, Z)$ and $\{f(x_1, x_2) : x_2 \in \Omega'_2\}$ is bounded, hence $f(x_1, .) \in \mathscr{A}(\Omega'_2, Z)$, for all $x_1 \in \Omega_1$.*

a) *If $\mathbb{C} \times X_2$ is a Baire space: $f(., x_2) \in \mathscr{G}(\Omega_1, Z)$ for all $x_2 \in N_2$ implies $f(., x_2) \in \mathscr{G}(\Omega_1, Z)$ for all $x_2 \in \Omega_2$.*
b) *If X_1, X_2 are Fréchet spaces: $f(., x_2) \in \mathscr{A}(\Omega_1, Z)$ for all $x_2 \in N_2$ implies $f \in \mathscr{A}(\Omega_1 \times \Omega_2, Z)$.*

Proof. Both results are properties of the $z' \circ f$, $z' \in Z'$, so we may replace Z by $\mathbb{C}$ and apply the Theorem. By Proposition 1.3.2, some connected component of Ω'_2 may be substituted for Ω'_2.

a) The proof of Theorem 3.2.4.a gives f analytic on $U \times \Omega'_2$; then f is $\mathscr{G}$-analytic on $U \times \Omega_2$ by Proposition 2.3.9, locally bounded on $U \times \Omega'_2$, hence analytic on $U \times \Omega_2$ by Theorem 2.4.4.b.

b) f is $\mathscr{G}$-analytic on $\Omega_1 \times \Omega_2$ and, by the proof of Theorem 3.2.4.b, each connected component of Ω_1 contains a nonempty open set ω_1 such that f is bounded on $\omega_1 \times \Omega'_2$. $\square$

The fact that $\mathscr{G}$-analyticity enjoys the Hartogs property (Prop. 2.3.8), but not the composition property in general, whereas analyticity (in the sense of Fréchet) enjoys the second one (Prop. 3.1.10) but not the first one in general, is incentive to define some sort of analyticity having both properties without any restriction. A relevant definition, developed by Silva [Sil], goes back to Luigi Fantappiè [Fa].

Definition 3.2.6. *Let X and Z be s.c.l.c. spaces, Ω an open set in X. A map $f : \Omega \to Z$ is $\mathscr{LF}$-analytic if $h \in \mathscr{A}(U, X)$, $h(U) \subset \Omega$, where U is an open set in $\mathbb{C}$, imply $f \circ h \in \mathscr{A}(U, Z)$.*

f is $\mathscr{LF}$-analytic $\Omega \to Z$ if and only if $z' \circ f$ is $\mathscr{LF}$-analytic $\Omega \to \mathbb{C}$ for all $z' \in Z'$; the set of $\mathscr{LF}$-analytic functions $\Omega \to \mathbb{C}$ is a ring and the set of $\mathscr{LF}$-analytic maps $\Omega \to Z$ a module over this ring. $\mathscr{LF}$-analyticity also has the sheaf property.

Theorem 3.2.7. a) *The class of $\mathscr{LF}$-analytic maps enjoys both the composition and the Hartogs properties without any restriction.*

b) *Analytic implies $\mathscr{LF}$-analytic, and this in turn implies $\mathscr{G}$-analytic, without any converse statement in general, on account of a); nevertheless:*

c) *Let X be metrizable: then any $\mathscr{LF}$-analytic linear map $X \to Z$ is continuous; more generally, an m-linear map $\lambda: X^m \to Z$ is continuous if each linear map $X \to Z$ obtained from λ by assigning values to all variables but one is $\mathscr{LF}$-analytic.*

d) *If X is a metrizable Baire space: any $\mathscr{LF}$-analytic map $X \to Z$ is analytic.*

From a) and d) follows a slight modification of Corollary 3.2.2, with $X_1 \times X_2$ a metrizable Baire space instead of X_1 and X_2 Fréchet spaces.

Proof. a) For the composition property: let Γ be open in another s.c.l.c. space Y, $f : \Omega \to \Gamma$, $g: \Gamma \to Z$; if f and g are $\mathscr{LF}$-analytic, then $h \in \mathscr{A}(U, X)$, $h(U) \subset \Omega$, imply first $f \circ h \in \mathscr{A}(U, Y)$, $f \circ h(U) \subset \Gamma$, and secondly $(g \circ f) \circ h = g \circ (f \circ h) \in \mathscr{A}(U, Z)$.

For the Hartogs property: let X_1, X_2 be s.c.l.c. spaces, Ω open in $X = X_1 \times X_2$, f a map $\Omega \to \mathbb{C}$ which is separately $\mathscr{LF}$-analytic, i.e. $f(., x_2)$ $\mathscr{LF}$-analytic $\Omega_1(x_2) \to \mathbb{C}$ for all $x_2 \in X_2$, $f(x_1, .)$ $\mathscr{LF}$-analytic $\Omega_2(x_1) \to \mathbb{C}$ for all $x_1 \in X_1$. A map $h \in \mathscr{A}(U, X)$ such that $h(U) \subset \Omega$ is a couple of maps

$h_1 \in \mathscr{A}(U, X_1), h_2 \in \mathscr{A}(U, X_2)$ such that $(h_1(\zeta), h_2(\zeta)) \in \Omega$ for all $\zeta \in U$; given $\alpha \in U$, let $\Delta(\alpha, \rho) \subset U$ be such that $\zeta_1, \zeta_2 \in \Delta(\alpha, \rho)$ implies $(h_1(\zeta_1), h_2(\zeta_2)) \in \Omega : (\zeta_1, \zeta_2) \mapsto f[h_1(\zeta_1), h_2(\zeta_2)]$ is separately analytic, hence analytic, on $\Delta(\alpha, \rho) \times \Delta(\alpha, \rho)$. Then $f \circ h$ is analytic on $\Delta(\alpha, \rho)$.

b) If $f \in \mathscr{A}(\Omega, Z): h \in \mathscr{A}(U, X), h(U) \subset \Omega$, imply $f \circ h \in \mathscr{A}(U, Z)$ since the class of analytic maps has the composition property.

If f is only $\mathscr{LF}$-analytic: $f \circ h \in \mathscr{A}(U, Z)$, whenever h is the restriction of an affine function $\mathbb{C} \to X$ to an open set U such that $h(U) \subset \Omega$, which means f $\mathscr{G}$-analytic.

c) Since X is metrizable, we may: take $Z = \mathbb{C}$ in the proofs of c) and d) (Prop. 2.4.6); assume the distance d on X given by the formula in 1.1.A and test the continuity at a point x of a map $X \to \mathbb{C}$ by considering only sequences $(x_k)_{k \in \mathbb{N}}$ tending to x such that $2^k d(x_k, x_{k-1}) \leqslant 1$ for all $k \in \mathbb{N}^*$, from which follows that $2^k p(x_k - x_{k-1})$ is bounded for all $p \in \operatorname{csn}(X)$.

Let λ be an $\mathscr{LF}$-analytic linear map $X \to \mathbb{C}$ and $(x_k)_{k \in \mathbb{N}}$ such a sequence. Since

$$h(\zeta) = x_0 + \sum_{k=1}^{\infty} (x_k - x_{k-1})\zeta^k$$

is an analytic function of $\zeta \in \Delta(0, 2)$, so is $\lambda[h(\zeta)] = \sum_{k=0}^{\infty} \alpha_k \zeta^k, \alpha_k \in \mathbb{C}$. Taking $\zeta = 0$ gives $\alpha_0 = \lambda(x_0)$; then

$$\lambda\left[\sum_{k=1}^{\infty} (x_k - x_{k-1})\zeta^{k-1}\right] = \sum_{k=1}^{\infty} \alpha_k \zeta^{k-1}$$

holds for $0 < |\zeta| < 2$, therefore for all $\zeta \in \Delta(0, 2)$ since the left hand member is analytic there. Taking $\zeta = 0$ again gives $\alpha_1 = \lambda(x_1 - x_0)$; and so on indefinitely. Thus we have $\alpha_k = \lambda(x_k - x_{k-1})$ for all $k \in \mathbb{N}^*$, and $\lambda(x_k) = \alpha_0 + \alpha_1 + \cdots + \alpha_k$ tends to $\sum_{k=0}^{\infty} \alpha_k = \lambda[h(1)] = \lambda(x)$ as $k \to \infty$.

Now let λ be an m-linear map $X^m \to \mathbb{C}$ with the property stated in c); in order to prove that λ is continuous at the point $(x^{(1)}, \ldots, x^{(m)})$, we choose sequences $(x_k^{(j)})_{k \in \mathbb{N}}$ $(j = 1, \ldots, m)$ tending to $x^{(j)}$, with the same property as (x_k) above, and set

$$h_j(\zeta_j) = x_0^{(j)} + \sum_{k=1}^{\infty} (x_k^{(j)} - x_{k-1}^{(j)})\zeta_j^k$$

$$= \sum_{k=0}^{\infty} (x_k^{(j)} - x_{k-1}^{(j)})\zeta_j^k$$

with $x_{-1}^{(j)} = 0$; then $\lambda[h_1(\zeta_1), \ldots, h_m(\zeta_m)]$ is separately analytic function of the $\zeta_j \in \Delta(0, 2)$, hence analytic, say $\lambda[h_1(\zeta_1), \ldots, h_m(\zeta_m)] = \sum_{k \in \mathbb{N}^m} \alpha_k \zeta^k$, and the relation

$$\alpha_k = \lambda[x^{(1)}_{k_1} - x^{(1)}_{k_1-1}, \ldots, x^{(m)}_{k_m} - x^{(m)}_{k_m-1}] \quad \text{for all } k \in \mathbb{N}^m$$

is easily obtained by an induction on m. Now

$$\lambda(x^{(1)}_{k_1}, \ldots, x^{(m)}_{k_m}) = \sum_{l \leqslant k} \lambda[x^{(1)}_{l_1} - x^{(1)}_{l_1-1}, \ldots, x^{(m)}_{l_m} - x^{(m)}_{l_m-1}] = \sum_{l \leqslant k} \alpha_l$$

(where $k, l \in \mathbb{N}^m$ and $l \leqslant k$ means $l_j \leqslant k_j$ for all $j \in \{1, \ldots, m\}$) tends to $\sum\limits_{k \in \mathbb{N}^m} \alpha_k = \lambda[h_1(1), \ldots, h_m(1)] = \lambda(x^{(1)}, \ldots, x^{(m)})$ as $k_1, \ldots, k_m \to \infty$.

d) Let f be $\mathscr{LF}$-analytic, hence $\mathscr{G}$-analytic, $\Omega \to \mathbb{C}$; since X is a Baire space, it is sufficient by Proposition 2.4.2.b, even redundant by Theorem 2.4.4.b, to prove $(\hat{D}^m_a f)$ continuous for all $a \in \Omega$, $m \in \mathbb{N}^*$. Moreover, by Theorem 2.2.9.c, each m-homogeneous polynomial map $(\hat{D}^m_a f)$ is related to an m-linear map $\lambda_m: X^m \to \mathbb{C}$ by the formulas

(1) $(\hat{D}^m_a f)(x) = \lambda_m(x, \ldots, x)$ (x written m times) and

(2) $\lambda_m(x_1, \ldots, x_m) = \dfrac{1}{2^m m!} \sum_\varepsilon (-1)^{(m-|\varepsilon|)/2} (\hat{D}^m_a f)(\varepsilon_1 x_1 + \cdots + \varepsilon_m x_m),$

where $\varepsilon = (\varepsilon_1, \ldots, \varepsilon_m) \in \{-1 + 1\}^m$. From (2) follows that, if $(\hat{D}^m_a f)$ is $\mathscr{LF}$-analytic, so is each linear map $X \to \mathbb{C}$ obtained from λ_m by assigning values to all variables but one; from (1), that λ_m continuous entails $(\hat{D}^m_a f)$ continuous. So we shall prove $(\hat{D}^m_a f) \mathscr{LF}$-analytic.

Let U be an open set in $\mathbb{C}$, $h \in \mathscr{A}(U, X)$: in order to prove that $(\hat{D}^m_a f) \circ h$ has a derivative at $\zeta_0 \in U$, we may, thanks to the m-homogeneity of $(\hat{D}^m_a f)$, assume $h(\zeta_0) \in \omega(a)$ (the largest balanced subset of $\Omega - a$). Then, for small enough $|\zeta - \zeta_0|$, we also have $h(\zeta) \in \omega(a)$ and formula (2) in Theorem 2.3.5:

$$\frac{1}{k!} (\hat{D}^m_a f) \circ h(\zeta) = \frac{1}{2\pi} \int_{-\pi}^{\pi} e^{-im\theta} f[a + e^{i\theta} h(\zeta)] \, d\theta;$$

$f[a + e^{i\theta} h(\zeta)]$ may be considered as a function of two complex variables ζ, θ, defined for small enough $|\zeta - \zeta_0|$ and $|\mathrm{Im}\,\theta|$, and separately analytic since f is $\mathscr{LF}$-analytic: then $(\zeta, \theta) \mapsto f[a + e^{i\theta} h(\zeta)]$ is continuous, and the integral an analytic function of ζ for small enough $|\zeta - \zeta_0|$.   □

3.3 Entire maps and functions

Again let X and Z be l.c. spaces, Z s.c.: an entire map is an $f \in \mathscr{A}(X, Z)$, an entire function an $f \in \mathscr{A}(X, \mathbb{C})$. We begin this section by considering the more general case of a balanced open set Ω in X and an $f \in \mathscr{A}(\Omega, Z)$:

the expansion

(1) $f(x) = f(0) + \sum_{k \in \mathbb{N}^*} \frac{1}{k!} (\hat{D}_0^k f)(x)$

is uniformly summable on compact subsets of Ω (Th. 3.1.5.a) and each $(\hat{D}_0^k f)$ is a continuous k-homogeneous polynomial map $X \to Z$.

Conversely, given a sequence $(\varphi_k)_{k \in \mathbb{N}^*}$ of such polynomial maps, the set where the series $\sum_{k \in \mathbb{N}^*} \varphi_k$ is summable is balanced (Ex. 1.2.2); therefore the question is: under what conditions are there a nonempty balanced open set Ω and $f \in \mathscr{A}(\Omega, Z)$ such that

(2) $f = f(0) + \sum_{k \in \mathbb{N}^*} \varphi_k$ or $\varphi_k = \frac{1}{k!}(\hat{D}_0^k f)$ for all $k \in \mathbb{N}^*$?

if such an Ω exists, for all $q \in \mathrm{csn}(Z)$ we have (i) $\sup_{k \in \mathbb{N}^*} q \circ \varphi_k(x) < \infty$ for all $x \in \Omega$ (Prop. 1.2.1.a), which implies $\limsup_{k \to \infty} [q \circ \varphi_k(x)]^{1/k} \leqslant 1$ for all $x \in \Omega$, actually (ii) $\limsup_{k \to \infty} [q \circ \varphi_k(x)]^{1/k} < 1$ for all $x \in \Omega$ since Ω is open, and conversely (ii) implies (i).

If there is an Ω satisfying (i) or (ii) for all $q \in \mathrm{csn}(Z)$, then (Prop. 2.3.6) the series $\sum_{k \in \mathbb{N}^*} \varphi_k$ is summable on Ω to an $f \in \mathscr{G}(\Omega, Z)$ such that (2) holds; but the counter-example in Remark 2.4.3(B) shows that in general $f \notin \mathscr{A}(\Omega, Z)$, unless X is a Baire space. In this case simple answers can be given.

Proposition 3.3.1. *Let X be a Baire space and $(\varphi_k)_{k \in \mathbb{N}^*}$ a sequence of continuous k-homogeneous polynomial maps $X \to Z$.*

a) *Let $\sum$ be the set of $x \in X$ such that $\{\varphi_k(x): k \in \mathbb{N}^*\}$ is bounded: if $\sum$ is a neighbourhood of the origin, then $\overset{\circ}{\sum}$ is the largest open balanced set Ω on which the series $\sum_{k \in \mathbb{N}^*} \varphi_k$ is summable to a map $\in \mathscr{A}(\Omega, Z)$.*

b) *In the special case $Z = \mathbb{C}$: $\overset{\circ}{\sum}$ is nonempty if and only if there is a $p \in \mathrm{csn}(X)$ such that $|\varphi_k| \leqslant p^k$ for all $k \in \mathbb{N}^*$.*

c) *The existence of $f \in \mathscr{A}(X, Z)$ satisfying (2), $\{\varphi_k(x): k \in \mathbb{N}^*\}$ bounded for all $x \in X$, $\lim_{k \to \infty} [q \circ \varphi_k(x)]^{1/k} = 0$ for all $x \in X$, $q \in \mathrm{csn}(Z)$, are three equivalent properties.*

Proof. a) $\sum$ is balanced, $\overset{\circ}{\sum}$ too provided $\overset{\circ}{\sum}$ contains the origin.
b) $\overset{\circ}{\sum} \neq \emptyset$ implies the existence of such a p (Th. 3.1.8.a), and conversely this existence implies $\overset{\circ}{\sum} \supset p^{-1}([0, 1[)$. $\square$
Without the Baire assumption, less simple criteria must be used.

Proposition 3.3.2. *Let* $(\varphi_k)_{k \in \mathbb{N}^*}$ *be a sequence of continuous k-homogeneous polynomial maps* $X \to Z$.

a) *A necessary and sufficient condition for the existence of* $f \in \mathscr{A}(X, Z)$ *satisfying* (2) *is that, given* $q \in \operatorname{csn}(Z)$ *and* $\varepsilon > 0$, *each* $x \in X$ *has a neighbourhood* ω *such that* $\sup_{\omega} (q \circ \varphi_k)^{1/k} \leqslant \varepsilon$ *for k large enough.*

b) *Consequently, let* $(\alpha_k)_{k \in \mathbb{N}^*}$ *be a sequence of complex numbers with* $\sup_{k \in \mathbb{N}^*} |\alpha_k|^{1/k} < \infty$: *if there exists* $f \in \mathscr{A}(X, Z)$ *satisfying* $\varphi_k = (\hat{D}_0^k f)/k!$ *for all* $k \in \mathbb{N}^*$, *there also exists* $g \in \mathscr{A}(X, Z)$ *satisfying* $\alpha_k \varphi_k = (\hat{D}_0^k g)/k!$ *for all* $k \in \mathbb{N}^*$.

c) *The condition* $\sup_K (q \circ \varphi_k)^{1/k} \to 0$ *as* $k \to \infty$, *for all* $q \in \operatorname{csn}(Z)$ *and* K *compact in* X, *is necessary for the existence of* $f \in \mathscr{A}(X, Z)$ *satisfying* (2), *and also sufficient if* X *is metrizable.*

Proof. a) Let (2) hold. Given $q \in \operatorname{csn}(Z)$, $x \in X$ and $\varepsilon > 0$, let $\lambda > \dfrac{1}{\varepsilon}$ and V be a balanced neighbourhood of the origin in X such that $q \circ f \leqslant M$ on $\lambda e^{i\theta}(x + V) = \lambda(e^{i\theta}x + V)$ for all $\theta \in \mathbb{R}$; hence follows, by the generalized inequalities of Cauchy, $q \circ \varphi_k \leqslant M$ on $\lambda(x + V)$ or $q \circ \varphi_k \leqslant \dfrac{M}{\lambda^k}$ on $\omega = x + V$ for all $k \in \mathbb{N}^*$.

Conversely, take $\varepsilon < 1$ in a). Then $\sup_{k \in \mathbb{N}^*} q \circ \varphi_k(x) < \infty$ for all $x \in X$, $q \in \operatorname{csn}(Z)$, hence (2) for $f = \sum_{k \in \mathbb{N}^*} \varphi_k \in \mathscr{G}(X, Z)$ by Proposition 2.3.6; moreover $\sup_{\omega} (q \circ \varphi_k) \leqslant \varepsilon^k$ for k large enough and finite for each k implies that, on ω, $q \circ f$ is the uniform limit of the sequence $q \circ (\varphi_1 + \cdots + \varphi_n)$, therefore bounded. If Z is a Banach space: on ω, f also is the uniform limit of the sequence $(\varphi_1 + \cdots + \varphi_n)$.

c) Let (2) hold: given $\varepsilon > 0$ and $q \in \operatorname{csn}(Z)$, each $x \in X$ has a neighbourhood ω such that $\sup_{\omega} (q \circ \varphi_k)^{1/k} \leqslant \varepsilon$ for k large enough; hence follows $\sup_K (q \circ \varphi_k)^{1/k} \leqslant \varepsilon$ for k large enough. Conversely, this property with $\varepsilon < 1$ implies $\{\varphi_k(x): k \in \mathbb{N}^*\}$ bounded for all $x \in X$ and $f(K)$ bounded, hence $f \in \mathscr{A}(X, Z)$ by Proposition 3.1.2 if X is metrizable. $\square$

Corollary 3.3.3. *Given a sequence* $(x'_k)_{k \in \mathbb{N}^*}$ *in* X', *let* $\varphi_k = x'^k_k$; *for the existence of an* $f \in \mathscr{A}(X, \mathbb{C})$ *satisfying* (2):

a) *a necessary and sufficient condition is* $\lim_{k \to \infty} \langle x, x'_k \rangle = 0$ *for all* $x \in X$ *with the equicontinuity of the sequence* (x'_k);

b) *the condition* $\sup_K |x'_k| \to 0$ *as* $k \to \infty$, *for each compact set* K *in* X, *is necessary, and also sufficient if* X *is metrizable.*

If X is a barrelled space, the condition $\lim_{k \to \infty} \langle x, x'_k \rangle = 0$ for all $x \in X$ implies that $x \mapsto \sup_{k \in \mathbb{N}^*} |\langle x, x'_k \rangle|$ is a continuous seminorm on X, i.e. the sequence (x'_k) equicontinuous.

Remark 3.3.4. In view of Proposition 3.3.2 and Corollary 3.3.3, the following questions seem natural enough.

(Q) If $f \in \mathscr{A}(X, Z)$ and the sequence $(\varphi_k)_{k \in \mathbb{N}^*}$ satisfy (2), given $q \in \mathrm{csn}(Z)$, does there also exist a neighbourhood V of the origin in X such that $\sup_V (q \circ \varphi_k)^{1/k} \to 0$ as $k \to \infty$?

(Q') In particular, if the sequence $(x'_k)_{k \in \mathbb{N}^*}$ in X' is equicontinuous and weak* convergent to the constant 0, does there exist a V such that $\sup_V |x'_k| \to 0$ as $k \to \infty$?

In the special case of an infinite dimensional Banach space X, (Q') can be reformulated as

(Q'') Does $\lim_{k \to \infty} \langle x, x'_k \rangle = 0$ for all $x \in X$ imply, not only $\sup_{k \in \mathbb{N}^*} \|x'_k\| < \infty$, but actually $\lim_{k \to \infty} \|x'_k\| = 0$?

and this last question has received several negative answers.

(A) The first one was given by Angus Taylor [Ta]$_1$ in the special case $X = l^p$ $(1 \leqslant p < \infty)$, the space of sequences $x = (\xi_1, \xi_2, \ldots)$ of complex numbers such that $\sum_{k=1}^{\infty} |\xi_k|^p = \|x\|^p < \infty$, with $\langle x, x'_k \rangle = \xi_k$ for all $k \in \mathbb{N}^*$.

The final negative answer is the Josefson-Nissenzweig theorem [Jo$_2$, Ni]: the adjoint space X' to any infinite dimensional Banach space X contains a sequence $(x'_k)_{k \in \mathbb{N}^*}$ such that $\lim_{k \to \infty} \langle x, x'_k \rangle = 0$ for all $x \in X$ but $\|x'_k\| = 1$ for all $k \in \mathbb{N}^*$. Then the entire function $f(x) = \sum_{k=1}^{\infty} \langle x, x'_k \rangle^k$ is bounded (by $r/1 - r$) for $\|x\| \leqslant r$ for all $r < 1$, but unbounded for $\|x\| \leqslant r$ for all $r > 1$, since $|f(x)| \leqslant M$ for $\|x\| \leqslant r$ would imply $|\langle x, x'_k \rangle|^k \leqslant M$ for $\|x\| \leqslant r$ by the generalized inequalities of Cauchy, hence $\|x'_k\| \leqslant M^{1/k}/r$ for all $k \in \mathbb{N}^*$.

So the negative answer to (Q'') is closely related to another striking fact: an entire function on an infinite dimensional normed space may have a finite radius of boundedness (see Ex. 3.3.6 below).

Since the proofs of the Josefson-Nissenzweig theorem are very difficult, specific proofs for two wide enough cases are not uninteresting.

(B) Let the infinite dimensional normed space X be separable, i.e. have a dense countable subset D. Then, the adjoint space X', with the strong topology, is an infinite dimensional Banach space, in which $\{x' \in X': \|x'\| = 1\}$ is complete but not compact, hence not precompact: one can find, first $\delta > 0$ and a sequence $(x'_n) \subset X'$ with $\|x'_n\| = 1$ for all n and $\|x'_m - x'_n\| \geq \delta$ whenever $m \neq n$, then, by the diagonal argument, a subsequence (x'_{n_k}) such that $\lim \langle x, x'_{n_k} \rangle$ exists for all $x \in D$, hence for all $x \in X$.

Since the sequence (x'_n) is equicontinuous, the subsequence (x'_{n_k}) is weak* convergent to an $x' \in X'$ and $(x'_{n_k} - x')$ to the constant 0, but $\lim_{k \to \infty} \|x'_{n_k} - x'\| = 0$ would contradict $\|x'_{n_k} - x'_{n_{k'}}\| \geq \delta$ whenever $k \neq k'$.

(C) Another specific argument can be used if X is a reflexive Banach space (see Th. 1.1.2). Having chosen the same (x'_n) as in (B) above, consider the closed subspace Y of X' spanned by the x'_n, which is reflexive too, and its adjoint space Y', both with the norm topology. Since the adjoint space to Y' is Y, it is separable, and Y' too by a classical property which we prove in (D) below for completeness; by the result of (B), the sequence (x'_n), as a sequence in the adjoint space to Y', contains a subsequence (x'_{n_k}) such that $\lim \langle y', x'_{n_k} \rangle$ exists for all $y' \in Y'$; since Y' can be algebraically identified with $X/\bigcap (\ker x'_n)$, an equivalent statement is the existence of $\lim \langle x, x'_{n_k} \rangle$ for all $x \in X$, and the argument goes on as in (B).

(D) Let again X be a normed space: if X' is separable for the norm topology, so is X (which, for example, shows that l^1 is not the adjoint space to l^∞). In fact, let $(x'_n)_{n \in \mathbb{N}}$ be a dense subset of $\{x' \in X': \|x'\| = 1\}$ and, for each n, let $x_n \in X$ satisfy $\|x_n\| = 1, |\langle x_n, x'_n \rangle| \geq \dfrac{1}{2}$: if the closed subspace of X spanned by the x_n were not X, there would exist an $x' \in X'$ with $\|x'\| = 1$, $\langle x_n, x' \rangle = 0$ for all n, hence $\|x' - x'_n\| \geq |\langle x_n, x' - x'_n \rangle| \geq \dfrac{1}{2}$ for all n.

Definition and Theorem 3.3.5. *Let* $p \in \mathrm{csn}(X)$, $q \in \mathrm{csn}(Z)$, $a \in X$, $f \in \mathscr{A}(X, Z)$ *and*

$$B(f, p, q) = \{r \geq 0 : q \circ f \text{ bounded on } a + p^{-1}([0, r])\}$$

$$C(f, p, q) = \left\{ r \geq 0 : q\left[f(x) - f(a) - \sum_{k=1}^{n} \frac{1}{k!} \hat{D}_a^k f(x) \right] \to 0 \right.$$

$$\left. \text{as } n \to \infty, \text{ uniformly for } p(x) \leq r \right\}.$$

a) *The upper bound of* $C(f, p, q)$, *known as the* (p, q)-*radius of uniform convergence of the expansion of f around a, is the number* $R(f, p, q)$ *given by*

(3) $s_k(f, p, q) = \sup \left\{ \dfrac{1}{k!} q \circ (\hat{D}_a^k f)(x): p(x) \leqslant 1 \right\}$ *for all* $k \in \mathbb{N}^*$

and

(4) $\dfrac{1}{R(f, p, q)} = \limsup_{k \to \infty} [s_k(f, p, q)]^{1/k}.$

The upper bound of $B(f, p, q)$, known as the (p, q)-radius of boundedness of f around a, is the same number $R(f, p, q)$ if $s_k(f, p, q) < \infty$ for all $k \in \mathbb{N}^*$, and 0 otherwise.

b) *Given $q \in \mathrm{csn}(Z)$, there always exist some $p \in \mathrm{csn}(X)$ such that f has a (p, q)-radius of boundedness > 0 around a given point $\in X$.*

Proof. a) Let a be the origin, $R(f, p, q)$ given by (3) and (4), $r > 0$.

By the generalized inequalities of Cauchy: $q \circ f \leqslant M$ on $p^{-1}([0, r])$ implies $s_k(f, p, q) \leqslant \dfrac{M}{r^k}$ for all $k \in \mathbb{N}^*$, $R(f, p, q) \geqslant r$. Conversely, let $p(x) \leqslant r < r + \varepsilon < R(f, p, q)$; then, for $k > n$, n large enough:

$$[s_k(f, p, q)]^{1/k} \leqslant \frac{1}{r + \varepsilon}, \frac{1}{k!} q \circ (\hat{D}_0^k f)(x) \leqslant \left(\frac{p(x)}{r + \varepsilon} \right)^k \leqslant \left(\frac{r}{r + \varepsilon} \right)^k,$$

hence $r \in C(f, p, q)$ and $r \in B(f, p, q)$ if moreover $s_k(f, p, q) < \infty$ for all $k \in \mathbb{N}^*$.

$r \in C(f, p, q)$ implies $\dfrac{1}{k!} q \circ (\hat{D}_0^k f) \leqslant 1$ on $p^{-1}([0, r])$ or $s_k(f, p, q) \leqslant \dfrac{1}{r^k}$ for k large enough, $R(f, p, q) \geqslant r$.

b) By Proposition 2.4.2.a: $q \circ f$ is bounded on some fundamental neighbourhood $a + p^{-1}([0, r])$ of any point $a \in X$.   □

The dependence on a of the (p, q)-radius around a, which involves plurisubharmonic functions, will be dealt with later on (Section 5.3).

Examples 3.3.6. (**A**) When X is an infinite dimensional normed space, part b) of the Theorem means that every entire map $X \to Z$ has a $(\| \quad \|, q)$-radius of boundedness > 0 for all $q \in \mathrm{csn}(Z)$, in particular every entire function $X \to \mathbb{C}$ has a $(\| \quad \|, | \quad |)$-radius of boundedness > 0, but this may be finite (Rem. 3.3.4(A)), in contrast with the finite dimensional case.

(**B**) When X is a l.c. space, $p \in \mathrm{csn}(X)$, two cases may occur. First let $p^{-1}(0)$ have an infinite codimension and $R \in \,]0, +\infty[$ be given: applying the Josefson-Nissenzweig theorem to the normed space $X/p^{-1}(0)$ (or the completed one), we get a sequence (x'_k) in X' weak* convergent to the constant 0 and such that $\sup\{|\langle x, x'_k\rangle|: p(x) \leqslant 1\} = \dfrac{1}{R}$ for all $k \in \mathbb{N}^*$; then the entire function $f(x) = \sum\limits_{k=1}^{\infty} \langle x, x'_k\rangle^k$ has a $(p, |\quad|)$-radius of boundedness around the origin equal to the given number R.

(C) If, on the contrary, $p^{-1}(0)$ has a finite codimension m, $p^{-1}(0)$ is the set of common zeros of m linearly independent elements x'_j of X', $j = 1, \ldots, m$, and since all norms on $X/p^{-1}(0)$ are equivalent, $\alpha \sup\limits_{1 \leqslant j \leqslant m} |x'_j| \leqslant p \leqslant \beta$. $\sup\limits_{1 \leqslant j \leqslant m} |x'_j|, \alpha > 0$.

Let f be an entire function with a $(p, |\quad|)$-radius of boundedness > 0 around the origin and λ_n the symmetric n-linear map $X^n \to \mathbb{C}$ associated by Theorem 2.2.9 to the n-homogeneous polynomial function $(\hat{D}^n_0 f)$:

$$\frac{1}{n!}|(\hat{D}^n_0 f)(x)| \leqslant s_n(f, p, |\quad|)\left(\beta \cdot \sup_{1 \leqslant j \leqslant m} |\langle x, x'_j\rangle|\right)^n \quad \text{for all } x \in X$$

implies

$$|\lambda_n(x_1, \ldots, x_n)| \leqslant s_n(f, p, |\quad|)\left(\beta n \sup_{\substack{1 \leqslant j \leqslant m \\ 1 \leqslant k \leqslant n}} |\langle x_k, x'_j\rangle|\right)^n$$

$$\text{for all } (x_1, \ldots, x_n) \in X^n.$$

Then $\langle x_1, x'_j\rangle = 0$ for all j implies $\lambda_n(rx_1, x_2, \ldots, x_n)$ bounded as $r \to \infty$ or $\lambda_n(x_1, \ldots, x_n) = 0$, and similarly for the other indeces k: $\lambda_n(x_1, \ldots, x_n)$ depends only on the $\langle x_k, x'_j\rangle$, $(\hat{D}^n_0 f)(x)$ depends only on the $x'_j(x)$, so does $f(x)$.

More precisely: since m points x_k can be chosen in X so that $\langle x_k, x'_j\rangle = \delta_{j,k}$, $f(x)$ is an analytic function of the m complex variables $\langle x, x'_j\rangle$. We conclude that, when $p^{-1}(0)$ has a finite codimension m, the entire functions which have a $(p, |\quad|)$-radius of boundedness > 0 are actually analytic functions of m complex variables, and their $(p, |\quad|)$-radius of boundedness is infinite.

(D) In both cases: given $q \in \mathrm{csn}(Z)$, an entire map $X \to Z$ may have a vanishing (p, q)-radius: for $x' \in X'$, the $(p, |\quad|)$-radius is > 0 if and only if x'/p is bounded, and then it is infinite.

The classical Liouville theorem is easily generalized as follows.

Proposition 3.3.7. *Let* $f \in \mathscr{G}(X, Z), n \in \mathbb{N}, n \leqslant \alpha < n + 1$; *let* $\limsup\limits_{|\zeta| \to \infty} \dfrac{q \circ f(\zeta x)}{|\zeta|^\alpha}$ $< \infty$ *for all* $x \in X$, $q \in \mathrm{csn}(Z)$: *then* f *is constant if* $n = 0$, *is a polynomial map of degree* $\leqslant n$ *if* $n \in \mathbb{N}^*$.

Proof. By the classical inequalities of Cauchy: for all $z' \in Z'$, $\zeta \mapsto z' \circ f(\zeta x)$ is a constant, namely $z' \circ f(0)$, if $n = 0$, is a polynomial of degree $\leqslant n$ if $n \in \mathbb{N}^*$, and this means $z' \circ (\hat{D}_0^k f) \equiv 0$ for all $k > n$.

Definition 3.3.8. *An entire map $f \in \mathscr{A}(X, Z)$ has the exponential type if*

$$t_q(x) = \limsup_{\mathbb{C} \ni \zeta \to \infty} \frac{\ln(q \circ f)(\zeta x)}{|\zeta|} < \infty \quad \text{for all } q \in \mathrm{csn}(Z), x \in X.$$

The next theorem shows that a weaker assumption is sufficient, and that exponential type implies boundedness on bounded sets.

Theorem 3.3.9. *Let Ω be an open neighbourhood of the origin in X, $f \in \mathscr{A}(\Omega, Z)$ and N a nonmeagre subset of Ω with the property: for all $x \in N$, $\zeta \mapsto f(\zeta x)$ has an extension $\in \mathscr{A}(\mathbb{C}, Z)$ satisfying $t_q(x) < \infty$ for all $q \in \mathrm{csn}(Z)$. Then f extends to a map $f \in \mathscr{A}(X, Z)$ of exponential type and, for every $q \in \mathrm{csn}(Z)$, there is a $p \in \mathrm{csn}(X)$ such that $q \circ f \leqslant q \circ f(0) + e^p - 1$ everywhere.*

Proof. First let $x \in N$ be given and $l = t_q(x)$: we claim that, for all $\varepsilon > 0$, $q \circ (\hat{D}_0^k f)(x) < (l + \varepsilon)^k$ *for large enough* k. If it were not so, we could find a subsequence (k_n) of the sequence $\mathbb{N}^*$ such that $q \circ (\hat{D}_0^{k_n} f)(x) \geqslant (l + \varepsilon)^{k_n}$ for each n; from this follows by the generalized inequalities of Cauchy, with

$$r_n = \frac{k_n}{l + \varepsilon}:$$

$$\sup_{|\zeta| = r_n} q \circ f(\zeta x) \geqslant \frac{1}{k_n!} q \circ (\hat{D}_0^{k_n} f)(r_n x) \geqslant \frac{k_n^{k_n}}{k_n!} \geqslant C \frac{e^{k_n}}{\sqrt{k_n}} = C' \frac{e^{(l + \varepsilon) r_n}}{\sqrt{r_n}}$$

where C, C' are constants. This contradicts the assumption since $r_n \to \infty$ with n.

Having proved our claim, we may use the Banach-Steinhaus theorem 3.1.8.a for the family $\mathscr{B}$ of k-homogeneous polynomial maps $(\hat{D}_0^k f)$, $k \in \mathbb{N}^*$: there is a $p \in \mathrm{csn}(X)$ such that $q \circ (\hat{D}_0^k f) \leqslant p^k$ for all $k \in \mathbb{N}^*$, hence follow $q \circ f \leqslant q \circ f(0) + \sum_{k \in \mathbb{N}^*} \frac{p^k}{k!}$ and $f = f(0) + \sum_{k \in \mathbb{N}^*} \frac{1}{k!} (\hat{D}_0^k f) \in \mathscr{A}(X, Z)$ by Proposition 3.3.2.a, since

$$\left[\frac{1}{k!} q \circ (\hat{D}_0^k f) \right]^{1/k} \leqslant \frac{p}{(k!)^{1/k}} \quad \text{for all } k \in \mathbb{N}^*. \quad \square$$

By similar arguments, but with a slightly different formulation, the result extends to more general types of growth: a map $f \in \mathscr{A}(X, Z)$ is said to have the *type of growth* τ if the assumption of the following theorem holds for all $q \in \mathrm{csn}(Z)$; then again B bounded in X implies $f(B)$ bounded in Z.

Theorem 3.3.10. *Let* $\tau(t_1, \ldots, t_m, u)$ *be a positive and separately increasing function of the positive real variables* $t_1, \ldots, t_m, u$; *let* $f \in \mathscr{A}(X, Z), q \in \mathrm{csn}(Z)$ *and a nonmeagre set* N *in* X *have the property: for each* $x \in N$, *there are positive numbers* $t_1, \ldots, t_m$ *such that* $q \circ f(\zeta x) \leqslant \tau(t_1, \ldots, t_m, |\zeta|)$ *for all* $\zeta \in \mathbb{C}$. *Then there also exist positive numbers* $\theta_1, \ldots, \theta_m$ *and* $p \in \mathrm{csn}(X)$ *such that* $q \circ f \leqslant 2\tau(\theta_1, \ldots, \theta_m, p)$ *everywhere.*
   Theorem 3.3.9 deals with the special case $\tau(t_1, t_2, u) = t_1 + e^{t_2 u}$.

Proof. Since the nonmeagre set N is contained in the union of the closed sets $\{x \in X : q \circ f(\zeta x) \leqslant \tau(t_1, \ldots, t_m, |\zeta|) \text{ for all } \zeta \in \mathbb{C}\}, (t_1, \ldots, t_m) \in \mathbb{N}^m$, there are $(\theta_1, \ldots, \theta_m) \in \mathbb{N}^m, a \in X$ and $p_0 \in \mathrm{csn}(X)$ such that $p_0(x - a) \leqslant 1$ implies $q \circ f(\zeta x) \leqslant \tau(\theta_1, \ldots, \theta_m, |\zeta|)$ for all $\zeta \in \mathbb{C}$ and therefore, by the generalized inequalities of Cauchy,

$$\frac{1}{k!} q \circ (\hat{D}_0^k f)(x) \leqslant \frac{1}{r^k} \tau(\theta_1, \ldots, \theta_m, r) \quad \text{for all } k \in \mathbb{N}^*, r > 0.$$

Then, by Proposition 2.2.11, for all $k \in \mathbb{N}^*, x \in X$ we have

$$\frac{1}{k!} q \circ (\hat{D}_0^k f)(x) \leqslant \left[\frac{p_0(x)}{r}\right]^k \tau(\theta_1, \ldots, \theta_m, r)$$

also for any $r > 0$, which makes it possible to take $r = 2p_0(x)$ for a given x; since $q \circ f(0) = q \circ f(0 . a) \leqslant \tau(\theta_1, \ldots, \theta_m, 0)$, we conclude that $q \circ f \leqslant 2\tau(\theta_1, \ldots, \theta_m, 2p_0)$ everywhere. $\square$

3.4 Bounding sets

This notion was introduced by H. Alexander [Al] in his Thesis and shortly afterwards deepened by S. Dineen [Di]$_2$.

Definition 3.4.1. *A set* B *in* X *is* bounding *if* $f(B)$ *is bounded in* $\mathbb{C}$ *for any entire function* f, *therefore also (Prop. 1.4.2) bounded in* Z *for any entire map* $f : X \to Z$.
   If B is bounding, its closure $\bar{B}$, its translated sets $a + B, a \in X$, are bounding too. Any relatively compact subset of X, a finite union of bounding sets, are bounding; bounding implies bounded since (Prop. 1.4.2) $X' \subset \mathscr{A}(X, \mathbb{C})$.

Now let X have an infinite dimension. When X is a normed space, an entire function $X \to \mathbb{C}$ may have an arbitrarily small radius of boundedness around the origin (Ex. 3.3.4.A): therefore an open nonempty subset of X cannot be bounding, a bounded subset may be not bounding. When X is not a normed space, an open nonempty subset of X cannot be bounded. Finally, when X is a s.r. space endowed with the weakened topology (see Rem. 1.4.3), the bounded subsets of X are relatively compact, hence bounding. Since B bounding implies $\bar{B}$ bounding, B bounding also implies B nowhere dense, i.e. the interior of $\bar{B}$ empty.

The following criterion is to be compared with Proposition 3.3.2.

Proposition 3.4.2. a) *A bounded set B is bounding if and only if* $\sup_{B} |\varphi_k|^{1/k} \to 0$ *as $k \to \infty$ whenever the continuous k-homogeneous polynomial maps φ_k: $X \to \mathbb{C}$ satisfy $\varphi_k = (\hat{D}_0^k f)/k!$ for all $k \in \mathbb{N}^*$ for some $f \in \mathscr{A}(X, \mathbb{C})$.*
b) *Consequently, if B is bounding, the balanced hull $\{\zeta x : x \in B, |\zeta| \leqslant 1\}$ of B is bounding too.*

Proof of a) Each continuous φ_k is bounded on some neighbourhood of the origin, hence bounded on the bounded set B; then $\sup_{B} |\varphi_k| \leqslant \left(\dfrac{1}{2}\right)^k$ for sufficiently large k implies $\sup_{B} \left| \sum_{k \in \mathbb{N}^*} \varphi_k \right| < \infty$.

Conversely, let $(\varphi_k)_{k \in \mathbb{N}^*}$ be a sequence of continuous m_k-homogeneous polynomial maps $X \to \mathbb{C} (0 < m_k < m_{k+1})$ such that

(∗)   the series $\sum_{k \in \mathbb{N}^*} \varphi_k$ is summable to some $f \in \mathscr{A}(X, \mathbb{C})$

but $|\varphi_k(x_k)| \geqslant \alpha_k = \beta^{m_k}, \beta > 0$, for some $x_k \in B$; by Proposition 3.3.2.b, we may take $\beta > 1$, and also multiply some φ_k by 0 or complex numbers of modulus 1, without removing property (∗): this we will do step by step.

First step: $|\varphi_1(x_1)| \geqslant \alpha_1$ and $\sum_{k \geqslant k_1} |\varphi_k(x_1)| \leqslant 1$ for some $k_1 > 1$; after multiplying φ_k by 0 for each $k \in]1, k_1[$, we have $|f(x_1)| \geqslant \alpha_1 - 1$, and this will remain after the next steps.

Second step: $|\varphi_{k_1}(x_{k_1})| \geqslant \alpha_{k_1}, |\varphi_1(x_{k_1}) + \varphi_{k_1}(x_{k_1})| \geqslant \alpha_{k_1}$ after multiplication of φ_{k_1} by a suitable number of modulus 1, and $\sum_{k \geqslant k_2} |\varphi_k(x_{k_1})| \leqslant 1$ for some $k_2 > k_1$; after multiplying φ_k by 0 for each $k \in]k_1, k_2[$, we have $|f(x_{k_1})| \geqslant \alpha_{k_1} - 1$, and this remains after the next steps.

$(n + 1)^{\text{th}}$ step: $|\varphi_{k_n}(x_{k_n})| \geqslant \alpha_{k_n}, |(\varphi_1 + \varphi_{k_1} + \cdots + \varphi_{k_n})(x_{k_n})| \geqslant \alpha_{k_n}$

after multiplication of φ_{k_n} by a suitable number of modulus 1, and so on.
Thus we have obtained an $f \in \mathscr{A}(X, \mathbb{C})$ unbounded on B. $\square$

Theorem 3.4.3. *Let $\mathscr{V} = (V_n)$ be an increasing sequence of open sets having X as their union: if B is bounding, there is an n such that $\sup\limits_{B} |f| \leqslant \sup\limits_{V_n} |f|$ for all $f \in \mathscr{A}(X, \mathbb{C})$, and therefore the closed convex hull of V_n contains B.*

Proof. For each increasing sequence $\mathscr{V}$ of nonempty open sets V_n having X as their union: $\mathscr{A}_{\mathscr{V}} = \left\{ f \in \mathscr{A}(X, \mathbb{C}): \sup\limits_{V_n} |f| < \infty \text{ for all } n \right\}$ is a linear subspace of $\mathscr{A}(X, \mathbb{C})$, and $(\mathscr{A}_{\mathscr{V}}, \pi_{\mathscr{V}})$ a Fréchet space with the norms $f \mapsto \sup\limits_{V_n} |f|$. We define a preorder relation on the set of all sequences $\mathscr{V}$ by setting $\mathscr{V} \prec \mathscr{W}$ if $V_n \supset W_n$ for sufficiently large n; the inductive limit (see (1.5(B)) $(\mathscr{A}(X, \mathbb{C}), \pi)$ of the Fréchet spaces $(\mathscr{A}_{\mathscr{V}}, \pi_{\mathscr{V}})$ is barrelled, and Hausdorff since $[f \mapsto |f(x)|] \in \pi$ for all $x \in X$.

Since B is a bounding set: $f \mapsto \sup\limits_{x \in B} |f(x)|$ is a lower semi-continuous seminorm on $(\mathscr{A}(X, \mathbb{C}), \pi)$, therefore a continuous one; this means that, for each $\mathscr{V} = (V_n)_{n \in \mathbb{N}^*}$ as above, there are an integer n and a number c for which $\sup\limits_{B} |f| \leqslant c \sup\limits_{V_n} |f|$ for all $f \in \mathscr{A}_{\mathscr{V}}$. But we claim that actually there are n and c for which this estimate holds for all $f \in \mathscr{A}(X, \mathbb{C})$.

If it were not so: for each $n \in \mathbb{N}^*$ there would exist $f_n \in \mathscr{A}(X, \mathbb{C})$ with $|f_n| \leqslant 1$ on V_n but $\sup\limits_{B} |f_n| \geqslant n$; then $|f_n|, |f_{n+1}|, \ldots$ are $\leqslant 1$ on V_n, each set $\{x \in X: |f_n(x)| \leqslant m \text{ for all } n \in \mathbb{N}^*\}$, $m \in \mathbb{N}^*$, contains V_1; denoting its interior by W_m, we get an increasing sequence $\mathscr{W}$ of open nonempty sets having X as their union. Since $\sup\limits_{W_m} |f_n| \leqslant m$ for all $m, n \in \mathbb{N}^*$, we have $f_n \in \mathscr{A}_{\mathscr{W}}$ for all $n \in \mathbb{N}^*$, but we cannot find an integer m and a number c such that $\sup\limits_{B} |f_n| \leqslant c \sup\limits_{W_m} |f_n|$ for all $n \in \mathbb{N}^*$.

Now $\sup\limits_{B} |f| \leqslant c \sup\limits_{V_n} |f|$ for all $f \in \mathscr{A}(X, \mathbb{C})$ implies $\sup\limits_{B} |f^k| \leqslant c \sup\limits_{V_n} |f^k|$ for all $k \in \mathbb{N}^*$, hence $\sup\limits_{B} |f| \leqslant \sup\limits_{V_n} |f|$ for all $f \in \mathscr{A}(X, \mathbb{C})$, in particular (with $f = \exp x'$) $\sup\limits_{B} (\mathscr{R}e\, x') \leqslant \sup\limits_{V_n} (\mathscr{R}e\, x')$ for all $x' \in X'$. $\square$

Corollary 3.4.4. *In a separable l.c. space, any bounding set is precompact.*

Proof. Let B be closed and bounding, $\{x_k: k \in \mathbb{N}^*\}$ dense in X. Given $p \in \mathrm{csn}(X)$ and $\varepsilon > 0$; X is the union of the open (p, ε)-balls $x_k + p^{-1}([0, \varepsilon[)$, $k \in \mathbb{N}^*$; by Theorem 3.4.3, for a suitable $n \in \mathbb{N}^*$, $V = \bigcup\limits_{1 \leqslant k \leqslant n} \{x_k + p^{-1}([0, \varepsilon[)\}$ has a closed convex hull [namely $K + p^{-1}([0, \varepsilon])$, where K is the (compact) convex hull of the finite set $\{x_1, \ldots, x_n\}$] containing B; then B can be covered by a finite number of $(p, 2\varepsilon)$-balls. $\square$

The question whether the assumption of separability could be avoided was first answered by S. Dineen [Di]$_1$. Given an infinite set A, $l^\infty(A)$ denotes the Banach space of all bounded maps $x: A \to \mathbb{C}$ with the sup norm; for each $\alpha \in A$, let $x_\alpha = 1_{\{\alpha\}}$: since $\alpha \neq \beta$ implies $\|x_\alpha - x_\beta\| = 1$, the set $B = \{x_\alpha : \alpha \in A\}$ is closed but not compact.

A first step in the proof that B is nevertheless bounding is the following proposition, where $St\, x$ denotes the support of x, namely $St\, x = x^{-1}(\mathbb{C}^*) = \{\alpha \in A : x(\alpha) \neq 0\}$.

Proposition 3.4.5. *Let φ be a continuous n-homogeneous polynomial map $l^\infty(A) \to \mathbb{C}$.*

a) *For any infinite partition $(S_i)_{i \in I}$ of A, the family of numbers $C_i = \sup\{|\varphi(x)|^2 : St\, x \subset S_i, \|x\| \leqslant 1\}$ is summable, $\sum_{i \in I} C_i \leqslant \sup_{\|x\| \leqslant 1} |\varphi(x)|^2 < \infty$.*
b) *Consequently: given $\varepsilon > 0$, any infinite subset of A contains another infinite subset S such that $St\, x \subset S$, $\|x\| \leqslant 1$ implies $|\varphi(x)| \leqslant \varepsilon$.*

Proof. a) Let λ be the symmetric n-linear map associated to φ by Theorem 2.2.9, and $M = \sup_{\|x\| \leqslant 1} |\varphi(x)|^2$; let $St\, x_j \subset S_{i(j)}$, $i(j) \neq i(j')$ whenever $j \neq j'$, $\|x_j\| \leqslant 1$ and $|\zeta_j| \leqslant 1$ for all $j \in \{1, \ldots, m\}$. Then $\|\zeta_1 x_1 + \cdots + \zeta_m x_m\| \leqslant 1$, $\psi(\zeta_1, \ldots, \zeta_m) = \lambda(\zeta_1 x_1 + \cdots + \zeta_m x_m, \ldots, \zeta_1 x_1 + \cdots + \zeta_m x_m)$ (where $\zeta_1 x_1 + \cdots + \zeta_m x_m$ is written n times) is an n-homogeneous polynomial in the m complex variables ζ_j; by the classical Gutzmer formula, the square moduli of the coefficients of this polynomial have a sum equal to the mean value of

$$|\psi(e^{i\theta_1}, \ldots, e^{i\theta_m})|^2 = |\varphi(e^{i\theta_1} x_1 + \cdots + e^{i\theta_m} x_m)|^2 \leqslant M.$$

But the coefficient of ζ_j^n is $\varphi(x_j)$ for all $j \in \{1, \ldots, m\}$, hence $\sum_{j=1}^m |\varphi(x_j)|^2 \leqslant M$, $C_{i(1)} + \cdots + C_{i(m)} \leqslant M$.
b) An infinite set admits an infinite partition into infinite subsets. $\square$

Theorem 3.4.6. *The set $B = \{x_\alpha : \alpha \in A\}$ is bounding in $l^\infty(A)$.*

Proof. Assume B not bounding; then, by Propositions 3.4.2.a and 3.3.2.b, we can find a sequence $\Phi = (\varphi_k)_{k \in \mathbb{N}^*}$ of continuous m_k-homogeneous $(0 < m_k < m_{k+1})$ polynomial maps $l^\infty(A) \to \mathbb{C}$ such that (i) the series $\sum_{k \in \mathbb{N}^*} \varphi_k$ is summable to some entire function $l^\infty(A) \to \mathbb{C}$; (ii) $\varphi_k(x_{\alpha_k}) = 1$ for some $\alpha_k \in A$, and all α_k are distinct. Let again λ_k be the continuous symmetric m_k-linear map associated to φ_k by formula (2) in Theorem 2.2.9.

1) By discarding a suitable amount of indeces k, we will step by step reach the following situation: for all $k \in \mathbb{N}^*$,

$$St\, x \subset \{\alpha_1, \ldots, \alpha_k\}, \quad St\, y \subset \{\alpha_{k+1}, \alpha_{k+2}, \ldots\}, \quad \|x\| \leqslant 1, \|y\| \leqslant 1$$

imply $|\lambda_k(x, \ldots, x, y, \ldots, y)| \leqslant \dfrac{2^{-m_k}}{m_k!}$ whenever y stands at least once among the m_k elements $x, \ldots x, y, \ldots, y$.

In the first step, we use Proposition 3.4.5.b, with $A_1 = \{\alpha_2, \alpha_3, \ldots\}$ instead of A, successively for the following continuous polynomial maps $l^\infty(A_1) \to \mathbb{C}$: $y \mapsto \lambda_1(x_{\alpha_1}, \ldots, x_{\alpha_1}, y)$ is 1-homogeneous, $y \mapsto \lambda_1(x_{\alpha_1}, \ldots, x_{\alpha_1}, y, y)$ is 2-homogeneous, $\ldots$, $y \mapsto \lambda_1(x_{\alpha_1}, y, \ldots, y)$ is $(m_1 - 1)$-homogeneous, $y \mapsto \lambda_1(y, \ldots, y)$ is m_1-homogeneous. We thus get an infinite subset A_2 of A_1 such that, for $St\, y \subset A_2$, $\|y\| \leqslant 1$, each of these polynomials has a modulus $\leqslant \dfrac{2^{-m_1}}{m_1!}$. From now on, we keep only the index 1 and the indeces k for which $\alpha_k \in A_2$, the first of which we denote 2; this implies that the sequence Φ contains an m_2-homogeneous φ_2 such that $\varphi_2(x_{\alpha_2}) = 1$ and $m_2 > m_1$.

In the second step, we use Proposition 3.4.5.b, with $A_2 \backslash \{\alpha_2\}$ instead of A, successively for the following continuous polynomial maps $l^\infty(A_2 \backslash \{\alpha_2\}) \to \mathbb{C}$: the m_2 maps $y \mapsto \lambda_2(x_{\alpha_1}, \ldots, x_{\alpha_1}, y)$, $y \mapsto \lambda_2(x_{\alpha_1}, \ldots, x_{\alpha_1}, x_{\alpha_2}, y)$, $\ldots$, $y \mapsto \lambda_2(x_{\alpha_2}, \ldots, x_{\alpha_2}, y)$ are 1-homogeneous, the $(m_2 - 1)$ maps $y \mapsto \lambda_2(x_{\alpha_1}, \ldots, x_{\alpha_1}, y, y)$, $y \mapsto \lambda_2(x_{\alpha_1}, \ldots, x_{\alpha_1}, x_{\alpha_2}, y, y)$, $\ldots$, $y \mapsto \lambda_2(x_{\alpha_2}, \ldots, x_{\alpha_2}, y, y)$ are 2-homogeneous, $\ldots$, $y \mapsto \lambda_2(y, \ldots, y)$ is m_2-homogeneous. We thus get an infinite subset A_3 of $A_2 \backslash \{\alpha_2\}$ such that, for $St\, y \subset A_3$, $\|y\| \leqslant 1$, the l-homogeneous ones $(l = 1, 2, \ldots, m_2)$ have moduli $\leqslant \dfrac{2^{l - 2m_2}}{m_2!}$, from which follows that

$$St\, x \subset \{\alpha_1, \alpha_2\}, \quad St\, y \subset A_3, \quad \|x\| \leqslant 1, \quad \|y\| \leqslant 1$$

imply $|\lambda_2(x, \ldots, x, y, \ldots, y)| \leqslant \dfrac{2^{-m_2}}{m_2!}$ whenever y stands at least once. From now on, we keep only the indeces 1, 2 and those k for which $\alpha_k \in A_3$, the first of which we denote 3; this implies that an m_3-homogeneous $\varphi_3 \in \Phi$ satisfies $\varphi_3(x_{\alpha_3}) = 1$, and $m_3 > m_2$.

An indefinite repetition of the argument leads to the desired situation: with the notation $S_k = \{\alpha_1, \ldots, \alpha_k\}$, $S = \{\alpha_k : k \in \mathbb{N}^*\}$, for each $k \in \mathbb{N}^*$,

$$(*) \qquad St\, x \subset S_k, \quad St\, y \subset S \backslash S_k, \quad \|x\| \leqslant 1, \quad \|y\| \leqslant 1$$

$$\text{imply } |\varphi_k(x + y) - \varphi_k(x)| \leqslant 1/m_k!$$

since the expansion of $\lambda_k(x + y, \ldots, x + y) - \lambda_k(x, \ldots, x)$ is made up of $2^{m_k} - 1$ terms $\lambda_k(x, \ldots, x, y, \ldots, y)$ in which y appears at least once.

2) The new sequence $(\varphi_k)_{k \in \mathbb{N}^*}$ still satisfies (i), (ii) in the beginning of the proof (Prop. 3.3.2.b). Now let

$$\psi_k(x) = \varphi_k(\mathbf{1}_{S_k} x), \qquad x \in l^\infty(S):$$

this is a m_k-homogeneous polynomial map $l^\infty(S) \to \mathbb{C}$; for all $x \in l^\infty(S)$,

$$|\varphi_k(x)|^{1/m_k} \to 0 \text{ as } k \to \infty \text{ (Prop. 3.3.2.c) and } |\psi_k(x) - \varphi_k(x)|^{1/m_k} \leqslant \left(\frac{1}{m_k!}\right)^{1/m_k}$$

$\|x\| \to 0$ by (∗) above, hence $|\psi_k(x)|^{1/m_k} \to 0$ too. We will nevertheless find $z \in l^\infty(S)$ such that $|\psi_k(z)| \geqslant 1$ for all $k \in \mathbb{N}^*$, and thus reach a contradiction; let $z(\alpha_k) = \zeta_k$ for all $k \in \mathbb{N}^*$.

Suppose we have found complex numbers $\zeta_1, \ldots, \zeta_k$ with modulus 1 such that $|\psi_k(\zeta_1 x_{\alpha_1} + \cdots + \zeta_k x_{\alpha_k})| \geqslant 1$, which is obviously possible for $k = 1$, since $\psi_1(x_{\alpha_1}) = 1$. Then $\psi_{k+1}(\zeta_1 x_{\alpha_1} + \cdots + \zeta_k x_{\alpha_k} + \zeta x_{\alpha_{k+1}})$ is a polynomial in ζ, whose leading term is $\zeta^{m_{k+1}}$ and whose maximum modulus for $|\zeta| = 1$ is therefore $\geqslant 1$. Thus we obtain a sequence $(\zeta_k)_{k \in \mathbb{N}^*}$ with $|\zeta_k| = 1$ for all $k \in \mathbb{N}^*$, hence a point $z \in l^\infty(S)$ such that $\psi_k(z) = \psi_k(\mathbf{1}_{S_k} z) = \psi_k(\zeta_1 x_{\alpha_1} + \cdots + \zeta_k x_{\alpha_k})$ has a modulus $\geqslant 1$ for all $k \in \mathbb{N}^*$. □

Remark 3.4.7. Putting Corollary 3.4.4 and Theorem 3.4.6 together, we see that, if Y is a subspace, or a closed subspace, of the space X, the entire functions on Y do not all admit entire extensions to X: there is no Hahn-Banach theorem for entire functions [Di]₃.

The space $X = l^\infty(\mathbb{N})$ and its subspace $Y = c_0(\mathbb{N})$ $\Big($whose elements are the maps $\mathbb{N} \ni n \mapsto \xi_n \in \mathbb{C}$ such that $\lim_{n \to \infty} \xi_n = 0$$\Big)$ are a counter-example: the set B considered in Theorem 3.4.6 is bounding in X but not in Y since Y is separable but B not precompact; therefore an $f \in \mathscr{A}(Y, \mathbb{C})$ which is unbounded on B admits no entire extension to X.

Exercises

3.1.1. By a classical theorem of Rado ([He]$_3$, Coroll. 15.15): if U is an open set in $\mathbb{C}$, f continuous $U \to \mathbb{C}$ and $f|_{f^{-1}(\mathbb{C}^*)}$ analytic, then f is analytic. Extend this to a continuous $f: \Omega \to Z$, where Ω is open in the l.c. space X and Z s.c.l.c. For a generalization, see Exercise 4.5.17.b.

3.2.1. Let Ω be a balanced convex open set in a Baire space X, q its gauge, $q(a) = 1$, Z a Banach space; let $f \in \mathscr{A}(\Omega, Z)$ and $\alpha > 0$ have the property: for every boundary point y of Ω sufficiently near a, the function $\Delta(0,1) \ni \zeta \mapsto f(\zeta y)$ has an analytic extension g_y to $\Delta(0,1) \cup \Delta(1,\alpha)$. Show that f has an analytic extension to the union of Ω and a suitable neighbourhood of a.
Hint: Choosing $x' \in X'$ such that $|x'| \leqslant q$ and $x'(a) = 1$, extend $f[\zeta(a+z)]$ from $\{(z,\zeta) \in (\ker x') \times \mathbb{C}: |\zeta| < 1/q(a+z)\}$ to a neighbourhood of $(0,1)$ in $(\ker x') \times \mathbb{C}$.

3.3.5. [Hir]. Let Z be a Banach space and $f \in \mathscr{A}(X, Z)$ have a $(p, \|\quad\|)$-radius of boundedness $R \in \,]0, +\infty[$ around the origin; with $a = 0$ in the notation of 3.3.5, let the subsequence (n_k) be such that $[s_{n_k}(f, p, \|\quad\|)]^{1/n_k} \to \dfrac{1}{R}$. Given $\lambda \in \,]0, 1[$,

prove:
a) by means of Proposition 1.4.2, the existence of $z_0' \in Z'$ such that

$$\limsup_{k \to \infty} \, [s_{n_k}(z_0' \circ f, p, |\quad|)]^{1/n_k} > \frac{\lambda}{R};$$

b) that $\left\{ z' \in Z': s_{n_k}(z' \circ f, p, |\quad|) \leqslant \left(\dfrac{\lambda^2}{R}\right)^{n_k} \text{ for all } k \geqslant k_0 \right\}$ has an empty interior in the Banach space Z';

c) that $\{z' \in Z': R(z' \circ f, p, |\quad|) > R\}$ is meagre in Z'.

Chapter 4
Plurisubharmonic functions

Summary

4.1.	Plurisubharmonic functions on an open set in a l.c. space.
Def. 4.1.6.	Submedian functions.
Th. 4.1.7.	The upper regularized u^* of a supremum u of submedian functions [Nov]$_1$.
4.2.	The finite dimensional case.
Coroll. 4.2.3.	Nearly subharmonic functions [Sz].
Th. 4.2.5.	The convergence theorem in the 1-dimensional case.
Th. 4.3.1.	v p.s.h. and f analytic imply $v \circ f$ p.s.h.
Def. 4.3.3.	Construction of classes of submedian functions [Lel]$_1$.
Def. 4.3.4.	Generalized polydiscs [Coe].
Th. 4.3.6.	Mean value inequalities with respect to a generalized polydisc [Coe].
Def. 4.4.1.	Pluriharmonic functions.
Prop. 4.4.4.	The Caratheodory pseudodistance.
Def. 4.5.1.	Unipolar sets.
Th. 4.5.2.3.	Extensions from $\Omega \backslash P$ to Ω, where P is a closed unipolar subset of Ω.
Th. 4.5.6.	A convergence theorem under the assumption u^* pluriharmonic.
Def. 4.5.8.	Pluripolar and strictly pluripolar sets.
Prop. 4.5.10.	Countable unions of pluripolar sets in a Fréchet space [Coe, Nov$_3$].
Prop. 4.5.12.	A characterization of strictly pluripolar sets [Lel]$_6$.
Th. 4.5.13.	An inverse function theorem [Lel]$_6$.
Th. 4.5.16.	Extensions from $\Omega \backslash P$ to Ω, where P is a closed pluripolar subset of Ω.
Def. 4.6.1.	Thin sets. Def. 4.6.2. The fine topology.
Th. 4.6.5.	A fine maximum principle [Coe].
Th. 4.6.7.	A generalization of the Perron envelopes [Coe].

Chapter 4
Plurisubharmonic functions

4.1 Plurisubharmonic functions on an open set Ω in a l.c. space X

For subharmonic functions, more generally for potential theory, on the complex plane $\mathbb{C}$, we again refer to [Brel]$_3$.

Definition 4.1.1. *A function* $u: \Omega \to [-\infty, +\infty[$ *is plurihypoharmonic (p.h.h.) if*: (i) *u is upper semicontinuous*; (ii) *for every complex line E in X, $u|_{E \cap \Omega}$ is hypoharmonic, i.e. subharmonic or* $\equiv -\infty$ *on each connected component of* $E \cap \Omega$.

Since (i) implies $u|_{E \cap \Omega}$ upper semicontinuous, (ii) is just a mean value (*MV*) inequality: for $a \in \Omega$, $b \in X$, whenever $\Omega \supset \{a + \zeta b : |\zeta| \leqslant 1\}$, then

$$u(a) \leqslant MV\,u(a + e^{i\theta}b) = \frac{1}{2\pi} \int_{-\pi}^{\pi} u(a + e^{i\theta}b)\, d\theta.$$

If $\zeta \mapsto u(a + \zeta b)$ is subharmonic on the disc $\Delta(0, R)$, then $[\theta \mapsto u(a + re^{i\theta}b)] \in \mathscr{L}^1(d\theta)$ for all $r \in]0, R[$, $MV\,u(a + re^{i\theta}b)$ and $\sup_{\theta \in \mathbb{R}} u(a + re^{i\theta}b)$ are convex increasing functions of $\ln r$ and tend to $u(a)$ as $r \to 0$. From this convexity follows the Liouville theorem for subharmonic functions on $\mathbb{C}$: such a function is either constant or unbounded from above. Then a p.h.h. function on X is also either constant or unbounded from above.

Proposition 4.1.2. a) *If u is p.h.h. on Ω and $u^{-1}(-\infty)$ has an interior point $a \in \Omega$, then $u \equiv -\infty$ on the connected component of Ω which contains a.* b) (*The maximum principle*). *If u is p.h.h. on Ω and $a \in \Omega$ is such that $u(a) = \sup_{\Omega} u$, then $u \equiv u(a)$ on the connected component of Ω which contains a.* c) *If u is the supremum of a finite family of p.h.h. functions on Ω, more generally the supremum of any such family provided that otherwise $u < +\infty$ and u is upper semicontinuous, then u is p.h.h. on Ω.* d) *The limit or infimum of a decreasing sequence, more generally of a lower directed family, of p.h.h. functions on Ω, is p.h.h. on Ω.*

Proof. a) If v is hypoharmonic on an open disc, $v^{-1}(-\infty)$ either is the disc or has no interior point; then $u^{-1}(-\infty)$ a neighbourhood of a implies

$u^{-1}(-\infty) \supset a + \omega(a)$ [where $\omega(a)$ is the largest balanced subset of $\Omega - a$] and therefore, by an argument already used (for Th. 2.4.4.b), the interior of $u^{-1}(-\infty)$ is open and closed in Ω.
b) By the maximum principle for subharmonic functions, $u = u(a)$ on $a + \omega(a)$; then the set $\left\{ x \in \Omega: u(x) = \sup_{\Omega} u \right\}$ is open and closed in Ω. □

Definition 4.1.3. *A p.h.h. function u is plurisubharmonic (p.s.h.) on Ω, $u \in \mathscr{P}(\Omega)$, if $u^{-1}(-\infty)$ has no interior point or, equivalently, contains no connected component of Ω.*

In other words, if Ω is connected, a p.h.h. function on Ω either $\in \mathscr{P}(\Omega)$ or is the constant $-\infty$.

Proposition 4.1.4. a) *$\mathscr{P}(\Omega)$ is a convex cone of functions.*
b) *Both p.h.h. and p.s.h. functions enjoy the following sheaf property: if Ω is the union of open sets Ω_i, $i \in I$, then a function $u: \Omega \to \overline{\mathbb{R}}$ is p.h.h. (resp.: p.s.h.) on Ω if and only if $u|_{\Omega_i}$ is p.h.h. (resp.: p.s.h.) on Ω_i for all $i \in I$.*

Proof of a). If u and $v \in \mathscr{P}(\Omega)$, $u + v$ is obviously p.h.h. Moreover, each connected component of Ω contains some point a such that $u(a) > -\infty$, $a + \omega(a)$ some point b such that $v(b) > -\infty$; then both functions $u[a + \zeta(b-a)], v[a + \zeta(b-a)]$ of ζ are subharmonic on an open disc containing $\overline{\Delta}(0, 1)$, $(u + v)[a + \zeta(b-a)]$ too. □

Examples 4.1.5. a) $f \in \mathscr{A}(\Omega, \mathbb{C})$ implies $|f|^\alpha \in \mathscr{P}(\Omega)$ for all $\alpha > 0$ and $\ln|f|$ p.h.h. on Ω. Note that, for $f \in \mathscr{G}(\Omega, \mathbb{C})$, $|f|^\alpha$ or $\ln|f|$ upper semicontinuous implies f locally bounded, hence $f \in \mathscr{A}(\Omega, \mathbb{C})$.
b) $p \in \text{csn}(X)$ implies $p^\alpha \in \mathscr{P}(X)$ for all $\alpha > 0$ (as a finite continuous supremum of $|x'|^\alpha$, $x' \in X'$) and $\ln p \in \mathscr{P}(X)$ unless $p \equiv 0$. Note that, for any seminorm p on X, p^α or $\ln p$ upper semicontinuous at the origin implies p continuous.
c) More generally: if $f \in \mathscr{A}(\Omega, Z)$, Z s.c.l.c., and $q \in \text{csn}(Z)$, then $q \circ f \in \mathscr{P}(\Omega)$ and $\ln(q \circ f)$ is p.h.h. on Ω. A special case of this, but without any topology on Z, was considered in Proposition 2.2.3.
d) If χ is an increasing convex function $]-\infty, \lambda[\to \mathbb{R}$, $\lambda \leq +\infty$, and $\chi(-\infty) = \lim_{t \to -\infty} \chi(t)$, then $u \in \mathscr{P}(\Omega)$, $u < \lambda$ imply $\chi \circ u \in \mathscr{P}(\Omega)$, as a finite upper semicontinuous supremum of functions $\alpha u + \beta$, $\alpha \in \mathbb{R}_+$, $\beta \in \mathbb{R}$.
e) Let v be a positive measure, $\int dv = 1$, on the circumference $T = \{t \in \mathbb{C}: |t| = 1\}$. If $u: \Omega \to [-\infty, +\infty[$ is upper semicontinuous and satisfies (∗) $u(a) \leq \int u(a + tb) \, dv(t)$ whenever $\Omega \supset \{a + \zeta b: |\zeta| \leq 1\}$, then u is p.h.h.

on Ω. In fact: since $MVu(a + te^{i\theta}b)$ does not depend on $t \in T$, we may write:

$$MVu(a + e^{i\theta}b) = \frac{1}{2\pi} \int \int u(a + te^{i\theta}b) \, dv(t) \, d\theta \geqslant \frac{1}{2\pi} \int u(a) \, d\theta = u(a).$$

Any convex function u satisfies $(*)$ with $v = \frac{1}{2}(\delta_1 + \delta_{-1})$, where δ_t is the Dirac measure with support t; for a real convex function u, the assumption of upper semicontinuity can be weakened as follows.

f) Let Ω be convex, $u: \Omega \to \mathbb{R}$ convex and locally bounded from above; then u is continuous and p.s.h. In fact: let $p \in \mathrm{csn}(X)$ be such that $\Omega \supset a + p^{-1}([0, 1])$ and $u \leqslant M$ on $a + p^{-1}([0, 1])$; then u convex implies $u \geqslant 2u(a) - M$ on $a + p^{-1}([0, 1])$ and $|u - u(a)| \leqslant \alpha[M - u(a)]$ on $a + p^{-1}([0, \alpha])$ for all $\alpha \in {]0, 1]}$.

g) So it turns out that a locally bounded from above supremum of real convex functions is continuous and p.s.h. On the contrary, a locally bounded from above supremum of p.s.h. functions is not always upper semicontinuous: for instance, if $f \in \mathscr{A}(\Omega, \mathbb{C})$ and $\omega = \{x \in \Omega: |f(x)| < 1\}$, then $\sup_{0 < \alpha < 1} |f|^\alpha = \lim_{\alpha \to 0} |f|^\alpha$ is lower, and not upper, semicontinuous on ω. Therefore we have to define a wider class than $\mathscr{P}(\Omega)$ containing the locally bounded from above suprema of functions $\in \mathscr{P}(\Omega)$, and this can be done in several ways, e.g. Definitions 4.1.6. and 4.3.3. below.

From now on, we denote by u^* the *upper regularized* of any function $u: \Omega \to \bar{\mathbb{R}}$, namely the smallest upper semicontinuous function $\geqslant u$:

$$u^*(x) = \limsup_{y \to x} u(y), \qquad x \in \Omega;$$

we also set $E_u = \{x \in \Omega: u(x) < u^*(x)\}$. By the definition of u^*: for $(\lambda, \mu) \in \mathbb{R}^2$, the set $\{x \in \Omega: u(x) \leqslant \lambda < \mu \leqslant u^*(x)\}$ has no interior point; from this follows E_u meagre in Ω in the following two cases:

h) if u is the supremum of a family $\mathscr{V}$ of *continuous* functions; in fact, $x \in E_u$ if and only if there are $(\lambda, \mu) \in \mathbb{Q}^2$ such that $u(x) \leqslant \lambda < \mu \leqslant u^*(x)$, and the set of such $x \in \Omega$ is closed in Ω;

h') if $u = \limsup_{\mathscr{F}} v$, where $\mathscr{F}$ is a filter on $\mathscr{V}$ with a countable basis; in fact $x \in E_u$ if and only if there are A_n in the basis of $\mathscr{F}$ and $(\lambda, \mu) \in \mathbb{Q}^2$ such that $\sup_{v \in A_n} v(x) \leqslant \lambda < \mu \leqslant u^*(x)$, and again the set of such $x \in \Omega$ is closed in Ω.

A family $\mathscr{V}$ of functions $v: \Omega \to \bar{\mathbb{R}}$ is said to be *locally bounded from above* if $u = \sup_{v \in \mathscr{V}} v$ is locally bounded from above, i.e. if $u^* < +\infty$ at each point $\in \Omega$.

Definition 4.1.6. *A function* $u: \Omega \to [-\infty, +\infty[$ *is submedian if*: (i) *u is locally bounded from above;* (ii) *for* $(a, b) \in X^2$, *whenever* $\Omega \supset \{a + \zeta b: |\zeta| \leqslant 1\}$, $[\theta \mapsto u(a + e^{i\theta}b)]$ *is* $(d\theta)$-*measurable, and* $u(a) \leqslant MV u(a + e^{i\theta}b)$.

The class of submedian functions on Ω is a convex cone. The limit of a decreasing sequence of submedian functions, the supremum of a finite, more generally countable and locally bounded from above, family of submedian functions, are again submedian. Among the submedian functions appear the p.h.h.ones and the suprema of countable locally bounded from above families of p.h.h. functions; further classes of submedian functions will appear in Definition 4.3.3.

Theorem 4.1.7. a) *Any submedian function u on Ω has a* p.h.h. *upper regularized u^*; if $u^{-1}(-\infty)$ has an interior point a, then $u \equiv -\infty$ on the connected component of Ω which contains a.*
b) *If $\mathscr{V}$ is a locally bounded from above family of submedian functions on Ω, then $u = \sup_{v \in \mathscr{V}} v$ has a* p.h.h. *upper regularized u^*.*
c) *If moreover a filter $\mathscr{F}$ on $\mathscr{V}$ has a countable basis, then: $u = \limsup_{\mathscr{F}} v =$*
$\inf_{A \in \mathscr{F}} \left(\sup_{v \in A} v \right)$ *has a* p.h.h. *upper regularized u^*; in particular, if a sequence (v_n) of submedian functions on Ω is locally bounded from above, then $u = \limsup v_n$ has a* p.h.h. *upper regularized u^*.*

Proof. First let w be an upper semicontinuous function $\Omega \to [-\infty, +\infty[$; given the compact set: $K = \{a + \zeta b: |\zeta| \leqslant 1\} \subset \Omega$, let $\cup$ be an open neighbourhood of the origin in X such that $K + \cup \subset \Omega$ and $w \leqslant M$ on $K + \cup$; we claim that $x \mapsto MV w(x + e^{i\theta}b)$ is an upper semicontinuous function on $a + U$. In fact: since the open set $K + U$ is completely regular, $(M - w)|_{K+U}$ is the supremum of the upper directed family of all real positive continuous functions $c \leqslant M - w$ on $K + U$; then, for $x \in a + U$, $M - MV w(x + e^{i\theta}b)$ is the supremum of the continuous functions $MV c(x + e^{i\theta}b)$.

The notation of this preliminary will be used in the proofs of a), b), c), but for u^* instead of w.

a) For $x \in a + U$: $u(x) \leqslant MV u(x + e^{i\theta}b) \leqslant MV u^*(x + e^{i\theta}b)$, hence $u^*(a) \leqslant MV u^*(a + e^{i\theta}b)$ whenever $\Omega \supset \{a + \zeta b: |\zeta| \leqslant 1\}$. The last statement in a) follows from the fact that $u^{-1}(-\infty)$ and $(u^*)^{-1}(-\infty)$ have the same interior.
b) For $x \in a + U$: $v(x) \leqslant MV v(x + e^{i\theta}b) \leqslant MV u^*(x + e^{i\theta}b)$ for all $v \in \mathscr{V}$ entails $u(x) \leqslant MV u^*(x + e^{i\theta}b)$, with the same conclusion as in a).
c) Here we shall use the notation $\overline{MV} w$ for the infimum of the mean values of all $d\theta$-measurable functions $\geqslant w$.

Let $(A_n)_{n \in \mathbb{N}}$ be a countable basis of $\mathscr{F}$, $A_{n+1} \subset A_n$ for all $n \in \mathbb{N}$, and $u_n = \sup_{v \in A_n} v$. For $x \in a + U$: $M - v(x) \geqslant \overline{MV}[M - u_n(x + e^{i\theta}b)]$ for all $v \in A_n$ entails $M - u_n(x) \geqslant \overline{MV}[M - u_n(x + e^{i\theta}b)]$ and, since $(M - u)$ is the limit of the increasing sequence $(M - u_n)$:

$$M - u(x) \geqslant \overline{MV}[M - u(x + e^{i\theta}b)] \geqslant MV[M - u^*(x + e^{i\theta}b)],$$

with the same conclusion as in a) and b).   □

In particular, $\mathscr{V}$ may be a locally bounded from above family of p.h.h. functions on Ω; in this case it turns out that $\sup_{v \in \mathscr{V}} v$ and $\limsup_{\mathscr{F}} v$ (provided that $\mathscr{F}$ has a countable basis) are submedian, even if $\mathscr{V}$ is uncountable. But this we shall be able to prove only after a thorough investigation of the one dimensional case, leading to the *convergence theorem* 4.2.6.

Corollary 4.1.8. *Let $\mathscr{V}$ be a locally bounded from above family of functions $v: \Omega \to [-\infty, +\infty[$: if each v has a p.h.h. upper regularized v^*, so has $u = \sup_{v \in \mathscr{V}} v$, and $u^* = \left(\sup_{v \in \mathscr{V}} v^* \right)^*$.*

Proof: Let $u' = \sup_{v \in \mathscr{V}} v^*$: $u \leqslant u'$ entails $u^* \leqslant u'^*$, and $v^* \leqslant u^*$ for all $v \in \mathscr{V}$ entails $u' \leqslant u^*$, hence $u'^* \leqslant u^*$. Now u'^* is p.h.h. by Theorem 4.1.7.b. for the family $(v^*)_{v \in \mathscr{V}}$.

Propositon 4.1.9. *Let E be a topological space with a countable fundamental family $\{\omega_1, \omega_2, \omega_3, \ldots\}$ of open sets. Any family $\mathscr{F}$ of functions $E \to \bar{\mathbb{R}}$ contains a countable subfamily $\mathscr{F}_0$ with the same upper semicontinuous majorants as $\mathscr{F}$.*

This is known as the Choquet topological lemma [Br Ch]; we shall use it below (see 4.2.D.) for the special case in which E is an open set in the complex plane.

Proof. Since there are homeomorphisms between $\bar{\mathbb{R}}$ and $[0, 1]$, the functions $f \in \mathscr{F}$ may be assumed to have their values in $[0, 1]$. By setting $U_1 = \omega_1$, $U_2 = \omega_2$, $U_3 = \omega_1$, $U_4 = \omega_2$, $U_5 = \omega_3$, $U_6 = \omega_1$ and so on, we again have a countable fundamental family of open sets, among which each ω in the initial family is found an infinity of times.

Let $s = \sup_{f \in \mathscr{F}} f$: for each $n \in \mathbb{N}^*$ there are an $x_n \in U_n$ such that $s(x_n) \geqslant \left(\sup_{U_n} s \right) - \frac{1}{2n}$ and then a function $f_n \in \mathscr{F}$ such that

$$f_n(x_n) \geqslant s(x_n) - \frac{1}{2n} \geqslant \left(\sup_{U_n} s\right) - \frac{1}{n}.$$

We claim that $\mathscr{F}_0 = \{f_n : n \in \mathbb{N}^*\}$ has the required property, i.e. s and $s_0 = \sup_{n \in \mathbb{N}^*} f_n$ have the same upper semicontinuous majorants.

In fact: given an upper semicontinuous function $g \geqslant s_0$, a point $a \in E$ and $\varepsilon > 0$, we can choose $n \in \mathbb{N}^*$ so that $\frac{1}{n} \leqslant \frac{\varepsilon}{2}$, $U_n \ni a$ and $\sup_{U_n} g \leqslant g(a) + \frac{\varepsilon}{2}$; then we have $s(a) \leqslant \sup_{U_n} s \leqslant \frac{1}{n} + \sup_{U_n} f_n \leqslant \frac{\varepsilon}{2} + \sup_{U_n} s_0 \leqslant g(a) + \varepsilon$, and conclude $g \geqslant s$.   □

4.2 The finite dimensional case

In parts **A, B, C** of this section, U denotes an open set in $\mathbb{C}^m$; we use the same notation $x = \sum_{j=1}^m \xi_j e_j$, $\|x\|^2 = \sum_{j=1}^m |\xi_j|^2$ as in §2.1. For the results we do not prove, we refer to $[\mathrm{He}]_1$, chapter I.

(A) Any function $u \in \mathscr{P}(U)$ is subharmonic; therefore u is locally integrable and $u^{-1}(-\infty)$ a null set for the Lebesgue measure dx on $\mathbb{C}^m$ (i.e. on $\mathbb{R}^{2m}$), which gives sense to the distributions (on U)

$$\Delta u = 4 \sum_{j=1}^m \frac{\partial^2 u}{\partial \xi_j \partial \overline{\xi}_j}, \qquad T_\lambda u = \sum_{j,k=1}^m \frac{\partial^2 u}{\partial \xi_j \partial \overline{\xi}_k} \lambda_j \overline{\lambda}_k,$$

$$\lambda = (\lambda_1, \ldots, \lambda_m) \in \mathbb{C}^m.$$

u subharmonic (resp. p.s.h.) on U implies that Δu (resp.: each $T_\lambda u$) is a positive Radon measure on U; conversely, a class of functions $u \in L^1_{\mathrm{loc}}(U)$ such that $\Delta u \geqslant 0$ (resp.: $T_\lambda u \geqslant 0$ for all $\lambda \in \mathbb{C}^m$) contains a unique function subharmonic (resp. p.s.h.) on U: uniqueness follows from the fact that, if u is subharmonic on U and $a \in U$, $u(a)$ is the limit as $r \to 0$ of the mean value of u over the ball with centre a, radius r.

In the special case of a continuous (resp.: twice continuously $\mathbb{R}$-differentiable) function $u : u$ is subharmonic if and only if the distribution (resp.: function) Δu is positive; u is p.s.h. if and only if each distribution (resp.: function) $T_\lambda u$ is positive.

(B) Let $\chi \in \mathscr{C}^\infty(\mathbb{R}^{2m})$, $\chi \geqslant 0$, $\chi(x)$ depending only on $\|x\|$, i.e. $\chi(x) = \psi(\|x\|)$, $\chi(x) = 0$ for $\|x\| \geqslant 1$ and $\int \chi(x)\,dx = 1$; let $\chi_n(x) = n^{2m}\chi(nx)$ and U_n be the

set of centres of the closed balls with radius $\dfrac{1}{n}$ *contained in U. If u is*
subharmonic (resp.: p.s.h.) on U, then $u * \chi_n \in \mathscr{C}^\infty(U_n)$ and $u * \chi_n$ is subharmonic (resp.: p.s.h.) on U_n since $\Delta(u * \chi_n) \geqslant 0$ [resp.: $T_\lambda(u * \chi_n) \geqslant 0$ for
all $\lambda \in \mathbb{C}^m$].

Moreover the sequence $(u * \chi_n)$ decreases to u. In fact, for $x \in U_n$:

$$u * \chi_n(x) = \int \ldots \underset{\|y\| \leqslant 1}{(2m)} \ldots \int u\left(x - \frac{y}{n}\right) \chi(y)\, dy$$

$$= 2m\omega_{2m} \int_0^1 \mu u\left(x, \frac{r}{n}\right) \psi(r) r^{2m-1}\, dr,$$

where ω_{2m} is the Lebesgue measure of the unit ball in $\mathbb{R}^{2m}$ and $\mu u\left(x, \dfrac{r}{n}\right)$
the mean value of u over the sphere $\|y - x\| = \dfrac{r}{n}$, which decreases to $u(x)$
since u is subharmonic.

Thus we see that any p.s.h. function is the limit of a decreasing sequence
of infinitely $\mathbb{R}$-differentiable p.s.h. functions. The following consequence is
most important, since it links p.s.h. functions to analytic functions.

Theorem 4.2.1. *If $f \in \mathscr{A}(U, \mathbb{C}^p)$ and v is p.s.h. on an open set $V(\text{in } \mathbb{C}^p)$
containing $f(U)$, then $v \circ f$ is p.h.h. on U.*

Proof. In order to prove that $v \circ f$ has property (ii) in Definition 4.1.1., let
U be a connected open set in $\mathbb{C}$ and ζ the complex variable. If v is twice
continuously $\mathbb{R}$-differentiable, then

$$\frac{\partial^2(v \circ f)}{\partial \zeta \partial \bar{\zeta}} = \sum_{j,k=1}^{p} \left(\frac{\partial^2 v}{\partial \xi_j \partial \bar{\xi}_k} \circ f\right) f_j' \overline{f_k'} \geqslant 0$$

with the notation $f = (f_1, \ldots, f_p)$. In the general case, since v is the limit of
a decreasing sequence (v_n) of infinitely $\mathbb{R}$-differentiable p.s.h. functions,
$v \circ f$ is the limit of the decreasing sequence $(v_n \circ f)$.

(C) The *Poisson kernel P for the unit disc is defined, for $|\zeta| < 1 = |e^{i\varphi}|$, by*

$$P(\zeta, e^{i\varphi}) = \frac{1 - |\zeta|^2}{|e^{i\varphi} - \zeta|^2} = \frac{d\theta}{d\varphi} \quad \text{if } e^{i\varphi} = \frac{\zeta + e^{i\theta}}{1 + \bar{\zeta}e^{i\theta}}.$$

Now let the open set U in $\mathbb{C}^m$ contain the compact set $\{x + \zeta_1 a_1 + \cdots + \zeta_m a_m : |\zeta_1| \leqslant 1, \ldots, |\zeta_m| \leqslant 1\}$; then for $|\zeta_1| < 1, \ldots, |\zeta_m| < 1$, every $u \in \mathscr{P}(U)$
satisfies the mean value inequality

(1) $u(x + \zeta_1 a_1 + \cdots + \zeta_m a_m)$

$$\leqslant \frac{1}{(2\pi)^m} \int \ldots (m) \ldots \int_{-\pi}^{\pi} \prod_{j=1}^{m} P(\zeta_j, e^{i\varphi_j}) u(x + e^{i\varphi_1} a_1 + \cdots + e^{i\varphi_m} a_m)\, d\varphi_1 \ldots d\varphi_m$$

or

(2) $u(x + \zeta_1 a_1 + \cdots + \zeta_m a_m)$

$$\leqslant \frac{1}{(2\pi)^m} \int \ldots (m) \ldots \int_{-\pi}^{\pi} u\left(x + \sum_{j=1}^{m} \frac{\zeta_j + e^{i\theta_j}}{1 + \bar{\zeta}_j e^{i\theta_j}} a_j\right) d\theta_1 \ldots d\theta_m,$$

where the m-uple integral is finite whenever $a_1, \ldots, a_m$ are linearly independent.

In fact: since $\zeta \mapsto u(x + e^{i\varphi_1} a_1 + \cdots + e^{i\varphi_{m-1}} a_{m-1} + \zeta a_m)$ is hypoharmonic on an open set containing $\bar{\Delta}(0, 1)$, the integral with respect to φ_m of the integrand in (1) is larger than or equal to:

$$2\pi \prod_{j=1}^{m-1} P(\zeta_j, e^{i\varphi_j}) u(x + e^{i\varphi_1} a_1 + \cdots + e^{i\varphi_{m-1}} a_{m-1} + \zeta_m a_m);$$

from the *Harnack inequality* $\dfrac{1 - |\zeta|}{1 + |\zeta|} \leqslant P(\zeta, e^{i\varphi}) \leqslant \dfrac{1 + |\zeta|}{1 - |\zeta|}$ follows that, if the m-uple integral in (1) were infinite for one point $(\zeta_1, \ldots, \zeta_m) \in [\Delta(0, 1)]^m$, it would be infinite for any such point, hence $u \equiv -\infty$ on the set:

$$\{x + \zeta_1 a_1 + \cdots + \zeta_m a_m : |\zeta_1| < 1, \ldots, |\zeta_m| < 1\}$$

which is open when $a_1, \ldots, a_m$ are linearly independent.

(D) Parts D and E of this section deal with the special case $\Omega = U$, an open set in $\mathbb{C}$, $\mathscr{V} \subset \mathscr{P}(U)$. In this case, and this case only, $\mathscr{P}(U)$ is the set of subharmonic functions on U; let also $\mathscr{S}(U) = -\mathscr{P}(U)$ [resp.: $\mathscr{S}^+(U)$] be the set of superharmonic [resp.: positive superharmonic] functions on U. If U is bounded and $A \subset U$, R_1^A denotes the infimum of all functions $\in \mathscr{S}^+(U)$ which are ($\leqslant 1$ on U and) equal to 1 on A; the lower regularized $(R_1^A)^* \in \mathscr{S}^+(U)$.

A set P in $\mathbb{C}$ is *polar* if and only if, on some (or every) open set $U \supset P$, there is u subharmonic (which may be chosen $\leqslant 0$ if U is bounded) such that $P \subset u^{-1}(-\infty)$. When a bounded open set $U \supset P$, P is polar if and only if $(R_1^P)^* \equiv 0$ ([He]$_1$, Th. 1 in I.3.3). Such a set is not only a null set for the Lebesgue measure on the plane: for any closed disc $\bar{\Delta}(\alpha, r)$, $\{\theta \in \mathbb{R} : \alpha + re^{i\theta} \in P\}$ is also a null set for the measure $d\theta$, since $[\theta \mapsto u(\alpha + re^{i\theta})] \in \mathscr{L}^1(d\theta)$ for every u subharmonic on an open set containing $\bar{\Delta}(\alpha, r)$.

A compact set K in $\mathbb{C}$ is *nonpolar* if and only if there is a positive nonzero measure μ carried by K such that, for one (or every) bounded open connected set $U \supset K: \xi \mapsto \int G_U(\xi, \eta) \, d\mu(\eta)$ is a bounded potential on U. Here G_U is the *Green function* of U, for instance

(3) $$G_U(\xi, \eta) = \ln \left| \frac{R^2 - \xi \bar{\eta}}{R(\xi - \eta)} \right| \quad \text{if } U = \{ \zeta \in \mathbb{C} : |\zeta| < R \},$$

which for an arbitrary U, bounded or not, can be defined as follows: for all $\beta \in U$, $G_U(\zeta, \beta)$ is the smallest positive hyperharmonic function of $\zeta \in U$ obtained by addition of $\ln(1/|\zeta - \beta|)$ to a harmonic function (or the constant $+\infty$); that it be the constant $+\infty$ does not depend on β since it occurs if and only if $\mathbb{C} \setminus U$ is polar. G_U is a continuous function $U^2 \to]0, +\infty]$ and increases with U; if U is the union of an increasing sequence (U_n), G_U is the limit of the increasing sequence (G_{U_n}). This makes it possible to prove the symmetry of G_U, i.e. $G_U(\beta, \alpha) = G_U(\alpha, \beta)$ for all $(\alpha, \beta) \in U^2$, by an argument ([Ts], Th. I.16) assuming that U is *regular* (for the Dirichlet problem), e.g. U bounded with a smooth boundary.

Given μ such that $\int G_U(\xi, \eta) \, d\mu(\eta)$ is bounded: by the Lusin theorem in measure theory, K contains another compact set L such that $v = \mu|_L$ is not zero and $\xi \mapsto \int G_U(\xi, \eta) \, d\mu(\eta)$ is continuous on L, therefore $\xi \mapsto \int G_U(\xi, \eta) \, dv(\eta)$ continuous on L, and also on U by the Evans-Vasilesco theorem [Ev, Va]. So K is nonpolar if and only if there is a positive nonzero measure v carried by K such that, for one (or every) bounded open set $U \supset K: \xi \to \int G_U(\xi, \eta) \, dv(\eta)$ is a bounded and continuous potential on U.

Let $-w \in \mathscr{S}^+(U)$ and $u = \sup_{n \in \mathbb{N}^*} \dfrac{w}{n}$; then $u(\xi) = 0$ if $w(\xi) > -\infty$, $u(\xi) = -\infty$ if $w(\xi) = -\infty$, and $u^* \equiv 0$ since $w^{-1}(-\infty)$ has no interior point; $E_u = \{\xi \in U : u(\xi) < u^*(\xi)\} = w^{-1}(-\infty)$. This example shows that, for a locally bounded from above family $\mathscr{V} \subset \mathscr{P}(U)$ and $u = \sup_{v \in \mathscr{V}} v$, E_u polar is the best possible *convergence theorem* in the one dimensional case; it can be obtained as follows in two steps: Proposition 4.2.2. and Theorem 4.2.5.

Proposition 4.2.2. *Let U be an open set in $\mathbb{C}$. If $\mathscr{V} \subset \mathscr{P}(U)$ is locally bounded from above, then $u = \sup_{v \in \mathscr{V}} v$ has an upper regularized $u^* \in \mathscr{P}(U)$ and every compact subset of $E_u = \{\xi \in U : u(\xi) < u^*(\xi)\}$ is polar.*

Proof. By the topological lemma 4.1.9., we can find an increasing sequence $(v_n)_{n \in \mathbb{N}} \subset \mathscr{P}(U)$ such that $u' = \sup v_n$ has the same upper regularized u^*, and $u' \leqslant u \leqslant u^*$; then it is enough to show that every compact set $K \subset U$ on which $u' < u'^*$ is polar. Since a countable union of polar sets is again polar, we may assume $K \subset \Delta(\alpha, r)$, $r < \rho$, $\bar{\Delta}(\alpha, \rho) \subset U$; let $u^* \leqslant M$ on $\bar{\Delta}(\alpha, \rho)$, $\int G_\Delta(., \eta) \, dv_n(\eta)$ be the potential in $\Delta = \Delta(\alpha, \rho)$ obtained by balayage on $\Delta(\alpha, r)$ of the positive superharmonic function $M - v_n$, i.e. the lower regularized of the infimum of all functions $\in \mathscr{S}^+(\Delta)$ which are $\geqslant M - v_n$ on $\Delta(\alpha, r)$.

Since the positive measure v_n is carried by $\bar{\Delta}(\alpha, r)$, choosing $\xi_0 \in \Delta$ such that $v_0(\xi_0) > -\infty$, we have

$$\left[\inf_{|\eta - \alpha| \leqslant r} G_\Delta(\xi_0, \eta) \right] \int dv_n \leqslant M - v_0(\xi_0) \quad \text{for all } n \in \mathbb{N},$$

hence $\|v_n\|$ bounded, and $\bar{\Delta}(\alpha, r)$ carries another positive measure v such that, modulo the extraction of a suitable subsequence: $\int c \, dv = \lim \int c \, dv_n$ for every finite continuous function c on Δ. Hence follow, for $\xi \in \Delta(\alpha, r)$, since $G_\Delta(., .)$ is lower semicontinuous:

$$M - u'(\xi) = \lim \int G_\Delta(\xi, \eta) \, dv_n(\eta) \geqslant \int G_\Delta(\xi, \eta) \, dv(\eta)$$

and

$$M - u'^*(\xi) \geqslant \int G_\Delta(\xi, \eta) \, dv(\eta);$$

then $u'(\xi) < u'^*(\xi)$ for all $\xi \in K$ entails

$$\lim \int G_\Delta(\xi, \eta) \, dv_n(\eta) > \int G_\Delta(\xi, \eta) \, dv(\eta) \quad \text{for all } \xi \in K.$$

But, if K were nonpolar, there would exist a positive non zero measure μ carried by K such that $\eta \mapsto \int G_\Delta(\xi, \eta) \, d\mu(\xi)$ is finite and continuous, hence

$$\lim \iint G_\Delta(\xi, \eta) \, d\mu(\xi) dv_n(\eta) = \iint G_\Delta(\xi, \eta) \, d\mu(\xi) dv(\eta),$$

a contradiction with the strict inequality above.   □

Corollary 4.2.3. *Let U be an open set in $\mathbb{C}$ and $\mathscr{N}_1(U)$, $\mathscr{N}_2(U)$ the following classes of functions $U \to [-\infty, +\infty[$, both of which may be termed "nearly subharmonic": $u \in \mathscr{N}_1(U)$ [resp.: $\mathscr{N}_2(U)$] if u has an upper regularized $u^* \in \mathscr{P}(U)$ and E_u is a null set for the Lebesgue measure on the plane [resp.: whenever $U \supset \bar{\Delta}(\alpha, r)$, $\{\theta \in \mathbb{R}: \alpha + re^{i\theta} \in E_u\}$ is a null set for the measure $d\theta$]. Then $\mathscr{N}_1(U)$ [resp.: $\mathscr{N}_2(U)$] contains the supremum of any locally bounded from above family $\mathscr{V} \subset \mathscr{N}_1(U)$ [resp.: $\mathscr{N}_2(U)$].*

In order to avoid repetition, we postpone other properties of those classes until Theorem 4.2.7.

Proof. $u = \sup_{v \in \mathscr{V}} v$ has (Coroll. 4.1.8.) an upper regularized $u^* \in \mathscr{P}(U)$, which (Prop. 4.1.9.) is also the upper regularized of $u' = \sup v_n$ for a suitably chosen sequence $(v_n) \subset \mathscr{V}$ and finally, by corollary 4.1.8. again, the upper regularized of $u'' = \sup v_n^*$. To sum up: $u' \leqslant u$, $u' \leqslant u''$, $u^* = u'^* = u''^*$.

Now $u(\zeta) < u^*(\zeta)$ requires $u'(\zeta) < u''^*(\zeta)$, therefore either $u'(\zeta) < u''(\zeta)$, $v_n(\zeta) < v_n^*(\zeta)$ for some n, or $u''(\zeta) < u''^*(\zeta)$; in other words, E_u is contained in the union of $E_{u''}$ and the E_{v_n}. By the assumption, each E_{v_n} is a null set for the Lebesgue measure on the plane [resp.: its intersection with the circumference of any closed disc contained in U is a null set for the measure $d\theta$]; on the other hand, $E_{u''}$ is a Borel set, every compact subset of which is polar (Prop. 4.2.2.), therefore $E_{u''}$ has the same property as the E_{v_n}.   □

(E) Let U be moreover bounded, A any subset of U, R_1^A the infimum of all functions $\in \mathscr{S}^+(U)$ which are ($\leqslant 1$ on U and) equal to 1 on A: this is a nearly superharmonic function, and vanishes a.e. if and only if A is polar; in other words (with the notation $d\zeta$ for the Lebesgue measure on the plane) $\varphi^*(A) = \int_U R_1^A \, d\zeta$ vanishes if and only if A is polar.

Now $\varphi^*(A)$ is the outer capacity of A for the Choquet [Ch] capacity on U defined as follows: for a compact set $K \subset U$, $\varphi(K) = \int R_1^K \, d\zeta$; for an open set $\omega \subset U$, $\varphi(\omega) = \sup_{K \subset \omega} \varphi(K)$; for any set $A \subset U$,

$$\varphi_*(A) = \sup_{K \subset A} \varphi(K) \qquad \text{and} \qquad \varphi^*(A) = \inf_{\omega \supset A} \varphi(\omega)$$

are the inner and outer capacities respectively. By Choquet's theory, every $K_{\sigma\delta}$ set A is capacitable, i.e. satisfies $\varphi_*(A) = \varphi^*(A)$, and therefore is polar if and only if its compact subsets are polar.

This makes it possible to improve Proposition 4.2.2. into Theorem 4.2.5.

Definition 4.2.4. *Let U be an open set in $\mathbb{C}$; a function $u: U \to [-\infty, +\infty[$ is quasisubharmonic, $u \in Q(U)$, if u has an upper regularized $u^* \in \mathscr{P}(U)$ and E_u is a polar set.*

Theorem 4.2.5. *Let U be an open set in $\mathbb{C}$; if $\mathscr{V} \subset \mathscr{P}(U)$ is locally bounded from above, then $u = \sup_{v \in \mathscr{V}} v \in Q(U)$.*

Proof. Resuming the proof of Proposition 4.2.2. with the same notation, we now have to show that $A_k = \left\{ \zeta \in U : u'(\zeta) \leqslant u^*(\zeta) - \dfrac{1}{k} \right\}$ is polar for all $k \in \mathbb{N}^*$. Since u' is the limit of an increasing sequence of upper semi-continuous functions v_n and u^* the limit of a decreasing sequence of continuous functions $c_n > u^*$, A_k is the intersection of the open sets $\left\{ \zeta \in U : v_n(\zeta) < c_n(\zeta) - \dfrac{1}{k} \right\}$; then A_k is a $K_{\sigma\delta}$, and it is enough to check that every compact subset of A_k is polar, which was done in the proof of Proposition 4.2.2.   □

Corollary 4.2.6. $Q(U)$ *contains the supremum of any locally bounded from above family* $\mathcal{V} \subset Q(U)$.

The proof of Corollary 4.2.3. fits exactly. So the three classes $\mathcal{N}_1(U)$, $\mathcal{N}_2(U)$, $Q(U)$ of nearly or quasi-subharmonic functions have in common the property to be preserved by the operation "supremum of a locally bounded from above family", which does not preserve $\mathcal{P}(U)$. Further common properties are listed in the following theorem and its corollary. Note that $\mathcal{P}(U) \subset Q(U) \subset \mathcal{N}_1(U) \cap \mathcal{N}_2(U)$.

Theorem 4.2.7. *Let* U *be an open set in* $\mathbb{C}$. a) *Given* $\alpha \in U$: *for all* $u \in \mathcal{N}_1(U)$ [*resp.*: $\mathcal{N}_2(U)$], $u^*(\alpha)$ *is the limit as* $r \to 0$ *of the mean value of* u *over the disc* $\Delta(\alpha, r)$ *or* $\bar{\Delta}(\alpha, r)$ [*resp.*: *the circumference of this disc*], *which is a convex increasing function of* $\ln r$. b) *Each class* $\mathcal{P}(U)$, $Q(U)$, $\mathcal{N}_1(U)$, $\mathcal{N}_2(U)$ *is a convex cone of functions; if the* u_k *are taken in any one of them, and* $\lambda_k > 0$ *for all* k, *then* $u = \sum_{k=1}^{n} \lambda_k u_k$ *implies* $u^* = \sum_{k=1}^{n} \lambda_k u_k^*$. c) *If* U *is connected, and the decreasing sequence* (u_n) *taken in any one class, then* $u = \lim u_n$ *either belongs to the same class or is the constant* $-\infty$, *and* $u^* = \lim u_n^*$.

Proof. a) By the definition of $\mathcal{N}_1(U)$, $\mathcal{N}_2(U)$, the mean value in question is the same for u and u^*, and $u^* \in \mathcal{P}(U)$.
b) We take the u_k in one of the last three classes; in any case $u' = \sum_{k=1}^{n} \lambda_k u_k^* \in \mathcal{P}(U)$, hence $u \leqslant u^* \leqslant u'$, with the following two consequences. First: whenever $U \supset \bar{\Delta}(\alpha, r)$, u^* has the same mean value as u and u' over the disc [if the u_k belong to $\mathcal{N}_1(U)$] or its circumference [if they belong to $\mathcal{N}_2(U)$]; this common mean value is $\geqslant u'(\alpha)$ since $u' \in \mathcal{P}(U)$, therefore $\geqslant u^*(\alpha)$, and $u^* \in \mathcal{P}(U)$; moreover this common mean value tends to $u^*(\alpha)$ and $u'(\alpha)$ as $r \to 0$, and $u^* = u'$. Secondly: $E_u \subset \bigcup_{k=1}^{n} E_{u_k}$.
c) We again take the u_n in one of the last three classes, in any case $u' = \lim u_n^*$ is hypoharmonic on U. If $u' \equiv -\infty$, then $u \equiv -\infty$, $u^* = u'$. If $u' \in \mathcal{P}(U)$, then $u \leqslant u^* \leqslant u'$, with the same consequences as in the proof of b). $\square$

Corollary 4.2.8. *Let* U *be a connected open set in* $\mathbb{C}$, μ *a positive measure on a compact metric space* T *and* v *a locally bounded from above function of* $(\zeta, t) \in U \times T$, *with the following two properties*: (i) *for all* $t \in T$, $v(.\,, t) \in \mathcal{P}(U)$ [*resp.*: $Q(U)$, $\mathcal{N}_1(U)$, $\mathcal{N}_2(U)$]; (ii) *for all* $\zeta \in U$, $v(\zeta, .)$ *is upper semicontinuous at* μ-*a.e. point* $t \in T$. *Then*: a) $u = \int v(.\,, t) \, d\mu(t) \in \mathcal{P}(U)$ [*resp.*: $Q(U)$, $\mathcal{N}_1(U)$, $\mathcal{N}_2(U)$] *or* $\equiv -\infty$; b) *if the upper regularized* v^* *of* v *with respect to* ζ *also has property* (ii) *above, then* $u^* = \int v^*(.\,, t) \, d\mu(t)$.

Proof. a) By the first assumption on v, $\sup_{t \in T} v(\zeta, t)$ is a locally bounded from above function of $\zeta \in U$. Consider once more, for each $n \in \mathbb{N}^*$, a finite partition of T into Borel sets T_n^j, $j = 1, \ldots, J(n)$, with diameters $\leqslant \frac{1}{n}$, and this time suppose moreover that each T_{n+1}^k is a subset of some T_n^j: on account of (i)

$$u_n(\zeta) = \sum_{j=1}^{J(n)} \mu(T_n^j) \sup_{t \in T_n^j} v(\zeta, t) \quad (0 \times -\infty = 0 \text{ as usual})$$

defines a decreasing sequence of functions $u_n \in Q(U)$ [resp.: $Q(U)$, $\mathcal{N}_1(U)$, $\mathcal{N}_2(U)$]; $u_n(\zeta)$ is the μ-integral of the function of t which, on each T_n^j, has the constant value $\sup_{t \in T_n^j} v(\zeta, t)$. On account of (ii), this sequence of μ-measurable functions of t decreases to $v(\zeta, t)$ μ-a.e., which makes $v(\zeta, .)$ μ-measurable and entails $u(\zeta) = \lim u_n(\zeta)$, $u \in Q(U)$ [resp.: $Q(U)$, $\mathcal{N}_1(U)$, $\mathcal{N}_2(U)$] or $\equiv -\infty$.

Finally, if $v(., t)$ is upper semicontinuous for all $t \in T$, u too is upper semicontinuous by the Fatou lemma.

b) Let $v^*(\zeta, t) = \limsup_{\zeta' \to \zeta} v(\zeta', t)$, $\zeta \in U$, and

$$u'(\zeta) = \int v^*(\zeta, t)\, d\mu(t), \, u_n'(\zeta) = \sum_{j=1}^{J(n)} \mu(T_n^j) \sup_{t \in T_n^j} v^*(\zeta, t).$$

By Corollary 4.1.8., $\zeta \mapsto \sup_{t \in T_n^j} v(\zeta, t)$ and $\zeta \mapsto \sup_{t \in T_n^j} v^*(\zeta, t)$ have the same upper regularized $\in \mathcal{P}(U)$, and therefore: if $v(., t) \in \mathcal{N}_1(U)$ for all $t \in T$, they have the same mean value over every closed disc $\bar{\Delta}(\alpha, r) \subset U$, and u_n, u_n' also have the same mean value over every such disc; similarly, if $v(., t) \in \mathcal{N}_2(U)$ for all $t \in T$, then u_n, u_n' have the same mean value over the circumference of every such disc. Given $\alpha \in U$, making $r \to 0$, we obtain $u_n^*(\alpha) = u_n'^*(\alpha)$; making $n \to \infty$, we obtain $u^*(\alpha) = u'^*(\alpha) = \int v^*(\alpha, t)\, d\mu(t)$ since u' is hypoharmonic. $\square$

Note that the proof and results remain unaltered if the assumption (i) is made only for μ-a.e.t, with $v(., t) \equiv -\infty$ for the other $t \in T$.

4.3 Back to the infinite dimensional case

Ω is again an open set in a l.c. space X. From Example 4.1.5.a follows that v p.s.h., $f\mathcal{G}$-analytic, do not in general imply $v \circ f$ p.s.h.: the following composition property, which generalizes Theorem 4.2.1., can hold for analytic maps only.

Theorem 4.3.1. *Let $f \in \mathscr{A}(\Omega, Z)$, Z s.c.l.c.: v p.s.h. on an open set Γ (in Z) containing $f(\Omega)$ implies $v \circ f$ p.h.h. on Ω.*

Proof. Let E be a complex line in X and U an open connected set in $\mathbb{C}$ such that $\{a + \zeta x : \zeta \in U\}$ is a connected component of $E \cap \Omega$.

First consider the special case where f is a continuous polynomial map φ: $\varphi(a + \zeta x) = \sum_{k=0}^{p} c_k \zeta^k$, $\zeta \in \mathbb{C}$. The open set $\{(\zeta_1, \ldots, \zeta_p) \in \mathbb{C}^p : c_0 + c_1 \zeta_1 + \cdots + c_p \zeta_p \in \Gamma\}$ has a connected component containing $(\zeta, \ldots, \zeta^p)$ for all $\zeta \in U$; on this connected component $v(c_0 + c_1 \zeta_1 + \cdots + c_p \zeta_p)$ is a p.s.h. function of $(\zeta_1, \ldots, \zeta_p)$ or $\equiv -\infty$; then, by Theorem 4.2.1., $\zeta \mapsto v \circ \varphi(a + \zeta x)$ is subharmonic on U or $\equiv -\infty$.

In the general case, we have to show that, if $|\zeta| \leq 1$ implies $a + \zeta x \in \Omega$, i.e. if $a \in \Omega$, $x \in \omega(a)$, then $v \circ f(a) \leq MV v \circ f(a + e^{i\theta}x)$; let $v \circ f(a) > -\infty$.

We know that the Taylor expansion around 0 of $\zeta \mapsto f(a + \zeta x)$:

$$f(a + \zeta x) = f(a) + \sum_{k \in \mathbb{N}^*} \frac{\zeta^k}{k!} (\hat{D}_a^k f)(x),$$

is uniformly summable for $|\zeta| \leq 1$; then, for sufficiently large n:

$$\varphi_n(a + \zeta x) = f(a) + \sum_{k=1}^{n} \frac{\zeta^k}{k!} (\hat{D}_a^k f)(x) \in \Gamma \quad \text{for all } \zeta \in \bar{\Delta}(0, 1)$$

and, by the first part of the proof, $\zeta \mapsto v \circ \varphi_n(a + \zeta x)$ is subharmonic on the connected component containing $\bar{\Delta}(0, 1)$ of the open set $\{\zeta \in \mathbb{C} : \varphi_n(a + \zeta x) \in \Gamma\}$. From

$$v \circ \varphi_n(a) \leq MV\{v \circ \varphi_n(a + e^{i\theta}x)\} \leq MV\left(\sup_{m \geq n} v \circ \varphi_m(a + e^{i\theta}x)\right)$$

follows, since v is upper semicontinuous:

$$v \circ f(a) \leq MV\left(\limsup_{n \to \infty} v \circ \varphi_n(a + e^{i\theta}x)\right) \leq MV\{v \circ f(a + e^{i\theta}x)\}$$

$\square$

Proposition 4.3.2. *Let μ be a positive measure on a compact metric space T and $v(., t) \in \mathscr{P}(\Omega)$ for all $t \in T$. Then $u(x) = \int v(x, t) \, d\mu(t)$ is plurihypoharmonic on Ω under either assumption: a) v an upper semicontinuous function of $(x, t) \in \Omega \times T$;*

b) *X metrizable; v a locally bounded from above function of (x, t); for all $x \in \Omega$, $v(x, .)$ upper semicontinuous at μ-a.e. point $t \in T$.*

Proof. By Corollary 4.2.8., u has property (ii) in Definition 4.1.1.; so we only have to check (i). In case b), this is obtained by the Fatou lemma, as it was for Corollary 4.2.8.a, in the end of the proof. In case a), let $v \leqslant M$ on $V \times T$, V a neighbourhood of a in Ω; since $V \times T$ is a completely regular space, $(M - v)|_{V \times T}$ is the supremum of the upper directed family of all real positive continuous functions $c \leqslant M - v$ on $V \times T$; then, for $x \in V$, $M\mu(1) - u(x)$ is the supremum of the continuous functions $x \mapsto \int c(x, t)\, d\mu(t)$. □

Definition 4.3.3. a) *Let $\mathscr{P}^0(\Omega) = \mathscr{P}(\Omega)$ and the classes $\mathscr{P}^n(\Omega)$, $n \in \mathbb{N}^*$, be inductively defined as follows: $u \in \mathscr{P}^{2n+1}(\Omega)$ if u is the limit of a locally bounded from above increasing sequence taken in $\mathscr{P}^{2n}(\Omega)$; $u \in \mathscr{P}^{2n+2}(\Omega)$ if u is the limit of a decreasing sequence taken in $\mathscr{P}^{2n+1}(\Omega)$, and $u^{-1}(-\infty)$ has an empty interior.*
b) *A function $u \colon \Omega \to [-\infty, +\infty[$ belongs to the class $Q(\Omega)$ if: (i) u is locally bounded from above and $u^{-1}(-\infty)$ has an empty interior; (ii) for every complex line E in X, $u|_{E \cap \Omega}$ is quasisubharmonic (Def. 4.2.4.) or $\equiv -\infty$ on each connected component of $E \cap \Omega$.*

$Q(\Omega)$ contains the supremum of *any* locally bounded from above family $\subset Q(\Omega)$ (Coroll. 4.2.6.) and the limit u of a decreasing sequence $\subset Q(\Omega)$, provided that $u^{-1}(-\infty)$ has an empty interior (Th. 4.2.7.c.). Therefore $\bigcup_{n \in \mathbb{N}} \mathscr{P}^n(\Omega) \subset Q(\Omega)$, and this would remain true if $\mathscr{P}^{2n+1}(\Omega)$ were defined as the class of suprema of all locally bounded from above families taken in $\mathscr{P}^{2n}(\Omega)$; but this more general case will require a separate treatment below (Coroll. 4.3.7. and 4.3.9.). All functions $\in Q(\Omega)$ are submedian, for the mean value inequality (ii) in Definition 4.1.6. follows from Theorem 4.2.7.a, and therefore (Th. 4.1.7.a) have upper regularized $\in \mathscr{P}(\Omega)$. By Theorem 4.1.7.c., if Ω is connected, the limit of a decreasing sequence taken in $Q(\Omega)$ [resp.: $\mathscr{P}^{2n+1}(\Omega)$], either belongs to $Q(\Omega)$ [resp.: $\mathscr{P}^{2n+2}(\Omega)$], or is the constant $-\infty$; therefore, given a filter $\mathscr{F}$, with a countable basis, on a locally bounded from above family $V \subset Q(\Omega)$, $\limsup_{\mathscr{F}} v$ either belongs to $Q(\Omega)$ or is the constant $-\infty$.

A finite family taken in $Q(\Omega)$ [resp.: $\mathscr{P}^n(\Omega)$] has its supremum in $Q(\Omega)$ [resp.: $\mathscr{P}^n(\Omega)$, by induction]; thus a countable locally bounded from above family taken in $\mathscr{P}^{2n}(\Omega)$ also has its supremum in $\mathscr{P}^{2n+1}(\Omega)$. $Q(\Omega)$, and each $\mathscr{P}^n(\Omega)$ by induction, are convex cones of functions: for $u, v \in Q(\Omega)$, the fact that $(u + v)^{-1}(-\infty)$ has an empty interior is obtained by the argument used for $u, v \in \mathscr{P}(\Omega)$ (Prop. 4.1.4.a).

We now aim at some generalization to $u \in Q(\Omega)$ of the relations between u and its upper regularized u^* which appear in Theorem 4.2.7. for $u \in Q(U)$, U open in $\mathbb{C}$. The following method, essentially due to Coeuré [Coe], rests on a skilful extension of inequalities (1), (2) in 4.2(C).

Definition 4.3.4. *In a q.c.l.c. space X, let $\mathscr{A} = (a_j)_{j \in J}$ be a summable family of vectors: the generalized polydisc (centred at the origin) generated by the a_j is the compact set*

$$K(\mathscr{A}) = \left\{ \sum_{j \in J} \zeta_j a_j \colon |\zeta_j| \leqslant 1 \quad \text{for all } j \in J \right\}$$

and its distinguished boundary the compact set

$$K^*(\mathscr{A}) = \left\{ \sum_{j \in J} \tau_j a_j \colon |\tau_j| = 1 \quad \text{for all } j \in J \right\}.$$

In both cases, compactness follows from the continuity of the map $[\bar{\Delta}(0, 1)]^J \ni (\zeta_j)_{j \in J} \mapsto \sum_{j \in J} \zeta_j a_j$ (see Ex. 1.2.2.). For $|\tau_j| = 1$, we shall write $\tau_j = e^{i\theta_j}$ and denote by $d\tau$ or $\bigotimes_{j \in J} \dfrac{d\theta_j}{2\pi}$ the unique Radon measure on the compact (for the product topology of $\mathbb{C}^J$) set $\mathbb{T}$ of $\tau = (\tau_j)_{j \in J} = (e^{i\theta_j})_{j \in J}$ such that, whenever a continuous function f on $\mathbb{T}$ depends only on a finite number of variables $\tau_{j_1}, \ldots, \tau_{j_m}$, i.e. $f(\tau) = g(\tau_{j_1}, \ldots, \tau_{j_m})$, then

$$\int f \, d\tau = \frac{1}{(2\pi)^m} \int \ldots (m) \ldots \int_{-\pi}^{\pi} g(e^{i\theta_{j_1}}, \ldots, e^{i\theta_{j_m}}) \, d\theta_{j_1} \ldots d\theta_{j_m}; \quad \tau(1) = 1.$$

Proposition 4.3.5. *Let Ω contain the generalized polydisc $x + K(\mathscr{A})$ (with centre x): for all $u \in \bigcup_{n \in \mathbb{N}} \mathscr{P}^n(\Omega)$, for all $r \in {]0, 1]}$ we have*

$$u(x) \leqslant \int u\left(x + r \sum_{j \in J} \tau_j a_j \right) d\tau = MV u[x + rK^*(\mathscr{A})],$$

which reads: mean value of u over the distinguished boundary of $x + rK(\mathscr{A})$; this is a convex increasing function of $\ln r$, which obviously tends to $u(x)$ as $r \to 0$ if $u \in \mathscr{P}(\Omega)$.

Proof. Since $\mathscr{P}^n(\Omega)$ proceeds from $\mathscr{P}(\Omega)$ by n successive limits of monotone sequences, we may consider only the case $u \in \mathscr{P}(\Omega)$. Then, for all $b \in K(\mathscr{A})$, $MV u(x + re^{i\varphi}b)$ is a convex increasing function of $\ln r$ and $\geqslant u(x)$, and

therefore $\dfrac{1}{2\pi}\displaystyle\iint u\left(x + re^{i\varphi}\sum_{j\in J}\tau_j a_j\right)d\tau\,d\varphi$ has the same properties; but this
is the right hand member of our mean value inequality if we remark that
$\int f(e^{i\varphi}\tau)\,d\tau$ does not depend on the real variable φ: first for an f continuous
on $\mathbb{T}$ (since such an f is the uniform limit of continuous functions depending
only on a finite number of variables); hence for an upper semicontinuous
f. $\square$

Theorem 4.3.6. *Let* Ω *contain the generalized polydisc* $x + K(\mathscr{A})$ *and* $u \in$
$\displaystyle\bigcup_{n\in\mathbb{N}}\mathscr{P}^n(\Omega)$. *Then:* a) *if* $|\zeta_j| < 1$ *for all* $j \in J$,

(1) $\quad u\left(x + \displaystyle\sum_{j\in J}\zeta_j a_j\right) \leqslant \displaystyle\int u\left(x + \sum_{j\in J}\frac{\zeta_j + e^{i\theta_j}}{1 + \overline{\zeta}_j e^{i\theta_j}}a_j\right)\bigotimes_{j\in J}\frac{d\theta_j}{2\pi};$

b) *if moreover the family* $(\zeta_j)_{j\in J}$ *of complex numbers is summable (which
implies* $\zeta_j \neq 0$ *only for a countable set of indeces j), the infinite product*
$\displaystyle\prod_{j\in J} P(\zeta_j, e^{i\varphi_j})$ *of Poisson kernels makes sense and formula* (1) *may be rewritten
as*

(2) $\quad u\left(x + \displaystyle\sum_{j\in J}\zeta_j a_j\right) \leqslant \displaystyle\int \prod_{j\in J} P(\zeta_j, e^{i\varphi_j})u\left(x + \sum_{j\in J}e^{i\varphi_j}a_j\right)\bigotimes_{j\in J}\frac{d\varphi_j}{2\pi}.$

Proof. a) Again we assume $u \in \mathscr{P}(\Omega)$. We dealt with the special case $\zeta_j = 0$
for all $j \in J$ in Proposition 4.3.5., but without fully using the assumption
$u \in \mathscr{P}(\Omega)$: only the upper semicontinuity of u: $x + K(\mathscr{A}) \to [-\infty, +\infty[$
and, for every $b \in K(\mathscr{A})$, the hypoharmonicity of $\zeta \mapsto u(x + \zeta b)$ on an open
set containing $\overline{\Delta}(0, 1)$.
   Now let $|\zeta_j| < 1$ for all $j \in J$ and $\lambda \in]0, 1[$ be given; let U be the set of
$\tau \in \mathbb{C}^J$ with $\|\tau\| = \sup_{j\in J}|\tau_j| < \dfrac{1}{\lambda}$, an open set in $l^\infty(J)$ but not in $\mathbb{C}^J$. For
$\tau \in U$:

$$v(\tau) = u\left(x + \sum_{j\in J}\frac{\zeta_j + \lambda\tau_j}{1 + \overline{\zeta}_j\lambda\tau_j}a_j\right)$$

makes sense and v is upper semicontinuous for both topologies of U;
furthermore, the map $U \ni \tau \mapsto x + \displaystyle\sum_{j\in J}\frac{\zeta_j + \lambda\tau_j}{1 + \overline{\zeta}_j\lambda\tau_j}a_j$, whose image lies in Ω,
belongs to $\mathscr{A}(U, X)$ as a uniform limit of analytic maps: here we choose
the topology which makes U open. For this topology, v is p.s.h. or $\equiv -\infty$
on U (Th. 4.3.1.) and, for each $\beta \in U$, $\zeta \mapsto v(\zeta\beta)$ is hypoharmonic on an
open set containing $\overline{\Delta}(0, 1)$.

Having checked this, we now consider the product topology, which makes $\mathbb{C}^J$ a complete l.c. space and $\mathscr{E} = (e_j)$ $\left(\text{here we write } \tau = \sum_{j \in J} \tau_j e_j\right)$ a summable family. Since $K(\mathscr{E}) = \left\{\sum_{j \in J} \tau_j e_j : \|\tau\| \leqslant 1\right\}$ is contained in U, by Proposition 4.3.5. we have

$$u\left(x + \sum_{j \in J} \zeta_j a_j\right) = v(0) \leqslant \int v(\tau)\,d\tau = \int u\left(x + \sum_{j \in J} \frac{\zeta_j + \lambda\tau_j}{1 + \bar{\zeta}_j\lambda\tau_j} a_j\right)d\tau$$

and finally, by the Fatou lemma, since u is upper semicontinuous:

$$u\left(x + \sum_{j \in J} \zeta_j a_j\right) \leqslant \int u\left(x + \sum_{j \in J} \frac{\zeta_j + \tau_j}{1 + \bar{\zeta}_j\tau_j} a_j\right)d\tau.$$

b) As we did in 1.2(A), we denote by $\mathscr{F}$ the family of finite subsets of J. Given the ζ_j with $|\zeta_j| < 1$ for all $j \in J$ and $\sum_{j \in J} |\zeta_j| < \infty$, the Harnack inequality (see 4.2(C)) implies the uniform summability, for $\tau' = (e^{i\varphi_j})_{j \in J} \in \mathbb{T}$, of the family of real functions $\tau' \mapsto \ln P(\zeta_j, e^{i\varphi_j})$, $j \in J$, therefore the existence of a real positive continuous function R on the compact set $\mathbb{T}$ with the property: for all $\varepsilon > 0 \; \exists\, F_0 \in \mathscr{F}$ such that $F_0 \subset F \in \mathscr{F}$ implies

$$\left| R(\tau') - \prod_{j \in F} P(\zeta_j, e^{i\varphi_j}) \right| \leqslant \varepsilon \quad \text{for all } \tau' = (e^{i\varphi_j})_{j \in J} \in \mathbb{T}.$$

Note that the $(d\tau)$-integral of R over $\mathbb{T}$ is 1; we set $R(\tau') = \prod_{j \in J} P(\zeta_j, e^{i\varphi_j})$.

If a continuous function f on $\mathbb{T}$ depends only on a finite number of variables, say $f(\tau) = g(\tau_{j_1}, \ldots, \tau_{j_m})$, then:

$$\int \ldots (m) \ldots \int_{-\pi}^{\pi} g\left(\frac{\zeta_{j_1} + e^{i\theta_{j_1}}}{1 + \bar{\zeta}_{j_1} e^{i\theta_{j_1}}}, \ldots, \frac{\zeta_{j_m} + e^{i\theta_{j_m}}}{1 + \bar{\zeta}_{j_m} e^{i\theta_{j_m}}}\right) \bigotimes_{k=1}^{m} \frac{d\theta_{j_k}}{2\pi}$$

$$= \int \ldots (m) \ldots \int_{-\pi}^{\pi} \prod_{k=1}^{m} P(\zeta_{j_k}, e^{i\varphi_{j_k}}) g(e^{i\varphi_{j_1}}, \ldots, e^{i\varphi_{j_m}}) \bigotimes_{k=1}^{m} \frac{d\varphi_{j_k}}{2\pi},$$

where: on the left, $\bigotimes_{k=1}^{m} \frac{d\theta_{j_k}}{2\pi}$ may be replaced by $\bigotimes_{j \in J} \frac{d\theta_j}{2\pi}$; on the right,

$$\prod_{k=1}^{m} P(\zeta_{j_k}, e^{i\varphi_{j_k}}) \bigotimes_{k=1}^{m} \frac{d\varphi_{j_k}}{2\pi} \quad \text{by} \quad \prod_{j \in J} P(\zeta_j, e^{i\varphi_j}) \bigotimes_{j \in J} \frac{d\varphi_j}{2\pi}.$$

The equality between the right hand members of (1) and (2) follows: first for u continuous on $x + K(\mathscr{A})$ $\left[\text{since then } u\left(x + \sum_{j \in J} \tau_j a_j\right) \text{ is the uniform}\right.$

limit of continuous functions on $\mathbb{T}$ depending only on a finite number of variables $\Big]$; hence for u upper semicontinuous on $x + K(\mathscr{A})$. $\square$

Corollary 4.3.7. *Let* Ω *contain the generalized polydisc* $x + K(\mathscr{A})$ *and* $\mathscr{V} \subset \bigcup_{n \in \mathbb{N}} \mathscr{P}^n(\Omega)$ *be locally bounded from above; if either* $u = \sup_{v \in \mathscr{V}} v$ *or* $u = \limsup_{\mathscr{F}} v$, *where* $\mathscr{F}$ *is a filter on* $\mathscr{V}$ *with a countable basis, then formulas* (1), (2) *in Theorem* 4.3.6. *still hold as follows, with* $u \leqslant M$ *on* $x + K(\mathscr{A})$: a) *if* $|\zeta_j| < 1$ *for all* $j \in J$,

$$(1') \qquad M - u\Big(x + \sum_{j \in J} \zeta_j a_j\Big) \geqslant \bar{\int}\bigg[M - u\Big(x + \sum_{j \in J} \frac{\zeta_j + e^{i\theta_j}}{1 + \bar{\zeta}_j e^{i\theta_j}} a_j\Big)\bigg]\bigotimes_{j \in J} \frac{d\theta_j}{2\pi};$$

b) *if moreover the family* $(\zeta_j)_{j \in J}$ *of complex numbers is summable,*

$$(2') \qquad M - u\Big(x + \sum_{j \in J} \zeta_j a_j\Big)$$

$$\geqslant \bar{\int} \prod_{j \in J} P(\zeta_j, e^{i\varphi_j})\bigg[M - u\Big(x + \sum_{j \in J} e^{i\varphi_j} a_j\Big)\bigg]\bigotimes_{j \in J} \frac{d\varphi_j}{2\pi}.$$

Proof. If $u = \sup_{v \in \mathscr{V}} v$, then (1), (2) for each v imply (1'), (2') for u since $M - u = \inf_{v \in \mathscr{V}} (M - v)$. If (A_n) is a countable basis of $\mathscr{F}$, $A_{n+1} \subset A_n$ for all $n \in \mathbb{N}$, and $u_n = \sup_{v \in A_n} v$, then (1'), (2') for each u_n imply (1'), (2') for u since $(M - u)$ is the limit of the increasing sequence $(M - u_n)$. $\square$

Corollary 4.3.8. *Let* Ω *be an open set in a Fréchet space* X.

a) *Given* $u \in \bigcup_{n \in \mathbb{N}} \mathscr{P}^n(\Omega)$ *and* $x \in \Omega$, *one can find generalized polydiscs* $K(\mathscr{A})$ *such that*

$$(3) \qquad u^*(x) = \lim_{r \to 0} MVu[x + rK^*(\mathscr{A})].$$

b) *If* $u_k \in \bigcup_{n \in \mathbb{N}} \mathscr{P}^n(\Omega)$ *and* $\alpha_k > 0$ *for all* $k \in \{1, \ldots, q\}$, *then*

$$\Big(\sum_{k=1}^{q} \alpha_k u_k\Big)^* = \sum_{k=1}^{q} \alpha_k u_k^*.$$

c) *If* $(u_k)_{k \in \mathbb{N}^*}$ *is a decreasing sequence taken in* $\bigcup_{n \in \mathbb{N}} \mathscr{P}^n(\Omega)$, *then* $(\lim u_k)^* = \lim u_k^*$.

Proof a) Since u^* is upper semicontinuous and $\geqslant u$, formula (3) is equivalent to the inequality

(4) $u^*(x) \leqslant MVu[x + rK^*(\mathscr{A})]$ whenever $\Omega \supset x + rK(\mathscr{A})$

which we shall now obtain for suitably chosen $\mathscr{A}$.

If $u^*(x) = u(x)$, (4) is included in Proposition 4.3.5., and holds for all $\mathscr{A}$. If not, let $(c_v) \subset X \setminus \{0\}$ be a sequence tending to 0 such that $u^*(x) = \lim_{v \to \infty} u(x + c_v)$ and let $(p_n)_{n \in \mathbb{N}^*}$ be an increasing sequence of seminorms defining the topology of X. For each $n \in \mathbb{N}^*$, choose $v(n)$ so that $p_n(c_{v(n)}) \leqslant \dfrac{1}{n^2}$, which implies that $v(n) \to \infty$ with n; let $\mathscr{A}$ be any summable family of vectors $\in X$ including $a_n = \sqrt{n} c_{v(n)}$ for all $n \in \mathbb{N}^*$, and $r > 0$ be such that $\Omega \supset x + rK(\mathscr{A})$; finally, if necessary by addition of a suitable constant, let $u \leqslant 0$ on $x + rK(\mathscr{A})$.

By formula (2) in Theorem 4.3.6., with $r\mathscr{A}$ instead of $\mathscr{A}$, and the Harnack inequality, we have

$$u\left(x + \frac{ra_n}{r\sqrt{n}}\right) \leqslant \frac{1 - (1/r\sqrt{n})}{1 + (1/r\sqrt{n})} \int u\left(x + r \sum_{j \in J} \tau_j a_j\right) d\tau \quad \text{for } r\sqrt{n} > 1,$$

from which (4) follows as $n \to \infty$.

b) Let $u, v \in \bigcup_{n \in \mathbb{N}} \mathscr{P}^n(\Omega)$. Given $x \in \Omega$, we have just constructed summable families $\mathscr{A}, \mathscr{B}$ of vectors in X such that $u^*(x), v^*(x)$ are the limits as $r \to 0$ of the mean values of u, v over the distinguished boundary of the generalized polydisc $x + rK(\mathscr{A} \cup \mathscr{B})$. Now (Def. 4.3.3), $(u + v) \in Q(\Omega)$, $(u + v)^* \in \mathscr{P}(\Omega)$ and, by Proposition 4.3.5., $(u + v)^*(x)$ is the limit as $r \to 0$ of the mean value of $(u + v)^*$ over the same distinguished boundary; thus we have $u^*(x) + v^*(x) \leqslant (u + v)^*(x)$, while the inequality $u^* + v^* \geqslant (u + v)^*$ is obvious.

c) $u = \lim u_k \leqslant u^* \leqslant \lim u_k^*$, and u^* is p.h.h. since $u_k \in Q(\Omega)$ for all k. Given $x \in \Omega$: $u_k^*(x) = u_k(x)$ for an infinity of indeces k implies $u^*(x) = \lim u_k^*(x)$; let then, for each large enough k, $(c_v^k) \subset X \setminus \{0\}$ be a sequence tending to 0 such that $u_k^*(x) = \lim_{v \to \infty} u_k(x + c_v^k)$ and, for $n \geqslant k$, the index $v(n, k)$ be chosen so that $p_n(c_{v(n,k)}^k) \leqslant \dfrac{1}{k^2 n^2}$, which implies that $v(n, k) \to \infty$ with n for each k. Let $\mathscr{A}$ be any summable family of vectors $\in X$ including each $a_n^k = \sqrt{n} c_{v(n,k)}^k$, $r_0 > 0$ be such that $\Omega \supset x + r_0 K(\mathscr{A})$ and finally $u_1 \leqslant 0$ on $x + r_0 K(\mathscr{A})$; then, for each k, by the same argument as in a):

$$u_k^*(x) = \lim_{n \to \infty} u_k\left(x + \frac{ra_n^k}{r\sqrt{n}}\right) \leqslant \int u_k\left(x + r \sum_{j \in J} \tau_j a_j\right) d\tau, \, 0 < r \leqslant r_0.$$

Hence follows, for $0 < r \leqslant r_0$:

(5) $\lim u_k^*(x) \leqslant MV u[x + rK^*(\mathscr{A})] \leqslant MV u^*[x + rK^*(\mathscr{A})]$;

making $r \to 0$ and using the upper semicontinuity of u^*, we get $\lim u_k^*(x) \leqslant u^*(x)$. $\square$

An example due to Noverraz [Nov]$_1$ shows that statement c) may fail if the space X is not complete. Let Ω be the open unit ball in $X = c_{00}(\mathbb{N}^*)$ with the notation $x = (\xi_1, \xi_2, \ldots)$, $\|x\| = \sup_{p \in \mathbb{N}^*} |\xi_p|$: for all $n \in \mathbb{N}^*$, $u_n(x) = \lim_{k \to \infty} (|\xi_1|^{1/k} + \cdots + |\xi_n|^{1/k}) - n$ defines on Ω a decreasing sequence of functions $u_n \in \mathscr{P}^1(\Omega)$, and may be rewritten as: $u_n(x) = -\text{Card}\{p \leqslant n: \xi_p = 0\}$; therefore $u = \lim u_n$ is the constant $-\infty$, u^* too, but $u_n = 0$ on a dense open subset of Ω, namely $\{x \in \Omega: \xi_1 \ldots \xi_n \neq 0\}$, hence $u_n^* \equiv 0$ for all $n \in \mathbb{N}^*$.

Corollary 4.3.9. *Let Ω be an open set in a Fréchet space X.*

a) *Let $\mathscr{V} \subset \bigcup_{n \in \mathbb{N}} \mathscr{P}^n(\Omega)$ be locally bounded from above; if either $u = \sup_{v \in \mathscr{V}} v$ or $u = \lim_{\mathscr{F}} \sup v$, where $\mathscr{F}$ is a filter on $\mathscr{V}$ with a countable basis, and $x \in \Omega$, one can find generalized polydiscs $K(\mathscr{A})$ such that, with $u \leqslant M$ on $x + r_0 K(\mathscr{A}) \subset \Omega$:*

(3') $M - u^*(x) = \lim_{r \to 0} \overline{MV}(M - u)[x + rK^*(\mathscr{A})]$.

b) *For each $k \in \{1, \ldots, q\}$, let $\alpha_k > 0$ and either $u_k = \sup_{v \in \mathscr{V}_k} v$ or $u_k = \lim_{\mathscr{F}_k} \sup v$, with $\mathscr{V}_k$ and $\mathscr{F}_k$ as in a); then $\left(\sum_{k=1}^q \alpha_k u_k\right)^* = \sum_{k=1}^q \alpha_k u_k^*$.*

c) *If $(u_k)_{k \in \mathbb{N}^*}$ is a decreasing sequence and each u_k as in b), then $(\lim u_k)^* = \lim u_k^*$.*

Proof. a) Since $M - u^*$ is lower semicontinuous and $\leqslant M - u$, formula (3') follows from

(4') $M - u^*(x) \geqslant \overline{MV}(M - u)[x + rK^*(\mathscr{A})]$ for all $r \leqslant r_0$,

which is obtained, for suitably chosen $\mathscr{A}$, by the argument used for Corollary 4.3.8.a., with (2') substituted for (2).

b) Given $x \in \Omega$, let $u_k \leqslant 0$ for $k \in \{1, \ldots, q\}$ on a generalized polydisc $x + r_0 K(\mathscr{A}) \subset \Omega$, chosen so that for each k:

$$- u_k^*(x) = \lim_{r \to 0} \overline{MV}(-u_k)[x + rK^*(\mathscr{A})]; \text{ since, for } r \leqslant r_0,$$

$$\sum_{k=1}^{q} \overline{MV}(-\alpha_k u_k)[x + rK^*(\mathscr{A})]$$

$$\geqslant \overline{MV}\left(-\sum_{k=1}^{q} \alpha_k u_k\right)[x + rK^*(\mathscr{A})]$$

$$\geqslant -MV\left(\sum_{k=1}^{q} \alpha_k u_k\right)^*[x + rK^*(\mathscr{A})],$$

we get $\displaystyle\sum_{k=1}^{q} \alpha_k u_k^* \leqslant \left(\sum_{k=1}^{q} \alpha_k u_k\right)^*$ by making $r \to 0$.

c) A similar argument leads to

$$-\lim u_k^*(x) \geqslant \overline{MV}(-u)[x + rK^*(\mathscr{A})] \geqslant -MV\,u^*[x + rK^*(\mathscr{A})]$$

instead of (5), hence $\lim u_k^* \leqslant u^*$.   □

Proposition 4.3.10. *Let again Ω be open in a Fréchet space X, let the sequence $(v_k) \subset \bigcup_{n \in \mathbb{N}} \mathscr{P}^n(\Omega)$ be locally bounded from above, $\limsup v_k \leqslant f$, f continuous on Ω, $f > -\infty$. Then, for every $\varepsilon > 0$ and every compact set $K \subset \Omega$, there is an integer k_0 such that $v_k \leqslant f + \varepsilon$ on K for all $k \geqslant k_0$.*

With this result, Noverraz [Nov]$_1$ generalizes the classical Hartogs lemma (Th. 1.6.13 in [Hö]) and obtains a refinement (Coroll. 4.3.11 below) of Theorem 3.1.5.c.

Proof. Let $u_k = \sup(v_k, v_{k+1}, \ldots)$: the assumption $\lim u_k \leqslant f$ implies $\lim u_k^* = (\lim u_k)^* \leqslant f$, hence $u_{k_0}^* \leqslant f + \varepsilon$ on K for some k_0, by the classical Dini lemma.   □

Corollary 4.3.11. *Let Ω be a connected open set in a Fréchet space X, Z a s.c.l.c. space, and $\mathscr{B} \subset \mathscr{A}(\Omega, Z)$ be such that $\sup\limits_{f \in \mathscr{B}} (q \circ f)$ is locally bounded for all $q \in \mathrm{csn}(Z)$; if a sequence $(f_n) \subset \mathscr{B}$ converges pointwise on some nonempty open subset of Ω, it actually converges uniformly on any compact subset of Ω.*

Proof. For all $q \in \mathrm{csn}(Z)$, we have to prove that $q \circ (f_n - f_{n'}) \to 0$ as n, $n' \to \infty$, uniformly on compact sets; this is obvious if there is an integer k such that $q \circ (f_n - f_{n'}) \equiv 0$ for n, $n' \geqslant k$. If it is not so, then $u_k = \sup\limits_{n, n' \geqslant k} \ln q \circ (f_n - f_{n'})$ defines a decreasing sequence taken in $\mathscr{P}^1(\Omega)$, $u = \lim u_k$ equals $-\infty$ on a nonempty open subset of Ω, therefore is the constant

$-\infty$; by Proposition 4.3.10, given $\varepsilon > 0$ and a compact set $K \subset \Omega$, there is an integer k_0 such that $u_k \leqslant \ln \varepsilon$ on K for all $k \geqslant k_0$. $\quad\square$

4.4 Analytic maps and pluriharmonic functions

Again Ω is an open set in a l.c. space X and Z at least s.c.l.c.

Definition 4.4.1. *A function* $u: \Omega \to \mathbb{R}$ *is pluriharmonic if* u *and* $-u$ *belong to* $\mathscr{P}(\Omega)$ *or, equivalently, if* u *is continuous and* $u|_{E \cap \Omega}$ *harmonic for every complex line* E *in* X.

By Theorem 4.3.1.: if $f \in \mathscr{A}(\Omega, Z)$, Z s.c.l.c., v pluriharmonic on an open set (in Z) containing $f(\Omega)$ implies $v \circ f$ pluriharmonic.

In the finite dimensional case: a pluriharmonic function u on an open set U in $\mathbb{C}^m$ is also harmonic, hence infinitely $\mathbb{R}$-differentiable and, with the notation of 4.2(A),

$$T_\lambda u = 0 \quad \text{for all } \lambda \in \mathbb{C}^m \quad \text{or} \quad \frac{\partial^2 u}{\partial \xi_j \partial \bar{\xi}_k} = 0 \quad \text{for all } j, k \in \{1, \ldots, m\};$$

conversely, a real continuous (resp.: twice continuously $\mathbb{R}$-differentiable) function u is pluriharmonic if each distribution (resp.: function) $\dfrac{\partial^2 u}{\partial \xi_j \partial \bar{\xi}_k}$ vanishes identically.

From this follows that $\displaystyle\sum_{j=1}^m \frac{\partial u}{\partial \xi_j} d\xi_j$ and its conjugate $\displaystyle\sum_{j=1}^m \frac{\partial u}{\partial \bar{\xi}_j} d\bar{\xi}_j$ are closed differentials; if U is star-shaped, by the Poincaré lemma there is another pluriharmonic function v (unique up to an additive constant) such that

$$i \, dv = \sum_{j=1}^m \left(\frac{\partial u}{\partial \xi_j} d\xi_j - \frac{\partial u}{\partial \bar{\xi}_j} d\bar{\xi}_j \right), \quad u + iv \in \mathscr{A}(U, \mathbb{C}).$$

Proposition 4.4.2. a) *For all* $f \in \mathscr{A}(\Omega, \mathbb{C})$, $\mathscr{R}e\, f$ *and* $\mathscr{I}m\, f$ *are pluriharmonic on* Ω.
b) *Conversely, if* Ω *is star-shaped, every pluriharmonic function on* Ω *is the real part of an* $f \in \mathscr{A}(\Omega, \mathbb{C})$ *which is unique up to a pure imaginary additive constant.*

Proof of b). Let Ω be star-shaped with respect to the origin and u pluriharmonic on Ω. For every finite dimensional subspace F of X, $U = F \cap \Omega$ is star-shaped, therefore $u|_{F \cap \Omega}$ is the real part of some $f_F \in \mathscr{A}(U, \mathbb{C})$, which is uniquely determined by $f_F(0) = 0$. From this uniqueness follows $f_{F_1} = f_{F_2}$ on $F_1 \cap F_2 \cap \Omega$, therefore the f_F are the restrictions to the $F \cap \Omega$ of a function $f \in \mathcal{G}(\Omega, \mathbb{C})$, since every complex line E in X is included in a 2-dimensional subspace of X; moreover e^f is locally bounded, hence continuous, and f is continuous too by Proposition 2.4.7.b. $\square$

Proposition 4.4.3. *Let Ω be connected.*

a) *Given $a, b \in \Omega$ and $M > 0$: $H_\Omega(a, b) = \sup \left\{ \left| \ln \dfrac{u(b)}{u(a)} \right| : u \text{ pluriharmonic} > 0 \text{ on } \Omega \right\}$ is finite and the pluriharmonic functions $u > 0$ on Ω which satisfy $u(a) \leqslant M$ are equicontinuous.*

b) *$\Omega' \subset \Omega$ entails $H_{\Omega'} \geqslant H_\Omega$ on Ω'^2; $f \in \mathcal{A}(\Omega, Z)$, Z s.c.l.c., Γ a connected open set (in Z) containing $f(\Omega)$ imply $H_\Gamma[f(a), f(b)] \leqslant H_\Omega(a, b)$ for all $(a, b) \in \Omega^2$.*

c) *The pseudodistance H_Ω is continuous on Ω; conversely, if $\sup\limits_\Omega p$ is finite, $p \in \mathrm{csn}(X)$, and $a \in \Omega$, then $p(x - a) \to 0$ as $H_\Omega(a, x) \to 0$.*

Proof. a) H_Ω satisfies the triangle inequality $H_\Omega(a, c) \leqslant H_\Omega(a, b) + H_\Omega(b, c)$. Given $a \in \Omega$, consider the set $\{b \in \Omega: H_\Omega(a, b) < \infty\}$, which will be open and closed in Ω if we prove $H_\Omega(a, b) < \infty$ for all $b \in a + \omega(a)$. In this case, denoting by R the radius of the open disc $\{\zeta \in \mathbb{C}: \zeta(b - a) \in \omega(a)\}$, we have $1 < R \leqslant \infty$ and

$$\frac{R - 1}{R + 1} \leqslant \frac{u(b)}{u(a)} \leqslant \frac{R + 1}{R - 1}$$

by the Harnack inequality; then $u(a) \leqslant M$ implies

$$|u(b) - u(a)| < \frac{2M}{r - 1} \quad \text{for all } b \in a + \frac{\omega(a)}{r}, r > 1.$$

c) H_Ω continuous follows from $H_\Omega(a, x) < \ln\dfrac{r + 1}{r - 1}, r > 1$, for all $x \in a + \dfrac{\omega(a)}{r}$. Now let $\Omega \subset a + p^{-1}([0, R[)$: if $x' \in X'$ satisfies $|x'| \leqslant p$, then $u(x) = R + \mathcal{R}e\langle x - a, x'\rangle$ defines a pluriharmonic function $u > 0$ on Ω; for all $\varepsilon > 0$, $H_\Omega(a, x) \leqslant \ln(1 + \varepsilon)$ entails $|u(x) - u(a)| \leqslant \varepsilon u(a)$ or $|\mathcal{R}e\langle x - a, x'\rangle| \leqslant \varepsilon R$ and, since p is a supremum of $|x'|$:

(1) $H_\Omega(a, x) \leqslant \ln(1 + \varepsilon) \Rightarrow p(x - a) \leqslant \varepsilon R.$ $\square$

  The *Caratheodory pseudodistance*, which we now define, proceeds from the same idea, with φ analytic, $|\varphi| < 1$, instead of u pluriharmonic, $u > 0$;

this version of the idea is more usual and we will use it in Theorem 4.4.5 (mainly in 6.2).

Proposition 4.4.4. *Let Ω be connected.*

a) *Given $a, b \in \Omega$:*

$$C_\Omega(a,b) = \sup\left\{\arg th\left|\frac{\varphi(a) - \varphi(b)}{1 - \overline{\varphi(a)}\varphi(b)}\right| : \varphi \in \mathscr{A}(\Omega, \mathbb{C}), |\varphi| < 1\right\}$$

is finite.
b) *$\Omega' \subset \Omega$ entails $C_{\Omega'} \geqslant C_\Omega$ on Ω'^2; $f \in \mathscr{A}(\Omega, Z)$, Z s.c.l.c., Γ a connected open set (in Z) containing $f(\Omega)$ imply $C_\Gamma[f(a), f(b)] \leqslant C_\Omega(a,b)$ for all $(a,b) \in \Omega^2$.*
c) *The pseudodistance C_Ω is continuous; conversely, if $\sup_\Omega p$ is finite, $p \in \mathrm{csn}(X)$, and $a \in \Omega$, then $p(x - a) \to 0$ as $C_\Omega(a,x) \to 0$.*

Proof. a) C_Ω satisfies the triangle inequality because, for $\alpha, \beta \in \Delta(0,1)$, $\arg th\left|\dfrac{\alpha - \beta}{1 - \bar{\alpha}\beta}\right|$ is the infimum of $\displaystyle\int_\alpha^\beta \frac{|d\zeta|}{1 - |\zeta|^2}$ along rectifiable curves in $\Delta(0,1)$.

For $b \in a + \omega(a)$, with the same radius R as in the proof of Proposition 4.4.3.a, we have $\left|\dfrac{\varphi(a) - \varphi(b)}{1 - \overline{\varphi(a)}\varphi(b)}\right| \leqslant \dfrac{1}{R}$ by the classical Pick lemma.
c) C_Ω continuous follows from $C_\Omega(a,x) < \arg th\dfrac{1}{r}$, $r > 1$, for all $x \in a + \dfrac{\omega(a)}{r}$. Now let $\Omega \subset a + p^{-1}([0, R[)$: if $x' \in X'$ satisfies $|x'| \leqslant p$, then $\varphi(x) = \dfrac{1}{R}\langle x - a, x'\rangle$ satisfies $|\varphi| < 1$, hence, for all $\varepsilon > 0$,

(2) $C_\Omega(a,x) \leqslant \arg th\,\varepsilon \Rightarrow p(x - a) \leqslant \varepsilon R.$ $\square$

Theorem 4.4.5. *Let A be a closed subset of Ω, $\mathring{A} = \emptyset$. If Z is complete, the following statements are equivalent:*

(1) *For every open set $\omega \subset \Omega$ with $\omega \cap A \neq \emptyset$, every function $\varphi \in \mathscr{A}(\omega\backslash A, \mathbb{C})$ satisfying $\limsup_{\omega\backslash A \ni x \to a} |\varphi(x)| < \infty$ for all $a \in \omega \cap A$ has a (unique) extension $\in \mathscr{A}(\omega, \mathbb{C})$.*
(2) *For every open set $\omega \subset \Omega$ with $\omega \cap A \neq \emptyset$, every map $f \in \mathscr{A}(\omega\backslash A, Z)$ satisfying $\limsup_{\omega\backslash A \ni x \to a} q \circ f(x) < \infty$ for all $a \in \omega \cap A$, $q \in \mathrm{csn}(Z)$, has a (unique) extension $\in \mathscr{A}(\omega, Z)$.*

Proof. We assume (1) and first remark that ω connected implies $\omega\backslash A$ connected: if, on the contrary, we had $\omega\backslash A = \omega_0 \cup \omega_1$ with ω_0, ω_1 open nonempty and $\omega_0 \cap \omega_1 = \emptyset$, the function equal to 0 on ω_0, 1 on ω_1, would admit a continuous extension to ω, implying disjoint closures of ω_0, ω_1 in ω.

Now let $f \in \mathscr{A}(\Omega\backslash A, Z)$ have the property $\limsup\limits_{\Omega\backslash A \ni x \to a} q \circ f(x) < \infty$ for all $a \in A$, $q \in \mathrm{csn}(Z)$: we show that f has a continuous extension $\tilde{f}$ to Ω. Given $a \in A$ and $q \in \mathrm{csn}(Z)$, let ω be a connected open neighbourhood of a in Ω such that $f(\omega\backslash A) \subset q^{-1}\left(\left[0, \dfrac{R}{2}\right[\right)$ and therefore

$$f(\omega\backslash A) \subset \Gamma(x) = f(x) + q^{-1}([0, R[) \quad \text{for all } x \in \omega\backslash A.$$

By the assumption (1) and the maximum modulus principle, every $\varphi \in \mathscr{A}(\omega\backslash A, \mathbb{C})$ with $|\varphi| < 1$ has an extension $\tilde{\varphi} \in \mathscr{A}(\omega, \mathbb{C})$ with $|\tilde{\varphi}| < 1$, hence $C_{\omega\backslash A} = C_\omega$ on $(\omega\backslash A)^2$; by Proposition 4.4.4, and formula (2) at the end of its proof, for $x, y \in \omega\backslash A$:

$$C_{\Gamma(x)}[f(x), f(y)] \leqslant C_{\omega\backslash A}(x, y) = C_\omega(x, y) \to 0 \quad \text{as } x, y \to a$$

and $C_\omega(x, y) \leqslant \arg th\, \varepsilon$, $\varepsilon > 0$, entails $q[f(x) - f(y)] \leqslant \varepsilon R$; then $\lim\limits_{\omega\backslash A \ni x \to a} f(x)$ exists in the complete space Z. Finally, by the assumption (1), each $z' \circ f$, $z' \in Z'$, has an analytic extension to Ω, which is $z' \circ \tilde{f}$. $\square$

4.5 Polar subsets

Polar subsets of an open set Ω in a l.c. space X have several possible definitions; among them, the following one, introduced by Noverraz [Nov]$_1$ under the name of "sets of class P", is closely related to the definitions (4.1.1. and 4.1.3) of p.h.h. and p.s.h. functions.

Definition 4.5.1. *A set $P \subset \Omega$ is unipolar in Ω if*: (i) *for every complex line E in X and every connected component U of $E \cap \Omega$, $P \cap U$ is either polar or U itself*; (ii) $\overset{\circ}{P} = \emptyset$ *or, equivalently, P does not include any connected component of Ω.*

The equivalence in (ii) is obtained by the same argument as the equivalence in Definition 4.1.3. If $u \in Q(\Omega)$ (Def. 4.3.3.b), $u^{-1}(-\infty)$ is unipolar in Ω; if $\varphi \in \mathscr{A}(\Omega, \mathbb{C})$, but also if $\varphi \in \mathscr{G}(\Omega, \mathbb{C})$, then $\varphi^{-1}(\alpha)$ is unipolar in Ω (see also Prop. 4.5.5.a) unless φ is the constant α on some connected component of Ω. A unipolar subset of Ω may be dense in Ω: such is $\varphi^{-1}(\alpha)$ for any non-continuous polynomial map $\varphi\colon X \to \mathbb{C}$ and any $\alpha \in \mathbb{C}$ (Prop. 2.4.8).

Any intersection of unipolar subsets of Ω is again such a set; so is a countable union of unipolar subsets of Ω, unless it includes a connected component of Ω, which may occur indeed (Prop. 4.5.10.a).

Theorem 4.5.2. *Let P be a closed unipolar subset of Ω.*

a) *Every function $\varphi \in \mathcal{A}(\Omega\backslash P, \mathbb{C})$ which satisfies $\lim\sup\limits_{\Omega\backslash P \ni x \to a} |\varphi(x)| < \infty$ for all $a \in P$ has a (unique) analytic extension to Ω; every function u pluriharmonic on $\Omega\backslash P$ which satisfies $\lim\sup\limits_{\Omega\backslash P \ni x \to a} |u(x)| < \infty$ for all $a \in P$ has a (unique) pluriharmonic extension to Ω.*

b) *Let Z be s.c.l.c. Every map $f \in \mathcal{A}(\Omega\backslash P, Z)$ which satisfies $\lim\sup\limits_{\Omega\backslash P \ni x \to a} q \circ f(x) < \infty$ for all $a \in P$, $q \in \mathrm{csn}(Z)$, has a (unique) analytic extension to Ω.*

c) *If Ω is connected, $\Omega\backslash P$ too.*

Proof. a) We have to prove that $\lim\limits_{\Omega\backslash P \ni x \to a} \varphi(x)$ or $\lim\limits_{\Omega\backslash P \ni x \to a} u(x)$ exists for all $a \in P$ and that the function equal to this limit on P, to φ or u on $\Omega\backslash P$, is analytic or pluriharmonic on Ω; we will get both results together by constructing, for every $a \in P$, a function $\tilde{\varphi}$ analytic or $\tilde{u}$ pluriharmonic on some open neighbourhood $a + V$ of a in Ω, such that $\tilde{\varphi} = \varphi$ or $\tilde{u} = u$ on $(a + V)\backslash P$.

Let $a + b \in [a + \omega(a)]\backslash P$: $\{\zeta \in \mathbb{C}: a + \zeta b \in P\}$ is a polar set, its complement in $\mathbb{C}$ contains circumferences with centre 0 and arbitrarily small radii ([Brel]$_3$, §VII.5); let then $r \in \,]0, 1]$ be such that $\Omega\backslash P$ contains the compact set $L = \{a + \zeta b: |\zeta| = r\}$ and V an open neighbourhood of the origin such that $\Omega\backslash P \supset L + V$, while $\Omega \supset K + V$, $K = \{a + \zeta b: |\zeta| \leqslant r\}$.

Now take $x \in a + V$: $P(x) = \{\zeta \in \mathbb{C}: |\zeta| \leqslant r, x + \zeta b \in P\}$ is a polar compact subset of $\Delta(0, r)$, $\varphi(x + \zeta b)$ or $u(x + \zeta b)$ a holomorphic or harmonic function on $\Delta(0, r')\backslash P(x)$ (where $r' > r$ depends on x), and is bounded for ζ in some neighbourhood of each point $\in P(x)$, therefore, by a theorem of Brelot [Brel]$_1$, has a holomorphic or harmonic exension to $\Delta(0, r')$. On the other hand:

$$\tilde{\varphi}(x) = MV\,\varphi(x + re^{i\theta}b) \quad \text{or} \quad \tilde{u}(x) = MV\,u(x + re^{i\theta}b)$$

makes sense since $\Omega\backslash P \supset L + V$ and defines (Th. 3.1.7.b or Prop. 4.3.2.a) $\tilde{\varphi}$ analytic or $\tilde{u}$ pluriharmonic on $a + V$; this mean value is also the value for $\zeta = 0$ of the above constructed extension, hence $\tilde{\varphi}(x) = \varphi(x)$ or $\tilde{u}(x) = u(x)$ if $x \in (a + V)\backslash P$.

b) If Z is complete, Theorem 4.4.5. may be used. But the above argument remains valid for $f \in \mathscr{A}(\Omega \backslash P, Z)$ under the weaker assumption Z s.c.: $\tilde{f}(x) = MVf(x + re^{i\theta}b)$ still makes sense and defines $\tilde{f} \in \mathscr{A}(a + V, Z)$; for all $z' \in Z'$, $z' \circ \tilde{f}(x) = MV z' \circ f(x + re^{i\theta}b)$ is also the value for $\zeta = 0$ of the extension of $z' \circ f$ obtained in a), therefore $z' \circ f(x) = z' \circ \tilde{f}(x)$ if $x \in (a + V) \backslash P$.

c) See the first paragraph of the proof of Theorem 4.4.5. $\square$

In the case of a complex valued function φ, the assumption $\limsup\limits_{\Omega \backslash P \ni x \to a} |\varphi(x)|$ $< \infty$ for all $a \in P$ can be considerably weakened as follows; the necessary complements of potential theory in $\mathbb{C}$ are proved below (Prop. 4.5.4.) for completeness.

Theorem 4.5.3. *Let again P be a closed unipolar subset of Ω and $\varphi \in \mathscr{A}(\Omega \backslash P, \mathbb{C})$; assume that each point $a \in P$ has an open neighbourhood ω in Ω such that $\varphi(\omega \backslash P) \cup \Delta(0, R)$ has a nonpolar complement in $\mathbb{C}$ for any radius R. Then φ has an extension $\in \mathscr{A}(\Omega, \mathbb{C})$.*

Proof. If we resume the construction made in the proof of Theorem 4.5.2.a (with ω instead of Ω), for a given $x \in a + V$ we now have $\varphi(x + \zeta b)$ (which we assume non constant) holomorphic on $\Delta(0, r') \backslash P(x)$ and such that the union of any disc and the open set $U(x) = \{\varphi(x + \zeta b) : \zeta \in \Delta(0, r') \backslash P(x)\}$ has a nonpolar complement in $\mathbb{C}$.

First we only use the fact that $\mathbb{C} \backslash U(x)$ is not polar: then (Prop. 4.5.4.b) there is a non constant bounded harmonic function h_0 on $U(x)$; the composed function $\zeta \mapsto h_0 \circ \varphi(x + \zeta b)$ is harmonic and bounded on $\Delta(0, r') \backslash P(x)$, therefore has a finite limit at each $\zeta_0 \in P(x)$. Given ζ_0, since h_0 is not constant, we can choose $w_0 \in U(x)$ such that this limit is not $h_0(w_0)$; then $\limsup\limits_{P(x) \ni \zeta \to \zeta_0} \left| \dfrac{1}{\varphi(x + \zeta b) - w_0} \right| < \infty$, $\dfrac{1}{\varphi(x + \zeta b) - w_0}$ has a holomorphic extension and $\varphi(x + \zeta b)$ a meromorphic extension to some open neighbourhood of ζ_0; repeating the argument for each $\zeta_0 \in P(x)$, we obtain a meromorphic extension of $\varphi(x + \zeta b)$ to $\Delta(0, r')$.

We shall now prove that this extension has a finite value at each $\zeta_0 \in P(x)$, and here we need the fact that the union of $U(x)$ and any disc has a nonpolar complement. Assume that ζ_0 is a pole of order k of the extension: then, for sufficiently small $|\zeta - \zeta_0|$, the extension may be written as $1/\psi^k(\zeta)$, where ψ maps biholomorphically some open neighbourhood δ of ζ_0 in $\Delta(0, r')$ onto some open disc $\Delta\left(0, \dfrac{1}{R}\right)$; on δ, $\dfrac{1}{\psi^k}$ assumes exactly k times each value $\in \mathbb{C} \backslash \overline{\Delta}(0, R^k)$, but each value $\in \mathbb{C} \backslash U(x) \backslash \overline{\Delta}(0, R^k)$ can be assumed only on $P(x) \cap \delta$, which is a polar set: $P(x) \cap \delta \subset u^{-1}(-\infty)$ for some $u \in \mathscr{P}(\delta)$.

Since the $1 - k$ map $1/\sqrt[k]{}$ is locally biholomorphic, $w \mapsto \sum\limits_{z^k = 1/w} u \circ \psi^{-1}(z)$
is subharmonic on $\mathbb{C} \backslash \bar{\Delta}(0, R^k)$, its value is $-\infty$ whenever $w \in \mathbb{C} \backslash U(x) \backslash \bar{\Delta}(0, R^k)$. $\square$

Proposition 4.5.4. a) *In a closed nonpolar set in* $\mathbb{C}$, *one can find two disjoint nonpolar compact sets.*

b) *There is a non constant bounded harmonic function on the open set U in $\mathbb{C}$ if and only if $\mathbb{C} \backslash U$ is not polar.*

Proof. a) Since a countable union of polar sets is again polar, we may start from a compact nonpolar set K, and K contains at least one point α such that the intersection of K with any disc centred in α is nonpolar; we claim that there are at least two such points (actually an arbitrary number of them, but 2 will serve our purpose).

If α is the only such point, $K \cap \Delta(\alpha, r)$ is nonpolar but $K \backslash \Delta(\alpha, r)$ polar for all $r > 0$. Now let $K \subset \Delta(\alpha, R)$ and φ be the Choquet capacity in $\Delta(\alpha, R)$ defined in 4.2(E):

$$\varphi[K \backslash \Delta(\alpha, r)] = 0 \quad \text{for all } r > 0 \text{ and } \varphi^*[K \cap \Delta(\alpha, r)] \leqslant \varphi[\Delta(\alpha, r)],$$

which tends to 0 with r since $R_1^{\Delta(\alpha, r)}(\zeta) = \inf\left[1, \dfrac{\ln(R/|\zeta - \alpha|)}{\ln(R/r)} \right]$. Hence follow $\varphi(K) = 0$ and K polar, a contradiction.

b) If $\mathbb{C} \backslash U$ is polar, every bounded harmonic function on U is constant since it has a harmonic extension to $\mathbb{C}$; if not, we use an argument of Noshiro [Nos]. We assume that U is connected and dense in $\mathbb{C}$, because

$$\zeta \mapsto \mathscr{R}e \frac{1}{\zeta - \beta}$$ has the desired properties if β belongs to the interior of $\mathbb{C} \backslash U$.

Let $K_j, j = 1$ or 2, be two disjoint nonpolar compact subsets of $\mathbb{C} \backslash U$ and G_j the Green function, with pole at infinity, of the open set $U_j = \mathbb{C} \backslash K_j$, which is connected since it includes U; we claim that $(G_1 - G_2)_{|U_1 \cap U_2}$ has the desired properties.

It is bounded: in fact, if the open disc Δ includes $K_1 \cup K_2$, $G_j(\zeta) - \ln|\zeta|$ has a finite limit as $\zeta \to \infty$ and $\sup\limits_{\Delta \cap U_j} G_j$ is finite, equal to the upper bound of G_j on the circumference of Δ.

It cannot be constant: in fact, the boundary of $U_1 \cap U_2$ is the union of $K_1 \subset U_2$ and $K_2 \subset U_1$; $m_1 = \inf\limits_{K_2} G_1 > 0$, but $\liminf\limits_{U_2 \ni \zeta \to \alpha_2} G_2(\zeta) \geqslant m_1$ cannot occur for all $\alpha_2 \in K_2$, and similarly $\liminf_{U_1 \ni \zeta \to \alpha_1} G_1(\zeta) \geq m_2 = \inf_{K_1} G_2$ cannot occur for all $\alpha_1 \in K_1$. $\square$

Proposition 4.5.5. a) *If* $\varphi \in \mathscr{G}(\Omega, \mathbb{C})$ *is not constant on any connected component of* Ω *and* P_0 *is polar in* $\mathbb{C}$, *then* $\varphi^{-1}(P_0)$ *is unipolar in* Ω.
b) *If* $\varphi \in \mathscr{A}(\Omega, \mathbb{C})$ *is not constant on any connected component of* Ω *and* P *is a closed unipolar subset of* Ω, *then* $\varphi(\Omega) \backslash \varphi(\Omega \backslash P)$ *is polar and, by* a), *may be any closed polar subset of the open set* $\varphi(\Omega)$.

Proof. a) Let E be a complex line in X and U a connected component of $E \cap \Omega$. If $\varphi|_U$ is constant, then $\varphi^{-1}(P_0) \cap U = \emptyset$ or U; if not, $P_0 \subset u^{-1}(-\infty)$, $u \in \mathscr{P}(\mathbb{C})$, implies $\varphi^{-1}(P_0) \cap U \subset v^{-1}(-\infty)$, $v = u \circ \varphi \in \mathscr{P}(U)$. Finally $\varphi^{-1}(P_0)$ cannot contain any connected component of Ω because φ is an open mapping (Prop. 2.4.7.a).
b) $F = \varphi(\Omega) \backslash \varphi(\Omega \backslash P)$ is a closed subset of $\varphi(\Omega)$; if F is not polar, we can successively find: $\alpha \in F$ such that the intersection of F with any disc centred in α is nonpolar; $a \in P$ such that $\varphi(a) = \alpha$; $a + b \in [a + \omega(a)] \backslash P$ with $\varphi(a + b) \neq \alpha$. For $|\zeta|$ sufficiently small: $\varphi(a + \zeta b) - \alpha = \psi^k(\zeta)$, where $k \in \mathbb{N}^*$, $\psi(0) = 0$ and ψ maps biholomorphically some open neighbourhood δ of 0 in $\Delta(0, 1)$ onto some open disc $\Delta(0, r)$, hence a contradiction by the argument used in the proof of Theorem 4.5.3. $\square$

The next theorem puts together:

a) a maximum principle for functions $\in Q(\Omega)$ (Def. 4.3.3.b).
b) a convergence theorem (see §4.2(D)) in which the main assumption is u^* (the upper regularized of u) pluriharmonic.

Theorem 4.5.6. *Let the open set* Ω *be connected.*

a) *If* $u \in Q(\Omega)$ *and* $a \in \Omega$ *are such that* $u(a) = \sup_{\Omega} u$, *then* $P = \{x \in \Omega: u(x) < u(a)\}$ *is unipolar in* Ω, *and* $E_u = \{x \in \Omega: u(x) < u^*(x)\} = P$.
b) *Let* $\mathscr{V} \subset \bigcup_{n \in \mathbb{N}} \mathscr{P}^n(\Omega)$ *be locally bounded from above; if* $u = \sup_{v \in \mathscr{V}} v$ *or* $u = \limsup_{\mathscr{F}} v$ *(where* $\mathscr{F}$ *is a filter on* $\mathscr{V}$ *with a countable basis) has a pluriharmonic upper regularized* u^*, *then* E_u *is either* Ω *or a unipolar subset of* Ω; *if* X *is a Fréchet space, the latter case only can occur.*

Proof. a) Let E be a complex line in X and U a connected component of $E \cap \Omega$ such that $P \cap U \neq U$: $u|_U$ is quasisubharmonic, $\leq u(a)$ on U but $= u(a)$ at some point in U, $(u|_U)^*$ is subharmonic, $\leq u(a)$ on U but $= u(a)$ at some point in U, therefore is the constant $u(a)$, and $P \cap U$ is polar, as the subset of U where $(u|_U) < (u|_U)^*$. By the same argument, u^* also is the constant $u(a)$.

b) $u - u^* \in Q(\Omega)$, $u - u^* \leqslant 0$ everywhere but $= 0$ at every point $\in \Omega \setminus E_u$. Now let X be a Fréchet space: then (Prop. 4.3.5 and Coroll. 4.3.9.a), given $x \in \Omega$, one can find generalized polydiscs $K(\mathscr{A}) \subset \Omega - x$ such that, if $u \leqslant M$ on $x + K(\mathscr{A})$:

$$MV(M - u^*)[x + K^*(\mathscr{A})]$$
$$= M - u^*(x) \geqslant \overline{MV}(M - u)[x + K^*(\mathscr{A})];$$

then the intersection of E_u with the distinguished boundary $x + K^*(\mathscr{A})$ has $(d\tau)$-measure 0.   □

Examples 4.5.7. (A) If X is not complete, $E_u = \Omega$ may occur. In fact, let (once more) $X = c_{00}(\mathbb{N}^*)$ with the notation $x = (\xi_1, \xi_2, \dots)$, $\|x\| = \sup_{n \in \mathbb{N}^*} |\xi_n|$: each function $u_n(x) = |\xi_1 \dots \xi_n|^{1/n^2}$, $n \in \mathbb{N}^*$, is p.s.h. and < 1 on the open unit ball Ω, $u_n(x) \leqslant \|x\|^{1/n}$; setting $m(x) = \text{Card }\{n \in \mathbb{N}^*: \xi_n \neq 0\}$, we also have $u(x) \leqslant \|x\|^{1/m(x)} < 1$ for all $x \in \Omega \setminus \{0\}$, $u(0) = 0$, for $u = \sup_{n \in \mathbb{N}^*} u_n$.

On the other hand, given $x \in \Omega$, let $y = (\eta_1, \dots, \eta_p, 0, 0, \dots)$ with $p > \sup\{n \in \mathbb{N}^*: \xi_n \neq 0\}$, $\eta_n = \xi_n$ whenever $\xi_n \neq 0$, $\eta_n = \varepsilon > 0$ whenever $n \leqslant p$ and $\xi_n = 0$: $\|y - x\| = \varepsilon$ and $u(y) \geqslant u_p(y) = [\mu(x)\varepsilon^{p-m(x)}]^{1/p^2}$, where $\mu(x)$ is the product of the nonzero $|\xi_n|$ if $x \neq 0$, $\mu(0) = 1$; for a given ε, the last member tends to 1 as $p \to \infty$, which implies that 1 is the upper bound of u on every neighbourhood of x, $u^*(x) = 1$ for all $x \in \Omega$.

For another counter-example, see Example 5.4.5(A).

(B) Let U be a domain of holomorphy in $\mathbb{C}^m$: the set of functions $\in \mathscr{A}(U, \mathbb{C})$ for which U is not a domain of existence is meagre and unipolar in $\mathscr{A}(U, \mathbb{C})$ endowed with the topology of compact convergence [Lel]₃.

Proof. We choose a norm on $\mathbb{C}^m$; for $a \in U$, $f \in \mathscr{A}(U, \mathbb{C})$, $k \in \mathbb{N}^*$, we set

$$r_a = \text{dist}(a, \complement U), \quad u_a^k(f) = \sup \left\{ \frac{1}{k!} |\hat{D}_a^k f(x)|: \|x\| = 1 \right\} \text{ and}$$

$$v_a(f) = \limsup_{k \to \infty} [u_a^k(f)]^{1/k}: \frac{1}{v_a(f)} \text{ is exactly the largest number } R \text{ such that}$$

the expansion $f(a) + \sum_{k \in \mathbb{N}^*} \frac{1}{k!} (\hat{D}_a^k f)(x)$ is summable for $\|x\| < R$; therefore $\frac{1}{v_a(f)} \geqslant r_a$ for all $f \in \mathscr{A}(U, \mathbb{C})$, and

$$E_a = \left\{ f \in \mathscr{A}(U,\mathbb{C}) \colon \frac{1}{v_a(f)} > r_a \right\}$$

is a proper subspace of $\mathscr{A}(U,\mathbb{C})$ since U is a domain of holomorphy. Now U is not a domain of existence for f if and only if $f \in E_a$ for some a in a countable dense subset of U; then we have to show that E_a is meagre and unipolar for all $a \in U$.

$u_a^k(f)$ depends continuously on the k-th order derivatives of f at the point a, hence also on $f \in \mathscr{A}(U,\mathbb{C})$; then $\ln u_a^k$ is p.s.h. on $\mathscr{A}(U,\mathbb{C})$ by Proposition 4.1.2.c; the sequence $\left(\frac{1}{k} \ln u_a^k \right)_{k \in \mathbb{N}^*}$ is locally bounded from above since

$$u_a^k(f) \leqslant \frac{1}{(\theta r_a)^k} \sup\{|f(a+x)| \colon \|x\| \leqslant \theta r_a\} \quad \text{for all } \theta \in \,]0,1[, k \in \mathbb{N}^*,$$

and $\ln v_a \in Q[\mathscr{A}(U,\mathbb{C})]$, $\ln v_a \leqslant -\ln r_a$ everywhere, $\ln v_a = -\ln r_a$ on the dense set $\mathscr{A}(U,\mathbb{C}) \backslash E_a : E_a$ is unipolar by Theorem 4.5.6.a, and meagre by Example 4.1.5.h'. □

Definition 4.5.8. *A set $P \subset \Omega$ is (strictly) pluripolar in Ω if there is a $w \in \mathscr{P}(\Omega)$ (and < 0) such that $P \subset w^{-1}(-\infty)$.*

Since $\mathscr{P}(\Omega)$ is a convex cone of functions, a finite union of (strictly) pluripolar subsets of Ω is again (strictly) pluripolar in Ω (for a countable union, see Proposition 4.5.10).

Thus a pluripolar subset of Ω is included in a unipolar one. If a closed pluripolar subset P of Ω is included in a closed unipolar one, then Theorem 4.5.2 and 4.5.3 remain valid for P. But the following counter-example of Coeuré [Coe] shows that a closed (even strictly) pluripolar subset of Ω is not always included in a closed unipolar one, which makes an independent theorem for closed pluripolar sets (Th. 4.5.16 below) useful.

For an example of a unipolar set (in a Fréchet space) which is not pluripolar, see Proposition 5.4.3 below. Also note that, by the Liouville theorem, $\emptyset$ is the only strictly pluripolar set in the whole space.

Proposition 4.5.9. *Let Ω be the open unit ball in an infinite dimensional Banach space X.*

a) *Given a sequence $(b_n)_{n \in \mathbb{N}^*} \subset X \backslash \{0\}$, the numbers $\varepsilon_n > 0$ can be chosen so that the family $\mathscr{A}$ of vectors $a_n = \varepsilon_n b_n$ is summable and the corresponding generalized polydisc $K(\mathscr{A})$ is a strictly pluripolar subset of Ω.*
b) *If moreover the b_n span a dense subspace of X (which may occur if and only if X is separable), there is no closed unipolar subset of Ω including $K(\mathscr{A})$.*

Proof. a) The condition $\sum_{n \in \mathbb{N}^*} \varepsilon_n \|b_n\| < 1$ ensures $\mathscr{A}$ summable and $K(\mathscr{A}) \subset$ Ω. Now let E_n be the finite dimensional subspace of X spanned by $b_1, \dots,$ b_n: since Ω is a Baire space, there is some point $c \in \Omega \setminus \bigcup_{n \in \mathbb{N}^*} E_n$; for each $n \in \mathbb{N}^*$, one can find $x_n' \in X'$, vanishing on E_n, with $\|x_n'\| = 1$ and $\langle c, x_n' \rangle \neq 0$, then $\lambda_n > 0$ so that $\sum_{n \in \mathbb{N}^*} \lambda_n \ln |\langle c, x_n' \rangle| > -\infty$; all this does not depend on the ε_n.

Then $w = \sum_{n \in \mathbb{N}^*} \lambda_n \ln |x_n'| \in \mathscr{P}(\Omega)$, $w < 0$ on Ω, $w^{-1}(-\infty) \supset E_n \cap \Omega$ for all $n \in \mathbb{N}^*$, and moreover $w^{-1}(-\infty) \supset K(\mathscr{A})$ for suitably chosen ε_n. In fact, for all $n \in \mathbb{N}^*$: $|\alpha_j| \leq 1$ for all $j \in \mathbb{N}^*$ implies

$$\left| \left\langle \sum_{j \in \mathbb{N}^*} \alpha_j a_j, x_n' \right\rangle \right| = \left| \left\langle \sum_{j > n} \alpha_j a_j, x_n' \right\rangle \right| \leq \sum_{j > n} \varepsilon_j \|b_j\|;$$

choose $r_n \in \,]0, 1[$ with $r_{n+1} < r_n$ and $r_{n+1} \leq e^{-1/\lambda_n}$, and set $\varepsilon_n \|b_n\| = r_n - r_{n+1}$ for all $n \in \mathbb{N}^*$; then $|\alpha_j| \leq 1$ for all $j \in \mathbb{N}^*$ implies

$$w\left(\sum_{j \in \mathbb{N}^*} \alpha_j a_j \right) \leq \sum_{n \in \mathbb{N}^*} \lambda_n \ln r_{n+1} = -\infty.$$

b) Let E be a complex line joining the origin and some point $x = \sum_{j=1}^n \alpha_j a_j \in$ $X \setminus \{0\}$: if a closed unipolar subset P of Ω includes $K(\mathscr{A})$, then $P \cap E$ includes the disc $\left\{ \zeta x : |\zeta| \leq \dfrac{1}{|\alpha_j|} \text{ for all } j = 1, \dots, n \right\}$, hence $P \supset E \cap \Omega$, $P = \Omega$ by the assumption b) since P is closed in Ω. □

Proposition 4.5.10. *Let Ω be connected and $u_k \in \mathscr{P}(\Omega)$, $u_k < 0$ on Ω, for all $k \in \mathbb{N}^*$.*

a) $\bigcup_{k \in \mathbb{N}^*} u_k^{-1}(-\infty)$ *is either Ω or a unipolar and strictly pluripolar subset of Ω.*
b) *If X is a Fréchet space, the latter case only can occur, and actually a countable union of pluripolar subsets of Ω has no interior point.*

Proof. a) If there is an $a \in \Omega$ such that $u_k(a) > -\infty$ for all $k \in \mathbb{N}^*$, we can choose the $\lambda_k > 0$ with $\lambda_k u_k(a) \geq -\dfrac{1}{k^2}$ for all $k \in \mathbb{N}^*$ and thus get a decreasing sequence $\left(\sum_{k=1}^n \lambda_k u_k \right) \subset \mathscr{P}(\Omega)$ whose limit $u \in \mathscr{P}(\Omega)$, $u < 0$, $u^{-1}(-\infty) \supset u_k^{-1}(-\infty)$ for all $k \in \mathbb{N}^*$.

If X is not complete, it may occur that $\bigcup_{k \in \mathbb{N}^*} u_k^{-1}(-\infty) = \Omega$.

In fact, let (once more) $X = c_{00}(\mathbb{N}^*))$ with the notation $x = (\xi_1, \xi_2, \ldots)$, $\|x\| = \sup_{n \in \mathbb{N}^*} |\xi_n|$: each function $u_n(x) = \ln|\xi_1 \ldots \xi_n|$ is p.s.h. and < 0 on the open unit ball Ω, which is the union of the strictly pluripolar sets $P_n = \{x \in \Omega: \xi_1 \ldots \xi_n = 0\}$.

b) We no longer assume $u_k < 0$, but $\Omega = \bigcup_{k \in \mathbb{N}^*} u_k^{-1}(-\infty)$. We choose an arbitrary point $x \in \Omega$: for each $j \in \mathbb{N}^*$, since $u_j^{-1}(-\infty)$ has no interior point, we can find $b_j \in \omega(x) \backslash \{0\}$ such that $u_j(x + b_j) > -\infty$, therefore $u_j(x + \zeta b_j) > -\infty$ for almost every $\zeta \in \bar{\Delta}(0, 1)$. Let the increasing sequence $(p_n)_{n \in \mathbb{N}^*}$ define the topology of X: for each $j \in \mathbb{N}^*$, we can also find $\lambda_j \in]0, 1]$ with $\lambda_j p_j(b_j) \leqslant \frac{1}{j^2}$, so that the family $\mathscr{A}$ of vectors $a_j = \lambda_j b_j$ is summable. When $\Omega \supset x + rK(\mathscr{A}), r > 0$, by formula (2) in Theorem 4.3.6, for $|\zeta| < 1$ and each $k \in \mathbb{N}^*$ we have

$$ u_k(x + \zeta r a_k) \leqslant \int P(\zeta, \tau_k) u_k\left(x + r \sum_{j \in \mathbb{N}^*} \tau_j a_j\right) d\tau; $$

since ζ can be chosen so that the left hand member $> -\infty$, each set $\left\{\tau \in \mathbb{T}: u_k\left(x + r \sum_{j \in \mathbb{N}^*} \tau_j a_j\right) = -\infty\right\}$ is a null set for the measure $d\tau$, and there is an $a \in x + rK^*(\mathscr{A})$ with $u_k(a) > -\infty$ for all $k \in \mathbb{N}^*$. $\quad\square$

Proposition 4.5.11. *Let Ω be connected and the supremum u of $\mathscr{V} \subset \mathscr{P}(\Omega)$ have a pluriharmonic upper regularized u^*. Then $E_u = \{x \in \Omega: u(x) < u^*(x)\}$ is either Ω or a strictly pluripolar subset of Ω and (Th. 4.5.6.b), if X is a Fréchet space, the latter case only can occur.*

Proof. Assume $E_u \neq \Omega$ and $\emptyset$, and let $u(x_0) = u^*(x_0)$, $x_0 \in \Omega$: for all $n \in \mathbb{N}^*$, there is $v_n \in \mathscr{V}$ such that $v_n(x_0) \geqslant u^*(x_0) - \frac{1}{n^2}$; then $w = \sum_{n \in \mathbb{N}^*} (v_n - u^*)$ is p.s.h. and < 0 on Ω, $E_u \subset w^{-1}(-\infty)$. For an example where $E_u = \Omega$, see 5.4.5(A). $\quad\square$

The next proposition repeats the beginning of §4.2(E). A function $v: \Omega \to]-\infty, +\infty]$ is said to be *plurisuper-* (resp.: *hyper-*) *harmonic* if $-v$ is p.s.h. (resp. p.h.h.) on Ω.

Proposition 4.5.12. *Let Ω be connected, A any subset of Ω, R_1^A the infimum of all positive plurisuperharmonic functions on Ω which are $\leqslant 1$ on Ω and $= 1$ on A: A is strictly pluripolar in Ω if and only if R_1^A vanishes somewhere in Ω, and then $\{x \in \Omega: R_1^A(x) > 0\}$ is unipolar in Ω.*

Note that R_1^A has a plurisuperharmonic lower regularized $(R_1^A)^*$ (Th. 4.1.7.b).

Proof. If A is strictly pluripolar in Ω, i.e. $A \subset u^{-1}(-\infty)$, $u \in \mathscr{P}(\Omega)$, $u < 0$ on Ω, then $R_1^A \leqslant \inf\left(1, -\dfrac{u}{n}\right)$ for all $n \in \mathbb{N}^*$, $R_1^A = 0$ on $\Omega \backslash u^{-1}(-\infty)$, a minimum for R_1^A (see Th. 4.5.6.a).

Conversely, if $R_1^A(x_0) = 0$, for each $n \in \mathbb{N}^*$ there is a positive plurisuperharmonic function v_n on Ω with $v_n = 1$ on A and $v_n(x_0) \leqslant \dfrac{1}{n^2}$: $v = \displaystyle\sum_{n \in \mathbb{N}^*} v_n$ is plurisuperharmonic on Ω and $A \subset v^{-1}(+\infty)$. $\square$

Every nondense subspace of the space X is pluripolar in X since, by the Hahn-Banach theorem $([\text{B0}]_2, \text{Chap. II}, \S 6, n^\circ 3)$, it is included in the set of zeroes of some $x' \in X' \backslash \{0\}$. From this follows, by Proposition 1.5.3, that all bounded sets are pluripolar in a space X which is the strict inductive limit of a sequence of complete spaces. But there also exist spaces in which all bounded sets are pluripolar, even if they are not included in nondense subspaces (Ex. 4.5.15 below); for their construction, the following *inverse function theorem* of Lelong $[\text{Lel}]_6$ will be useful.

Theorem 4.5.13. *Let the function u: $\Omega \times \mathbb{R}_+ \to [-\infty, +\infty[$ have the property that $[(x, z) \mapsto u(x, |z|)] \in \mathscr{P}(\Omega \times \mathbb{C})$, from which follows that, for all $x \in \Omega$, $u(x, r)$ is a convex increasing function of $\ln r$ and therefore either $u(x, r) = u(x, 0)$ for all r or $u(x, r) \to +\infty$ with r; also let $u(x, 0) < 0$ for all $x \in \Omega$.*

a) *A plurihyperharmonic function $\ln v(., s)$, and therefore (Ex. 4.1.5.d), for all $\alpha > 0$, a p.s.h. function $1/v^\alpha(., s)$, are defined on Ω, for all $s \in \mathbb{R}_+$, by*

$$\begin{cases} v(x, s) = +\infty & \text{if } u(x, r) = u(x, 0) \text{ for all } r \in \mathbb{R}_+, \\ u[x, v(x, s)] = s & \text{if } u(x, r) \to +\infty \text{ with } r. \end{cases}$$

b) *Let Ω be connected, $u(x, 1) \leqslant 0$ for all $x \in \Omega$ and $u(x, r) \to +\infty$ with r for at least one $x \in \Omega$; for every subset A of Ω which is not strictly pluripolar in Ω but satisfies $M(r) = \sup\limits_{x \in A} u(x, r) < \infty$ for all $r \in \mathbb{R}_+$ we have*

$$u(x, r) \leqslant M[r^{1/R_1^A(x)}] \quad \text{for all } r \in [1, +\infty[, x \in \Omega.$$

Proof. a) Given $s \in \mathbb{R}_+$, $\lambda \geqslant 0$, $\lambda < v(x_0, s)$, since $u(x_0, \lambda) < s$ and u is upper semicontinuous, for x sufficiently near x_0 we have $u(x, \lambda) < s$ and $\lambda < v(x, s)$; v is lower semicontinuous and now we have to prove $\ln v(a, s) \geqslant MV$

$\ln v(a + e^{i\theta}b, s)$ whenever $\Omega \supset \{a + \zeta b: |\zeta| \leqslant 1\}$; for this we use an argument of pseudoconvexity (see [Hö], §2.6), with a given $s \in \mathbb{R}_+$.

$\ln v(a + e^{i\theta}b, s)$ is the limit of a strictly increasing sequence of real continuous functions $c_n(e^{i\theta})$; let $f_n \in \mathscr{A}(\Delta, \mathbb{C})$, $\Delta = \Delta(0, 1)$, be such that $\mathscr{R}e\, f_n(\zeta) \to c_n(e^{i\theta})$ as $\Delta \ni \zeta \to e^{i\theta}$, for all $\theta \in \mathbb{R}$. Then, given the integer n, for $|\zeta| < 1$ and $1 - |\zeta|$ sufficiently small, we have

$$\ln v(a + \zeta b, s) \geqslant \mathscr{R}e\, f_n(\zeta) \text{ or } u[a + \zeta b, |\exp f_n(\zeta)|] \leqslant s;$$

since $(\zeta, z) \mapsto u(a + \zeta b, |z|)$ is p.s.h. on $\Delta \times \mathbb{C}$, the left hand member of the latter inequality is a subharmonic function of $\zeta \in \Delta$ by Theorem 4.2.1, and by the maximum principle both inequalities still hold for all $\zeta \in \Delta$. For $\zeta = 0$ the former one gives

$$\ln v(a, s) \geqslant \mathscr{R}e\, f_n(0) = MV\, c_n(e^{i\theta}).$$

b) Since the sets $\{x \in \Omega: u(x, r) = u(x, 0) \text{ for all } r \in \mathbb{R}_+\}$ and $\{x \in \Omega: v(x, s) = +\infty\}$ coincide for all $s \in \mathbb{R}_+$, $-\ln v(., s) \in \mathscr{P}(\Omega)$ for all $s \in \mathbb{R}_+$, the sets in question are strictly pluripolar in Ω and cannot include A: $M(r) \to +\infty$ with r; moreover M is a continuous increasing function on $\mathbb{R}_+$, $M(1) \leqslant 0$. Now, for all $s \in \mathbb{R}_+$, $\ln v(., s)$ is positive plurisuperharmonic on Ω and, for $x \in A$, satisfies $\ln v(x, s) \geqslant \ln \delta(s)$ if $\delta(s) \in [1, +\infty[$ satisfies $M[\delta(s)] = s$; then $\ln v(x, s) \geqslant R_1^A(x). \ln \delta(s)$ for all $x \in \Omega$.

Given $s \in \mathbb{R}_+$ and $x \in \Omega$, since $R_1^A(x) > 0$, the value $r = [\delta(s)]^{R_1^A(x)}$ satisfies $r \leqslant v(x, s)$ or

$$u(x, r) \leqslant s = M[\delta(s)] = M[r^{1/R_1^A(x)}];$$

finally, for a given $x \in \Omega$, $\delta(s)$ may be any number $\in [1, +\infty[$ and r too since $0 < R_1^A(x) \leqslant 1$. $\quad\square$

Corollary 4.5.14. *Let X be a Fréchet space whose topology is defined by an increasing sequence of seminorms p_n such that $\ln p_n(x) = u(x, t_n)$ for all $n \in \mathbb{N}^*$, where $(x, z) \mapsto u(x, |z|)$ is p.s.h. on $X \times \mathbb{C}$ and the sequence (t_n) increases strictly to $+\infty$. Then either X is a Banach space or any bounded set $A \subset X$ is strictly pluripolar in some open set containing A.*

Proof. We may assume that $\{x \in X: u(x, r) \text{ does not depend on } r\}$ is not X (otherwise X is a normed space), that $t_1 = 1$, A is bounded, i.e. $M_n = \sup_A \ln p_n < \infty$ for each $n \in \mathbb{N}^*$, finally that A is not strictly pluripolar in the open set $\Omega = p_1^{-1}([0, 1[)$.

Thus all assumptions of the theorem are fulfilled, including $M(r) < \infty$ since $M(r) \leqslant M_n$ for all $r \leqslant t_n$. Now let $(R_1^A)^*$ be the lower regularized of

$R_1^A, 0 \leqslant (R_1^A)^* \leqslant R_1^A \leqslant 1$, and $x \in \Omega$ be such that $(R_1^A)^*(x) > \lambda > 0$: by part b) of the theorem, we have

$$\ln p_n(x) \leqslant M[t_n^{1/R_1^A(x)}] \leqslant M[t_n^{1/(R_1^A)^*(x)}] \leqslant M(t_n^{1/\lambda}),$$

which proves that the open set $\{x \in \Omega : (R_1^A)^*(x) > \lambda\}$ is bounded, and X is a normed space unless this open set is empty for all $\lambda > 0$, i.e. $(R_1^A)^* \equiv 0$. Then, by Proposition 4.5.6.b, R_1^A vanishes on a dense subset of Ω and A is strictly pluripolar in Ω by Proposition 4.5.12. $\square$

Example 4.5.15. *In the Fréchet space* $X = \mathscr{A}(\mathbb{C}, \mathbb{C})$, *endowed with the topology of compact convergence, any bounded set* A *is strictly pluripolar in some open set containing* A.

Proof. X is not a normed space since any bounded set in X is relatively compact, hence its interior empty by the F. Riesz theorem ($[\text{Bo}]_2$, Chap. I, §2, Th. 3). So we only have to check that $[(x, z) \mapsto \sup_{\theta \in \mathbb{R}} \ln|x(ze^{i\theta})|] \in \mathscr{P}(X \times \mathbb{C})$: in fact, this supremum is a continuous function of $(x, z) \in X \times \mathbb{C}$ and, for given $(x, y) \in X^2, (\alpha, \beta) \in \mathbb{C}^2, \theta \in \mathbb{R}$,

$$x[(\alpha + \zeta\beta)e^{i\theta}] + \zeta y[(\alpha + \zeta\beta)e^{i\theta}] \text{ an analytic function of } \zeta \in \mathbb{C}.$$

Also note the existence of bounded sequences (x_n) in X such that any subspace containing the sequence is dense in X, e.g. $x_n(\zeta) = \left(\dfrac{\zeta}{n}\right)^n$, $n \in \mathbb{N}^*$, $x_0(\zeta) = 1$. $\square$

Theorem 4.5.16. *Let* P *be a closed pluripolar subset of* Ω. *Every function* $u \in \mathscr{P}(\Omega \backslash P)$ *which satisfies* $\limsup\limits_{\Omega \backslash P \ni x \to a} u(x) < \infty$ *for all* $a \in P$ *has a unique extension* $\in \mathscr{P}(\Omega)$.

Proof. We begin with the uniqueness, which is not so obvious as in Theorem 4.5.2. Let $P \subset w^{-1}(-\infty)$, $w \in \mathscr{P}(\Omega)$, $a \in P$, $b \in \omega(a)$, $w(a + b) > -\infty$: since $\zeta \mapsto w(a + \zeta b)$ is subharmonic on an open set including $\bar{\Delta}(0, 1)$, the set $\{\zeta \in \bar{\Delta}(0, 1) : a + \zeta b \in P\}$ is polar, and we can find a sequence of radii $r_k \to 0$ such that $|\zeta| = r_k$ implies $a + \zeta b \in \Omega \backslash P$; then the value at a of the desired extension has to be $\lim MV u(a + r_k e^{i\theta} b)$.

Now it is enough to construct, for every $a \in P$, a function $\tilde{u}$ p.s.h. on some open neighbourhood W of a in Ω, with $\tilde{u} = u$ on $W \backslash P$: it is enough because $P \cap W \cap W'$ is a closed pluripolar subset of $W \cap W'$, and p.s.h. functions have the sheaf property. Given w as above, we can choose W so that $\sup\limits_{W} w < \infty$, then a constant c such that $w + c \leqslant 0$ on W; in order to lighten the notation, we assume $w \leqslant 0$ on Ω and construct $\tilde{u}$ on Ω as follows.

For each $n \in \mathbb{N}^*$, let $u_n = u + \dfrac{w}{n}$ on $\Omega \backslash P$, $u_n = -\infty$ on P: this u_n is upper semicontinuous on Ω on account of the assumption made on u at all points $\in P$, and the inequality $u_n(a) \leqslant MV\, u_n(a + re^{i\theta}b)$ holds for sufficiently small radii r, for all $a \in \Omega$, $b \in X$; moreover the increasing sequence $(u_n) \subset \mathscr{P}(\Omega)$ is locally bounded from above, on account of the same assumption, hence $v = \lim u_n \in \mathscr{P}^1(\Omega)$ (Def. 4.3.3.a), $v^* \in \mathscr{P}(\Omega)$.

From $v = u$ on $\Omega \backslash w^{-1}(-\infty)$, $v \leqslant u$ on $w^{-1}(-\infty) \backslash P$, follows $v^* = u$ on $\Omega \backslash w^{-1}(-\infty)$. Now let $a \in w^{-1}(-\infty) \backslash P$: choosing $b \in \omega(a)$ with $w(a + b) > -\infty$, we can (as above) find a sequence of radii $r_k \to 0$ such that $|\zeta| = r_k$ implies $w(a + \zeta b) > -\infty$, $v^*(a + \zeta b) = u(a + \zeta b)$; then

$$v^*(a) = \lim MV\, v^*(a + r_k e^{i\theta} b)$$

$$= \lim MV\, u(a + r_k e^{i\theta} b) = u(a)$$

and $\tilde{u} = v^*$ is the desired extension. $\quad\square$

Corollary 4.5.17. *Let again P be a closed pluripolar subset of Ω.*

a) *Every function u pluriharmonic on Ω which satisfies $\limsup\limits_{\Omega \backslash P \ni x \to a} |u(x)| < \infty$ for all $a \in P$ has a (unique) pluriharmonic extension to Ω.*
b) *If Ω is connected, $\Omega \backslash P$ too.*
c) *Every function $\varphi \in \mathscr{A}(\Omega \backslash P, \mathbb{C})$ which satisfies $\limsup\limits_{\Omega \backslash P \ni x \to a} |\varphi(x)| < \infty$ for all $a \in P$ has a (unique) extension $\in \mathscr{A}(\Omega, \mathbb{C})$.*
d) *Let the l.c. space Z be complete: every map $f \in \mathscr{A}(\Omega \backslash P, Z)$ with satisfies $\limsup\limits_{\Omega \backslash P \ni x \to a} q \circ f(x) < \infty$ for all $a \in P$, $q \in \mathrm{csn}(Z)$, has a (unique) extension $\in \mathscr{A}(\Omega, Z)$.*

Proof. a) u and $-u$ have unique extensions $\in \mathscr{P}(\Omega)$, which take opposite values at each point $a \in P$, by the argument used in the first paragraph of the proof of Theorem 4.5.16.
b) See again the first paragraph of the proof of Theorem 4.4.5.
c) Since $u = \mathscr{R}e\, \varphi$ and $v = \mathscr{I}m\, \varphi$ have (unique) pluriharmonic extensions to Ω, φ has a continuous extension $\tilde{\varphi}$ to Ω; we shall prove that $\tilde{\varphi}$ is analytic on $a + \omega(a)$ for all $a \in P$.

Since $a + \omega(a)$ is star-shaped, by Proposition 4.4.2 the extension of u, restricted to $a + \omega(a)$, is the real part of an analytic function ψ; then $\tilde{\varphi} - \psi$ is locally constant, hence constant, on $[a + \omega(a)] \backslash P$, a connected open set.
d) A consequence of Theorem 4.4.5. $\quad\square$

4.6 A fine maximum principle

A fine maximum principle was introduced by Coeuré [Coe] as a generalization to the infinite dimension of a theorem of Brelot [Brel]$_2$ which we will present below (Th. 4.6.3) in the one dimensional case.

Definition 4.6.1. *A set T in X is thin at a point $a \notin T$ if there is some v p.s.h. on an open neighbourhood of a such that $\limsup\limits_{T \ni x \to a} v(x) < v(a)$; T is thin at a point $a \in T$ if $T\backslash\{a\}$ is thin at this point.*

Since $\sup\limits_{\emptyset} v$ is understood to be $-\infty$, $a \notin \bar{T}$ implies T thin at a. A finite union of sets each of which is thin at a is again thin at a: in fact, if T_1 and T_2 are thin at $a \notin T_1 \cup T_2$, v_1 and v_2 p.s.h. on an open neighbourhood of a and (for $i = 1, 2$) $\limsup\limits_{T_i \ni x \to a} v_i(x) < v_i(a)$, then $\limsup\limits_{x \to a} v_i(x) = v_i(a)$, $i = 1, 2$, implies $\limsup\limits_{T_1 \cup T_2 \ni x \to a} (v_1 + v_2)(x) < (v_1 + v_2)(a)$.

In the following subsections (A) to (D), several properties of thinness in the finite dimensional case are presented in a bounded open set U in $\mathbb{C}$, which will be chosen once for all.

(A) Given $\alpha \in U$ and v subharmonic on some open neighbourhood of α, one can find (by means of the positive measure Δv) $u \in \mathscr{P}(U)$ such that $v - u$ is harmonic on some open neighbourhood of α: therefore T is thin at $\alpha \notin T$ if and only if there is $u \in \mathscr{P}(U)$ such that $\limsup\limits_{T \ni \zeta \to \alpha} u(\zeta) < u(\alpha)$.

In other criteria appears R_v^A, defined for every set $A \subset U$ and every $v \in \mathscr{S}^+(U)$ as the infimum of all functions in $\mathscr{S}^+(U)$ which are ($\leqslant v$ on U and) equal to v on A: $-R_v^A \in Q(U)$ (Th. 4.2.5), R_v^A has a harmonic restriction to $U\backslash\bar{A}$ and a lower regularized $(R_v^A)^*$ which is the smallest function in $\mathscr{S}^+(U)$ equal to v on $A\backslash$(a polar set); $(R_v^A)^* = R_v^A$ if A is open.

A polar set P is thin at every point $\alpha \in \mathbb{C}$. In fact, if U includes P and α, for each $n \in \mathbb{N}^*$ there is $v_n \in \mathscr{S}^+(U)$ equal to $+\infty$ on $P\backslash\bar{\Delta}\left(\alpha, \dfrac{1}{n}\right)$ and, if necessary by balayage on $U\backslash\bar{\Delta}\left(\alpha, \dfrac{1}{n}\right)$, one may assume $v_n(\alpha) \leqslant \dfrac{1}{n^2}$; then $\sum\limits_{n \in \mathbb{N}^*} v_n \in \mathscr{S}^+(U)$ and equals $+\infty$ on $P\backslash\{\alpha\}$. From this follows that thinness is unaltered by addition or subtraction of a polar set; $(R_v^T)^*(\alpha) < v(\alpha)$ for one $v \in \mathscr{S}^+(U)$ implies T thin at α.

Let the countable family of closed discs $A_n = \bar{\Delta}(\alpha_n, r_n)$, $n \in \mathbb{N}^*$, contained in U, be such that any open subset of U is the union of a subfamily of open discs $\mathring{A}_n = \Delta(\alpha_n, r_n)$: each $v_n = R_1^{A_n}$ is continuous, therefore in $\mathscr{S}^+(U)$, and for all $\zeta \in U$

(*) $\qquad v_n(\zeta) \leqslant \inf\left[1, \dfrac{\ln(R_n/|\zeta - \alpha_n|)}{\ln(R_n/r_n)} \right]$ if $U \subset \Delta(\alpha_n, R_n)$.

If the set $T \subset U$ satisfies $(R_{v_n}^T)^*(\alpha) < v_n(\alpha)$ for some $n \in \mathbb{N}^*$, then T is thin at α. Conversely, if $\alpha \in \bar{T}\backslash T$, $v \in \mathscr{S}^+(U)$ and $\liminf_{T \ni \zeta \to \alpha} v(\zeta) > v(\alpha)$, one may assume $v(\alpha) = 1$, $v \geqslant \lambda > 1$ on $T \cap K$ for some compact neighbourhood K of α in U, hence $\dfrac{v}{v_n} \geqslant \lambda$ on $T \cap K$ for all $n \in \mathbb{N}^*$. On the other hand, if (A_{n_k}) is a decreasing sequence with $\bigcap_k A_{n_k} = \{\alpha\}$, by (*)$v_{n_k}$ decreases to 0 uniformly on the boundary of K; $v \geqslant \lambda v_{n_k}$ on this boundary and $A_{n_k} \subset K$ imply $v \geqslant \lambda v_{n_k}$ on $U \backslash K$ [for the subharmonic function equal to $(\lambda v_{n_k} - v)^+$ on $U\backslash K$, to 0 on K, must be $\equiv 0$], and $v \geqslant \lambda v_{n_k}$ on T implies $v_{n_k}(\alpha) = 1 = v(\alpha) > R_{v_{n_k}}^T(\alpha)$. Finally, if $\alpha \notin \bar{T}$, we have $v_n(\alpha) > R_{v_n}^T(\alpha)$ whenever A_n lies in the connected component of $U \backslash \bar{T}$ containing α, since $R_{v_n}^T$ is harmonic on this component, but not v_n.

Thus A is thin at α if and only if $(R_{v_n}^A)^*(\alpha) < v_n(\alpha)$ for some $n \in \mathbb{N}^*$, hence also if and only if $(R_w^A)^*(\alpha) < w(\alpha)$, where $w = \sum_{n \in \mathbb{N}^*} \dfrac{v_n}{n^2} \in \mathscr{S}^+(U)$: w may be termed a continuous universal test function for the thinness of any subset A of U at any point $\alpha \in U$. By the convergence theorem 4.2.5, the set $\{\alpha \in U : (R_w^A)^*(\alpha) < R_w^A(\alpha)\}$ is polar, hence also the set $\{\alpha \in A : (R_w^A)^*(\alpha) < w(\alpha)\}$: for every set A in $\mathbb{C}$, $\{\alpha \in A : A \text{ thin at } \alpha\}$ is polar, while $\{\alpha \in \mathbb{C} : A \text{ thin at } \alpha\}$ is an F_σ.

(B) If $A \subset U$ is the union of an increasing sequence (A_n), then $(R_v^A)^*$ is the limit of the increasing sequence $(R_v^{A_n})^*$ for all $v \in \mathscr{S}^+(U)$. If $v \in \mathscr{S}^+(U)$ is continuous (finite or not), for any set $A \subset U$, $(R_v^A)^*$ is the infimum of the R_v^ω, ω open containing $A\backslash$(a polar set): in fact, for all $\varepsilon > 0$, $\omega = \{\zeta \in U : (R_v^A)^*(\zeta) > v(\zeta) - \varepsilon\}$ is open, contains $A\backslash$(a polar set), and $(R_v^A)^* + \varepsilon > v$ on ω implies $(R_v^A)^* + \varepsilon \geqslant R_v^\omega$ on U.

Now let G be the Green function of U and $G_\alpha(\zeta) = G(\alpha, \zeta)$; from the properties just listed follows the important formula

(1) $\qquad (R_{G_\alpha}^A)^*(\beta) = (R_{G_\beta}^A)^*(\alpha)$ for all $A \subset U$, $\alpha \in U$, $\beta \in U$.

Proof of (1). a) If A is closed in U and $\alpha \in U' = U\backslash A$ with the Green function G', then $G_\alpha - G_\alpha' = h$ is the largest harmonic minorant of G_α on U'; if $v \in \mathscr{S}^+(U)$ is equal to G_α on A, then the subharmonic function equal to $(h - v)^+$ on U', to 0 on A, must be $\equiv 0$, for G_α is a potential on U, which means that any subharmonic function $\leqslant G_\alpha$ on U is $\leqslant 0$: $v \geqslant h$ on U', hence

$R^A_{G_\alpha} \geq h$ on U', $G_\alpha = G'_\alpha + (R^A_{G_\alpha})^*$ on U': for A closed in U, α, $\beta \in U \backslash A$, formula (1) follows from the symmetry of the Green functions of U and U' (see 4.2(D)).

b) If A is closed in U: formula (1) for $A \backslash D\left(\alpha, \dfrac{1}{n}\right) \backslash D\left(\beta, \dfrac{1}{n}\right)$ for all $n \in \mathbb{N}^*$

entails (1) for $A \backslash \{\alpha, \beta\}$, or A since $\{\alpha, \beta\}$ is a polar set. Any open subset ω of U is the union of an increasing sequence of closed subsets A_n of U and (1) for the A_n entails (1) for ω. Finally (1) for open sets ω entails (1) for any $A \subset U$ because G_α, G_β are continuous functions $U \to \bar{\mathbb{R}}_+$.   □

A consequence of formula (1) is the fact that $(R^A_{G_\alpha})^* < G_\alpha$ at one point in U implies A thin at α.

(C) From the relation $(R^A_{v_1 + v_2})^* = (R^A_{v_1})^* + (R^A_{v_2})^*$ for finite continuous v_1, $v_2 \in \mathscr{S}^+(U)$ ([He]$_R$, §10.A) follows, by the density of the $(v_1 - v_2)|_K$ in $\mathscr{C}(K)$ for any compact set $K \subset U$, the existence of a unique positive Radon measure μ^A_α on U such that

(2) $(R^A_v)^*(\alpha) = \int v \, d\mu^A_\alpha$ for all $v \in \mathscr{S}^+(U)$,

and particularly for all $\xi \in U$:

$$\int G(\xi, \eta) \, d\mu^A_\alpha(\eta) = \int G_\xi \, d\mu^A_\alpha = (R^A_{G_\xi})^*(\alpha) = (R^A_{G_\alpha})^*(\xi).$$

So the potential in U of the measure μ^A_α results from the balayage of G_α on A; consequently μ^A_α is carried by the boundary of A in U if $\alpha \notin \mathring{A}$, while $\mu^A_\alpha = \varepsilon_\alpha$ (the Dirac measure) if A is not thin at α, especially if $\alpha \in \mathring{A}$.

Since the functions $\in \mathscr{S}^+(U)$ equal to w, or to $(R^A_w)^*$, on $A \backslash$(a polar set), are the same, from (2) follows $\int (R^A_w)^* \, d\mu^A_\alpha = \int w \, d\mu^A_\alpha$ for all $w \in \mathscr{S}^+(U)$; with the universal function w constructed in (A), this means that the set of all points in U at which A is thin is a null set for μ^A_α. A deeper analysis ([He]$_R$, §28.B) proves that the set of all points in $U \backslash \{\alpha\}$ at which $U \backslash A$ is thin is also a null set for μ^A_α.

(D) The special case in which $B = U \backslash A$ is relatively compact in U will be useful. In this case, formula (2) gives $\int h \, d\mu^A_\alpha = h(\alpha)$ for any h harmonic ≥ 0 on U, in particular $\int d\mu^A_\alpha = 1$. On the other hand, given the bounded set B, let $\bar{B} \cup \{\alpha\} \subset U \subset U'$, where U' is again a bounded open set with Green function G': since $G'_\alpha - G_\alpha = h$ harmonic ≥ 0 on U, the restrictions to U of the functions $\in \mathscr{S}^+(U')$ which are equal to G'_α on $U' \backslash B \backslash$(a polar set) are exactly the $h + v$, $v \in \mathscr{S}^+(U)$, v equal to G_α on $U \backslash B \backslash$(a polar set); therefore the measure $\mu_\alpha^{U \backslash B}$ does not depend on the bounded open set U containing $\bar{B}$ and α.

We are now in a position to prove the fine maximum (or minimum) principle.

Definition 4.6.2. *The fine topology $\mathscr{F}_1$ on the complex plane is the coarsest one for which all subharmonic (or superharmonic) functions are continuous.*

Since $\zeta \mapsto |\zeta - \alpha|$ is subharmonic, $\mathscr{F}_1$ is finer than the initial topology $\mathscr{T}_1$. A set containing α is a fine neighbourhood of α if and only if its complement in $\mathbb{C}$ is thin at α. Given $A \subset U$ as in (**C**) above, the set of all points in $U \setminus A$ at which A is thin is the fine interior of $U \setminus A$, hence a new formulation of the last two statements in (**C**): the complement in U of the fine boundary of A in U is a null set for μ_α^A unless A is a fine neighbourhood of α.

A $(\mathscr{T}_1\text{-})$ connected open set ω cannot be thin at any $(\mathscr{T}_1\text{-})$ boundary point ζ_0 of ω [Brel]$_3$: both boundaries coincide, and the fine lim inf (or lim sup) at ζ_0 of a function $\omega \to \bar{\mathbb{R}}$ is actually a property of this function.

Theorem 4.6.3. *Let ω be a bounded connected open set in $\mathbb{C}$, $v \in \mathscr{S}^+(\omega)$, $M > 0$: if v has a fine lim inf $\geqslant M$ at every boundary point of ω, then $v \geqslant M$ on ω.*

Proof. In what follows, the bounded open set U which served in (**B**) and (**C**) above will be chosen containing $\bar{\omega}$. Assuming that, for some $\alpha \in \omega$,

$$v(\alpha) < m < M, \quad \text{we} \quad \text{set} \quad A = \{\zeta \in \omega : v(\zeta) \leqslant m\} \quad \text{and} \quad A_n = \Big\{\zeta \in A:$$

$\text{dist}(\zeta, \complement\omega) \geqslant \dfrac{1}{n}\Big\}$, $n \in \mathbb{N}^*$, whenever $A_n \ni \alpha$: these A_n are compact subsets of ω. As observed in (**D**) above, the measures $\mu_\alpha^{U \setminus A}$, $\mu_\alpha^{U \setminus A_n}$ (carried by $\bar{A}$) do not depend on U and may be denoted by $\mu_\alpha^{\complement A}$, $\mu_\alpha^{\complement A_n}$, their mass is 1; as to $\mu_\alpha^{\complement A_n}$, we may even replace U by ω, and this we do only in order to derive from (2) the inequalities

$$v(\alpha) \geqslant \int v \, d\mu_\alpha^{\complement A_n}, \, v(\alpha) - m \geqslant \int (v - m) \, d\mu_\alpha^{\complement A_n} \geqslant -m\mu_\alpha^{\complement A_n}(B_n),$$

where B_n is the set of boundary points ζ of A_n such that $\text{dist}(\zeta, \complement\omega) = \dfrac{1}{n}$

(in fact, for any other boundary point ζ of A_n, either $v(\zeta) = m$ or $\complement A_n$ is thin at ζ, and the set of all points at which $\complement A_n$ is thin is a null set for $\mu_\alpha^{\complement A_n}$). We will reach a contradiction by showing that $\mu_\alpha^{\complement A_n}(B_n) \to 0$ as $n \to \infty$.

Any compact subset of A is contained in some A_n, any open subset of U containing $U \setminus A$ also contains some $U \setminus A_n$; then any function in $\mathscr{S}^+(U)$ which is $> G_\alpha$ on $U \setminus A$ is also $> G_\alpha$ on some $U \setminus A_n$, the sequence $(R_{G_\alpha}^{U \setminus A_n})$ decreases to $R_{G_\alpha}^{U \setminus A}$ on U, to $(R_{G_\alpha}^{U \setminus A})^*$ on $U \setminus (\text{a polar set})$; in other words, the potential in U of the measure $\mu_\alpha^{\complement A_n}$ decreases to the potential of $\mu_\alpha^{\complement A}$ except on a polar set. Hence follow:

a) $\iint G(\xi, \eta) \, d\lambda(\xi) \, d\mu_\alpha^{\zeta A_n}(\eta)$ decreases to $\iint G(\xi, \eta) \, d\lambda(\xi) \, d\mu_\alpha^{\zeta A}(\eta)$ whenever the positive measure λ on U generates a bounded potential in U, since every λ-measurable polar set is a null set for such a λ (a consequence of Th. 4.5.2.a);

b) $\int c \, d\mu_\alpha^{\zeta A_n} \to \int c \, d\mu_\alpha^{\zeta A}$ for every finite continuous function c on $\bar{A}$, since this function c is the uniform limit of suitably chosen differences (restricted to $\bar{A}$) of continuous potentials in U.

Finally, at every point $\in \bar{A} \setminus \omega$, the assumption made on fine lim inf v implies that A is thin; then $\mu_\alpha^{\zeta A}(\bar{A} \setminus \omega) = 0$ and for all $\varepsilon > 0$ we can choose a positive continuous function c on $\bar{A}$, equal to 1 on $\bar{A} \setminus \omega$, with $\int c \, d\mu_\alpha^{\zeta A} \leq \varepsilon$; for sufficiently large n we have $c \geq 1 - \varepsilon$ on B_n and $\int c \, d\mu_\alpha^{\zeta A_n} \leq 2\varepsilon$, hence

$$\mu_\alpha^{\zeta A_n}(B_n) \leq \frac{2\varepsilon}{1 - \varepsilon},$$ which completes the proof. $\square$

Returning to an infinite dimensional l.c. space X, we first remark a property of the finitely open sets (see §2.3): whereas, by Definition 4.6.1, the complement of an open set ω is thin at every point in ω, this is no longer true if ω is only finitely open; Coeuré's thesis also contains the following counter-example.

Let the sequence $(a_k)_{k \in \mathbb{N}^*} \subset X$ tend to the origin and $a_1, \ldots, a_k$ be linearly independent for all $k \in \mathbb{N}^*$ (the existence of such a sequence was discussed in Rem. 2.3.11). The set $A = \bigcup_{k \in \mathbb{N}^*} \{\zeta a_k : |\zeta| \geq 1\}$ is not thin at the origin: in fact, if V is a balanced neighbourhood of the origin and $v \in \mathscr{P}(V)$, $v(0) \leq \sup_{|\zeta|=1} v(\zeta a_k)$ holds whenever $a_k \in V$, hence $\limsup_{A \ni x \to 0} v(x) = v(a)$. Yet $\varphi = X \setminus A$ is finitely open since any m-dimensional linear subspace of X contains at most m elements a_k.

This example shows that, by taking into account the finitely open sets in the following definition, we get an actually finer topology on X.

Definition 4.6.4. *The fine topology $\mathscr{F}$ on X is the coarsest one for which all finitely open sets are open and all p.s.h. functions continuous.*

A set $A \ni a$ is a fine neighbourhood of a if and only if there is a set T thin at a such that $A \cup T$ contains a finitely open set containing a. Then, for every complex line $E \ni a$ in X: the intersection of $A \cup T$ with E is a neighbourhood of a in E, while $E \cap T$ is thin at a in E, therefore $E \cap A$ a fine neighbourhood of a in E; the intersection with E of any $\mathscr{F}$-neighbourhood of a in X is an $\mathscr{F}_1$-neighbourhood of a in E.

If ω is open in X (for the initial topology $\mathscr{T}$): the fine interior of $X \setminus \omega$ is the union of its $\mathscr{T}$-interior and the $\mathscr{T}$-boundary points a of ω which are contained in some finitely open set φ such that $\omega \cap \varphi$ is thin at a; the set of the other $\mathcal{T}$-boundary points of ω is the *fine boundary of* ω, which we denote by $\Delta(\omega)$.

The following property of an open set ω makes the fine maximum principle valid in ω, hence its designation (FMP).

(FMP) Every point $x \in \omega$ lies on some complex line E such that the connected component of x in $E \cap \omega$ is bounded.

The topology of X admits a fundamental family of open sets with property (FMP), namely the $\omega = a + p^{-1}([0, 1[), a \in X, p \in \mathrm{csn}(X), p \not\equiv 0$.

Theorem 4.6.5. *Let the open set ω have property (FMP) and $u \in \mathscr{P}(\omega)$: if* $\limsup\limits_{\substack{\omega \ni x \to a \\ \mathcal{T}}} u(x) < +\infty$ *and* $\limsup\limits_{\substack{\omega \ni x \to a \\ \mathscr{F}}} u(x) \leqslant 0$ *for all $a \in \Delta(\omega)$, then $u \leqslant 0$ on ω.*

Proof. Given $x \in \omega$, let E be a complex line containing x such that the connected component $U \ni x$ of $E \cap \omega$ is bounded, and a any boundary point of U in E. Since U is not thin at a in E, $E \backslash U$ is not an $\mathscr{F}_1$-neighbourhood of a in E, neither is $E \backslash (E \cap \omega)$, $X \backslash \omega$ is not an $\mathscr{F}$-neighbourhood of a in X: $\Delta(\omega)$ contains the boundary of U in E.

From the first assumption made on $\Delta(\omega)$ follows $u_{|U}$ bounded from above by the standard maximum principle; from the second one, that $u_{|U}$ has an $\mathscr{F}_1$-lim sup $\leqslant 0$ at every boundary point of U, hence $u \leqslant 0$ on U by Theorem 4.6.3. $\quad\square$

Corollary 4.6.6. *Let the open set ω have property (FMP). Given a function* $f: \Delta(\omega) \to \bar{\mathbb{R}}$ *which is bounded from below, the infimum H_f^ω of all pluri-hyperharmonic functions w on ω which satisfy*

$$(3) \qquad \liminf_{\substack{\omega \ni x \to a \\ \mathcal{T}}} w(x) > -\infty \quad and \quad \liminf_{\substack{\omega \ni x \to a \\ \mathscr{F}}} w(x) \geqslant f(a) \quad for\ all\ a \in \Delta(\omega)$$

has a plurihyperharmonic lower regularized $(H_f^\omega)^*$.

Proof. By the theorem, all functions w are $\geqslant \inf\limits_{\Delta(\omega)} f$, hence the result by Theorem 4.1.7.b. $\quad\square$

The map $f \mapsto H_f^\omega$ is subadditive: $H_{f+g}^\omega \leqslant H_f^\omega + H_g^\omega$, and this implies $(H_{f+g}^\omega)^* \leqslant (H_f^\omega)^* + (H_g^\omega)^*$ by Corollary 4.3.9.b if X is a Fréchet space. In the following theorem, the data f are defined on an open set Ω containing $\bar{\omega}$, and H_f^ω is written for $H_{f|_{\Delta(\omega)}}^\omega$.

Theorem 4.6.7. *In a Fréchet space X, let the open set ω have property (FMP), and $\Omega \supset \bar{\omega}$ be another open set.*

a) *If $u \in \mathscr{P}(\Omega)$ and $f = u|_{\Delta(\omega)}$ is bounded from below: for every plurihyper-harmonic function w on ω, (3) is equivalent to*

(4) $\displaystyle\liminf_{\substack{\omega \ni x \to a \\ \mathscr{F}}} w(x) > -\infty$ *for all $a \in \Delta(\omega)$ and $w \geqslant u$ on ω;*

$H_u^\omega = (H_u^\omega)^*$ *is the smallest w satisfying (3) or (4).*
b) *If $\mathscr{V} \subset \mathscr{P}(\Omega)$ is locally bounded from above, $u = \displaystyle\sup_{v \in \mathscr{V}} v$ and $f = u|_{\Delta(\omega)}$ bounded from below: again (3) and (4) are equivalent, and $H_u^\omega = (H_u^\omega)^* = H_{u^*}^\omega = (H_{u^*}^\omega)^*$ is the smallest w satisfying (3) or (4).*

Proof. a) (3) implies $w \geqslant u$ on ω by Theorem 4.6.5; conversely, $w \geqslant u$ on ω implies, by the fine continuity of u:

$$\liminf_{\substack{\omega \ni x \to a \\ \mathscr{F}}} w(x) \geqslant \lim_{\substack{\omega \ni x \to a \\ \mathscr{F}}} u(x) = u(a) \quad \text{for all } a \in \Delta(\omega).$$

Since $(H_u^\omega)^* \geqslant \displaystyle\inf_{\Delta(\omega)} u$, the inequality $(H_u^\omega)^* \geqslant u$ will complete the proof of a).

In fact, given $x \in \omega$, ω contains generalized polydiscs $x + K(\mathscr{A})$ satisfying (4′) in Corollary 4.3.9 for H_u^ω:

$$(H_u^\omega)^*(x) \geqslant \overline{MV} H_u^\omega[x + K^*(\mathscr{A})] \geqslant MVu[x + K^*(\mathscr{A})] \geqslant u(x)$$

by Proposition 4.3.5.
b) Let $m = \displaystyle\inf_{\Delta(\omega)} u$: the inequality $(H_v^\omega)^* \geqslant v'$ obtained in a) for $v' = \sup(v, m)$ for all $v \in \mathscr{V}$ implies $(H_u^\omega)^* \geqslant u$ since $u \geqslant v'$ on $\Delta(\omega)$. Given $x \in \omega$, one can find generalized polydiscs $K(\mathscr{A})$ such that, with $u \leqslant M$ on $x + r_0 K(\mathscr{A}) \subset \omega$, (3′) in Corollary 4.3.9 holds:

$$M - u^*(x) = \lim_{r \to 0} \overline{MV}(M - u)[x + rK^*(\mathscr{A})]$$

$$\geqslant \lim_{r \to 0} MV(M - (H_u^\omega)^*)[x + rK^*(\mathscr{A})]$$

$$= M - (H_u^\omega)^*(x)$$

by Proposition 4.3.5 again. Thus $(H_u^\omega)^* \geqslant u^*$, which implies $(H_u^\omega)^* \geqslant H_{u^*}^\omega$ by a) for u^*, and then we may write

either $H_{u^*}^\omega \geqslant (H_{u^*}^\omega)^* \geqslant (H_u^\omega)^*$ or $H_{u^*}^\omega \geqslant H_u^\omega \geqslant (H_u^\omega)^*$.   □

Exercises

4.4.4. Show that, for $a, b \in \Omega$:

$$C_\Omega(a, b) = \sup \left\{ \arg th \left| \frac{\varphi(a) - \varphi(b)}{\varphi(a) + \overline{\varphi(b)}} \right| : \varphi \in \mathscr{A}(\Omega, \mathbb{C}), \operatorname{Re} \varphi > 0 \right\};$$

remarking that the additional condition $\varphi(a) - \varphi(b) \in \mathbb{R}$ leaves the supremum unaltered, compare H_Ω and C_Ω.

4.5.1. Show that unipolar sets have the sheaf property: if Ω is the union of open sets Ω_i, $i \in I$, a set $P \subset \Omega$ is unipolar in Ω if and only if $P \cap \Omega_i$ is unipolar in Ω_i for all $i \in I$.

4.5.7. Let $U \subset \mathbb{C}^m$ be the domain of existence of a bounded function $\in \mathscr{A}(U, \mathbb{C})$ and $\Omega = \{f \in \mathscr{A}(U, \mathbb{C}): |f| < 1\}$: show that the set of functions $\in \Omega$ for which U is not a domain of existence is strictly pluripolar in Ω endowed with the topology of uniform convergence.

4.5.8. Let $X = c_0(\mathbb{N}^*)$, the Banach space of complex sequences $x = (\xi_1, \xi_2, \dots)$ tending to 0, $P = \{x \in X: \xi_k = 0 \text{ for some } k\}$: show that P is pluripolar in X, and the intersection of P with any bounded open set in X is strictly pluripolar in this set.

4.5.12. Let Ω be a connected open set in X, A a subset of Ω which is not strictly pluripolar in Ω, (u_n) a sequence $\subset \mathscr{P}(\Omega)$: if $u_n < 0$ for all n and $u_n \to -\infty$ uniformly on A, then $u_n \to -\infty$ on every compact subset of Ω.

4.5.16 [Nov]$_1$. Let P be a (closed or not) unipolar or pluripolar subset of Ω; if u and v are p.s.h. (resp.: submedian) on Ω, then $u \leqslant v$ on $\Omega \backslash P$ entails $u \leqslant v$ (resp.: $u \leqslant v^*$, hence $u^* \leqslant v^*$) on Ω.

4.5.17. [Nov]$_1$. Let the open set Ω be connected. a) Let A be closed in Ω, the function $u: \Omega \to [-\infty, +\infty[$ upper semicontinuous, $A \subset u^{-1}(-\infty) \neq \Omega$: if $u|_{\Omega \backslash A} \in \mathscr{P}(\Omega \backslash A)$, then $u \in \mathscr{P}(\Omega)$ and A is pluripolar in Ω. *Hint*: consider the decreasing sequence $u_k = \sup(u, -k)$, $k \in \mathbb{N}$.
b) Let $f \in \mathscr{A}(\Omega, Z)$ (Z s.c.l.c.) be continuous, $f(\Omega)$ have a nonempty interior and B be closed and pluripolar in an open set containing $f(\Omega)$: if $f|_{\Omega \backslash f^{-1}(B)}$ is analytic, then f too is analytic (a generalization of the Rado theorem, exercise 3.1.1.).

Chapter 5
Problems involving plurisubharmonic functions

Summary

5.1.1.	Equivalent definitions of pseudoconvexity.
Prop. 5.1.3. and 5.1.4.	Pseudoconvexity of $u^{-1}([-\infty, \lambda[)$ for u p.s.h. and $\lambda \in \mathbb{R}$.
Th. 5.1.5.	On a pseudoconvex open set, any p.s.h. function is locally an infimum of finite continuous p.s.h. functions.
Prop. 5.1.6.	A pseudoconvex open set containing a p-ball is invariant under the translations $\in p^{-1}(0)$ [Di$_4$, Nov$_2$].
Prop. 5.2.2.	Domain of existence $\Rightarrow$ Domain of holomorphy $\Rightarrow$ Holomorphically convex $\Rightarrow$ Pseudoconvex.
Prop. 5.2.3.	Properties of convex domains.
Th. 5.2.4.	Spaces in which Holomorphically convex $\Rightarrow$ Domain of holomorphy or existence [Nov]$_2$.
Prop. 5.2.6.	Any Schauder basis of a Fréchet space is equischauder.
Th. 5.2.7.	Spaces in which Pseudoconvex $\Rightarrow$ Domain of existence [Di No Sch].
Prop. 5.3.1.	The p-radius of boundedness of a p.s.h. function.
Th. 5.4.2.	and Proposition 5.4.3. P.s.h. functions with a minimal growth [Lel]$_5$.
Th. 5.4.4.	The order of a p.s.h. function with non minimal growth [Lel]$_8$.
Th. 5.4.10.	Homogeneous p.s.h. functions [Lel]$_3$.
Th. 5.4.12.	A characterization of 1-homogeneous p.s.h. functions.
Th. 5.5.1.	The density number for a p.s.h. function at a given point [Lel]$_7$.
Th. 5.5.2.	The "minimum principle" of Kiselman [Ki]$_3$.
Prop. 5.5.5.	The density number at x is an upper semicontinuous function of x.
Th. 5.5.6.	The density number of $v \circ f$ for v p.s.h. and f analytic [Ki]$_3$.

Chapter 5
Problems involving plurisubharmonic functions

5.1 Pseudoconvexity in a l.c. space X

Definition and Theorem 5.1.1. *The open set Ω is pseudoconvex if it has the equivalent properties*

(1) *the convex hull in Ω, with respect to $\mathscr{P}(\Omega)$, of every compact subset K of Ω, namely*

$$\hat{K}_{\mathscr{P}(\Omega)} = \left\{ x \in \Omega \colon u(x) \leqslant \sup_K u \text{ for all } u \in \mathscr{P}(\Omega) \right\},$$

is such that $\hat{K}_{\mathscr{P}(\Omega)} + V \subset \Omega$ for some neighbourhood V of the origin in X;
(2) *for every compact subset K of Ω, Ω contains the closure of $\hat{K}_{\mathscr{P}(\Omega)}$;*
(3) *the distance of x to $X \backslash \Omega$ in the direction y, namely*

$$\delta(x, y) = \sup\{r > 0 \colon |\zeta| \leqslant r \text{ implies } x + \zeta y \in \Omega\},$$

is such that $\ln \delta$ is a plurihyperharmonic function of $(x, y) \in \Omega \times X$;
(4) *$\ln \delta(. , y)$ is plurihyperharmonic on Ω for all $y \in X$.*

Proof of their equivalence. The following remarks hold for any open set Ω: $\delta(x, y) > 0$ for all $(x, y) \in \Omega \times X$; $\delta(. , \alpha y) = \dfrac{1}{|\alpha|} \delta(. , y)$ for all $\alpha \in \mathbb{C}^*$; if $\delta(x_0, y_0) > r$, i.e. if $|\zeta| \leqslant r$ implies $x_0 + \zeta y_0 \in \Omega$, then $|\zeta| \leqslant r$ also implies $x + \zeta y \in \Omega$ for x, y sufficiently near x_0, y_0: $v = \ln \delta$ is lower semicontinuous on $\Omega \times X$; finally, if $\Omega \neq X$, v plurihyperharmonic or v plurisuperharmonic are equivalent properties since $b \in X \backslash \Omega$ implies $\delta(x, x - b) \leqslant 1$ for all $x \in \Omega$.

Since (1) obviously implies (2), we assume (2) and check the mean value inequality $v(a, b) \geqslant MV\, v(a + e^{i\theta}x, b + e^{i\theta}y)$ whenever $\Omega \supset \{a + \zeta x \colon |\zeta| \leqslant 1\}$. By an argument already used for Theorem 4.5.13.a, $v(a + e^{i\theta}x, b + e^{i\theta}y)$ is the limit of a strictly increasing sequence of real continuous functions $c_n(e^{i\theta})$; let $f_n \in \mathscr{A}(\Delta, \mathbb{C})$, $\Delta = \Delta(0, 1)$, be such that $\mathscr{R}e\, f_n(\zeta) \to c_n(e^{i\theta})$ as $\Delta \ni \zeta \to e^{i\theta}$, for all $\theta \in \mathbb{R}$. Then for each n there is a radius $\rho_n < 1$ such that $|\zeta| = \rho_n$ implies $v(a + \zeta x, b + \zeta y) > \mathscr{R}e \, f_n(\zeta)$ or $a + \zeta x + \alpha e^{f_n(\zeta)}(b + \zeta y)$ $\in \Omega$ for all $\alpha \in \bar{\Delta}(0, 1)$; if this remains true for $|\zeta| \leqslant \rho_n$, then $a + \alpha e^{f_n(0)}b \in \Omega$ for all $\alpha \in \bar{\Delta}(0, 1)$ or

$$v(a, b) > \mathscr{R}e \, f_n(0) = MV \, c_n(e^{i\theta}).$$

So the question is whether

$$A_n = \{\alpha \in \bar{\Delta}(0, 1): a + \zeta x + \alpha e^{f_n(\zeta)}(b + \zeta y) \in \Omega \text{ for all } \zeta \in \bar{\Delta}(0, \rho_n)\}$$

is the whole disc $\bar{\Delta}(0, 1)$. We know that $0 \in A_n$; we first check that A_n is an open subset of $\bar{\Delta}(0, 1)$, and this does not involve the assumption (2): in fact, if $\alpha_0 \in A_n$,

$$K_0 = \{a + \zeta x + \alpha_0 e^{f_n(\zeta)}(b + \zeta y): |\zeta| \leqslant \rho_n\}$$

is a compact subset of Ω, the origin in X has a neighbourhood V such that $K_0 + V \subset \Omega$, and this V contains $(\alpha - \alpha_0)e^{f_n(\zeta)}(b + \zeta y)$ for all $\zeta \in \bar{\Delta}(0, \rho_n)$ for sufficiently small $|\alpha - \alpha_0|$.

We now check that A_n is closed, and this rests on the assumption (2). In fact, by Theorem. 4.2.1.: for all $\alpha \in A_n$, $u \in \mathscr{P}(\Omega)$, $u[a + \zeta x + \alpha e^{f_n(\zeta)}(b + \zeta y)]$ is a hypoharmonic function of ζ on an open set containing $\bar{\Delta}(0, \rho_n)$, therefore $\leqslant \sup_K u$ for $|\zeta| \leqslant \rho_n$ if

$$K = \{a + \zeta x + \alpha e^{f_n(\zeta)}(b + \zeta y): |\alpha| \leqslant 1, |\zeta| = \rho_n\}.$$

Thus, for $|\zeta| \leqslant \rho_n$: $a + \zeta x + \alpha e^{f_n(\zeta)}(b + \zeta y)$ lies in $\hat{K}_{\mathscr{P}(\Omega)}$ if $\alpha \in A_n$, in the closure of $\hat{K}_{\mathscr{P}(\Omega)}$ if $\alpha \in \bar{A}_n$.

Since (3) obviously implies (4), we finally assume (4); let K be a compact subset of Ω and V a balanced open neighbourhood of the origin in X such that $K + V \subset \Omega$: for all $x \in K$, $y \in V$, $|\zeta| \leqslant 1$ implies $x + \zeta y \in \Omega$, i.e. $\delta(x, y) > 1$; then $a \in \hat{K}_{\mathscr{P}(\Omega)}$ implies $\delta(a, y) \geqslant \inf_{x \in K} \delta(x, y) \geqslant 1$ for all $y \in V$, and $\hat{K}_{\mathscr{P}(\Omega)} + \theta V \subset \Omega$ for all $\theta \in \,]0, 1[$. $\square$

Proposition 5.1.2. a) Ω *is pseudoconvex if and only if every connected component of Ω is pseudoconvex.* b) *Any convex open set is pseudoconvex (see also Prop. 5.2.3. below).* c) *The interior of the intersection of any family (resp. the union of an upper directed family) of pseudoconvex open sets Ω_i is again pseudoconvex.*

Proof. a) Obvious with property (4) in Definition 5.1.1.
b) Let K be a compact subset of the convex open set Ω and V a balanced convex open neighbourhood of the origin in X such that $K + V \subset \Omega$: $C = \{x \in \Omega: x + V \subset \Omega\}$ is convex, $\bar{C} = C$ because $y \in \bar{C}$ implies $y + rV \subset$

Ω for all $r \in {]}0,1[$; finally $C \supset \hat{K}_{\mathscr{P}(\Omega)}$ because $y \notin C$ implies the existence ([Bo]$_2$, Chap. II, §3, Prop. 4) of an $x' \in X'$ such that $\sup_{C} \mathscr{R}e \, x' < \mathscr{R}e \, x'(y)$, hence $\sup_{K} |e^{x'}| \leqslant \sup_{C} |e^{x'}| < |e^{x'(y)}|$, although $|e^{x'}| \in \mathscr{P}(\Omega)$.

c) For $x \in \Omega$, the interior of $\bigcap_{i \in I} \Omega_i : x \mapsto \delta(x,y)$ is the lower regularized of $x \mapsto \inf_{i \in I} \delta_i(x,y)$. The other statement requires the sheaf property (Prop. 4.1.4.b) of p.h.h. functions: for $x \in \Omega_j$, $\delta(x,y)$ is the supremum of the upper directed family $(\delta_i(x,y))_{\Omega_i \supset \Omega_j}$. $\quad\square$

Proposition 5.1.3. a) *Ω is pseudoconvex if and only if $E \cap \Omega$ is pseudoconvex for every 2-dimensional affine subspace E of X.*
b) *Ω is pseudoconvex if (and only if, of course) every boundary point of Ω has an open neighbourhood ω such that $\omega \cap \Omega$ is pseudoconvex.*
c) *If Ω is pseudoconvex and $u \in \mathscr{P}(\Omega)$, for all $\lambda \in \mathbb{R}$ the open set $\omega = u^{-1}([-\infty, \lambda[)$ is pseudoconvex too.*

Proof. a) Property (4) (in Def. 5.1.1.) for Ω and every $y \in X$ implies (4) for $E \cap \Omega$ and every y such that $E + y = E$. Conversely, let E contain a, $a + b$, $a + y$, Ω contain $\{a + \zeta b : |\zeta| \leqslant 1\}$ and $E \cap \Omega$ be pseudoconvex: $\ln \delta(a + \zeta b, y)$ is a hyperharmonic function of ζ on an open set containing $\bar{\Delta}(0,1)$, hence the mean value inequality for $\ln \delta(., y)$.

Note that the analogous statement for 1-dimensional affine subspaces would be void: any open set U in $\mathbb{C}$ is pseudoconvex since $-\ln d(\zeta, \complement U) = \sup_{\alpha \in \complement U} \ln \frac{1}{|\zeta - \alpha|}$, as a continuous supremum of harmonic functions, is a subharmonic function of $\zeta \in U$.
b) This is true in finite dimensional spaces ([Hö], Th. 2.6.10).
c) Let E be a 2-dimensional affine subspace of X and K a compact subset of $E \cap \omega$: the convex hull K' of K in $E \cap \Omega$, with respect to $\mathscr{P}(E \cap \Omega)$, is compact since $E \cap \Omega$ is pseudoconvex ([Hö], Th. 2.6.11) and moreover $\sup_{K'} u = \sup_{K} u < \lambda$ implies $K' \subset E \cap \omega$; but K' contains the convex hull of K in $E \cap \omega$ with respect to $\mathscr{P}(E \cap \omega)$; then $E \cap \omega$ is pseudoconvex too. $\quad\square$

Proposition 5.1.4. *Let $u \in \mathscr{P}(\Omega)$, $\lambda \in \mathbb{R}$: a connected component ω of the open set $u^{-1}([-\infty, \lambda[)$ is pseudoconvex if $\omega + V \subset \Omega$ for some neighbourhood V of the origin in X.*

Proof: Assuming $\omega \supset \{a + \zeta b : |\zeta| \leqslant 1\}$, we have to check the mean value inequality $v(a,y) \geqslant MV \, v(a + e^{i\theta}b, y)$ for $v = \ln \delta$, $\delta(x,y) = \sup\{r > 0:$

$|\zeta| \leqslant r$ implies $x + \zeta y \in \omega\}, x \in \omega$. By the argument used for Theorem 5.1.1, given a radius $\rho_n < 1$ such that $|\zeta| = \rho_n$ implies $v(a + \zeta b, y) > \mathcal{R}e\, f_n(\zeta)$ or $a + \zeta b + \alpha e^{f_n(\zeta)} y \in \omega$, for all $\alpha \in \bar{\Delta}(0, 1)$, we have to show that

$$A_n = \{\alpha \in \bar{\Delta}(0, 1): a + \zeta b + \alpha e^{f_n(\zeta)} y \in \omega \text{ for all } \zeta \in \bar{\Delta}(0, \rho_n)\}$$

is the whole disc $\bar{\Delta}(0, 1)$.

Let $|\beta| \leqslant k$, $k > 0$, imply $\beta e^{f_n(\zeta)} y \in V$ for all $\zeta \in \bar{\Delta}(0, \rho_n)$. Then $\alpha_0 \in A_n$, $|\alpha| \leqslant 1$, $|\alpha - \alpha_0| \leqslant k$ imply $\{a + \zeta b + \alpha e^{f_n(\zeta)} y: |\zeta| \leqslant \rho_n\}$ contained in Ω, hence in $u^{-1}([-\infty, \lambda[)$ since $u[a + \zeta b + \alpha e^{f_n(\zeta)} y]$ is a hypoharmonic function of ζ on an open set including $\bar{\Delta}(0, \rho_n)$, and finally in ω; so $\alpha_0 \in A_n$, $|\alpha - \alpha_0| \leqslant k$, $|\alpha| \leqslant 1$ entail $\alpha \in A_n$, and $0 \in A_n$ completes the proof. $\square$

Theorem 5.1.5. *Given a pseudoconvex open set Ω and $u \in \mathcal{P}(\Omega)$, one can find an upper directed family of pseudoconvex open sets Ω_p whose union is Ω and, on each Ω_p in this family, a finite continuous p.s.h. function u_p so that $u(x) = \inf\{u_p(x): x \in \Omega_p\}$ for all $x \in \Omega$.*

Proof: Let $\tilde{u} = u$ on Ω, $\tilde{u} = +\infty$ on $X \backslash \Omega$ if $\Omega \neq X$. The families (Ω_p), (u_p) will be indexed by a subset Π of $\mathrm{csn}(X)$ which, taking up an argument of [Fe Si] we define as follows: $p \in \mathrm{csn}(X)$ belongs to Π if and only if the set Ω_p of points $x \in \Omega$ such that

$$\gamma_p(x) = \inf_{y \in X} [p(x - y) + e^{-\tilde{u}(y)}] > 0$$

is not empty, and then $u_p = -\ln \gamma_p$ on Ω_p.

$\gamma_p(x)$ is always finite because $\tilde{u} \not\equiv -\infty$; $x \in \Omega_p$ and $p(x - x') < \gamma_p(x)$ imply $x' \in \Omega_p$ and $\gamma_p(x') \geqslant \gamma_p(x) - p(x - x')$, therefore Ω_p is open and γ_p uniformly continuous on Ω_p; $p \leqslant p'$ implies $\Omega_p \subset \Omega_{p'}$ and $\gamma_p \leqslant \gamma_{p'}$ on Ω_p. Now $\gamma_p(x) \leqslant e^{-u(x)}$ for all $x \in \Omega_p$; given $x \in \Omega$ and $\lambda \in \,]0, e^{-u(x)}[$, one can find $p \in \mathrm{csn}(X)$ and $\rho > 0$ such that $p(x - y) \leqslant \rho$ implies $y \in \Omega$ and $e^{-u(y)} > \lambda$, hence $x \in \Omega_{np}$ and $\gamma_{np}(x) \geqslant \inf(n\rho, \lambda)$ for all $n \in \mathbb{N}^*$: the theorem is proved except $u_p \in \mathcal{P}(\Omega_p)$, and Ω_p pseudoconvex.

For this we write

$$\gamma_p(x) = \inf\{p(x - y) + |\zeta|: y \in X, \zeta \in \mathbb{C}^*, |\zeta| \geqslant e^{-\tilde{u}(y)}\}$$

or, setting $y = x + \zeta z$:

(*) $\gamma_p(x) = \inf\{[1 + p(z)]|\zeta|: z \in X, \zeta \in \mathbb{C}^*, \tilde{u}(x + \zeta z) + \ln|\zeta| \geqslant 0\}.$

Since Ω is pseudoconvex in X, $\Omega \times \mathbb{C}$ is pseudoconvex in $X \times \mathbb{C}$ and also $\omega = \{(x, \alpha) \in \Omega \times \mathbb{C}: u(x) + \ln|\alpha| < 0\}$ by Proposition 5.1.3.c. Then, by property (4) in Definition 5.1.1., with $(x, 0)$ instead of x and $(z, 1)$ instead of y:

$$\delta(x,z) = \sup\{r > 0: |\zeta| \leqslant r \text{ implies } x + \zeta z \in \Omega, u(x + \zeta z) + \ln|\zeta| < 0\}$$
$$= \sup\{r > 0: 0 < |\zeta| \leqslant r \text{ implies } \tilde{u}(x + \zeta z) + \ln|\zeta| < 0\}$$

is such that $-\ln \delta(x,z)$ is a p.h.h. function of $x \in \Omega$ for all $z \in X$; for $x \in \Omega$, a comparison with (∗) gives

$$\gamma_p(x) = \inf_{z \in X} [1 + p(z)]\delta(x,z) = \inf\{\delta(x,z): z \in X, p(z) = 0 \text{ or } 1\};$$

u_p, a finite continuous supremum of p.h.h. functions, is p.s.h. on Ω_p.

Moreover, if $K \subset \Omega_p$ is compact and $a \in \hat{K}_{\mathscr{P}(\Omega)}$:

$$\delta(a,z) \geqslant \inf_{x \in K} \delta(x,z) \text{ for all } z \in X \text{ entails } \gamma_p(a) \geqslant \inf_{x \in K} \gamma_p(x) = \gamma > 0;$$

this proves $\hat{K}_{\mathscr{P}(\Omega)} + p^{-1}([0,\gamma[) \subset \Omega_p$, hence Ω_p pseudoconvex, and something more: for every finite dimensional subspace E of X, the pseudoconvex open sets $E \cap \Omega_p$, $E \cap \Omega$ have the approximation property of functions $\in \mathscr{A}(E \cap \Omega_p, \mathbb{C})$ by functions $\in \mathscr{A}(E \cap \Omega, \mathbb{C})$ (Th. 4.3.3 in [Hö]). □

If X is a normed space, the result gets much simpler: $\Omega_{\|\ \|} = \Omega$ and u is the limit of the decreasing sequence $(u_{n\|\ \|})$ of continuous p.s.h. functions on Ω; therefore $\hat{K}_{\mathscr{P}(\Omega)}$ is also the convex hull of K with respect to the continuous p.s.h. functions on Ω, a closed subset of Ω. This was pointed out by Mujica ([Mu], §38).

Proposition 5.1.6. *Let $\Omega \neq X$ be pseudoconvex and $p \in \operatorname{csn}(X)$ have the property: the open set Ω_p of points $x \in \Omega$ such that $\gamma_p(x) = \inf_{y \in X\backslash\Omega} p(x - y) > 0$ is not empty.*
a) Ω_p is pseudoconvex too, $-\ln\gamma_p$ is a finite continuous p.s.h. function on Ω_p and $\inf_{z \in X\backslash\Omega_p} p(x - z) = \gamma_p(x) > 0$ for all $x \in \Omega_p$.
b) Let moreover Ω be connected: then $\Omega = \Omega + p^{-1}(0)$, and Ω is the inverse image, under the canonical map $X \to X/p^{-1}(0)$, of a pseudoconvex open set in $X/p^{-1}(0)$ endowed with the quotient topology; Ω_p obviously has the same property.

Proof. a) For the first statements, take $\tilde{u} \equiv -\infty$ on Ω in the proof of Theorem 5.1.5. For the last one, for all $x \in \Omega_p$, $z \in \Omega\backslash\Omega_p$, $p(x - z) \geqslant p(x - y) - p(z - y) \geqslant \gamma_p(x) - p(z - y)$ for all $y \in X\backslash\Omega$.
b) Let $a \in \Omega_p$ and $\alpha = \gamma_p(a) > 0$: $p(x - a) < \alpha$ implies $x + p^{-1}(0) \subset \Omega$. Given $b \in \Omega$, there is a finite dimensional affine subspace E of X such that a and b lie in the same connected component U of $E \cap \Omega$; let V be the interior in U of $\{x \in U: \Omega \supset x + p^{-1}(0)\}$: $V \supset [a + p^{-1}([0,\alpha[)] \cap U$. We

claim that $x \in V$, $y \in U$, $\|y - x\| < \frac{1}{2} d(x, \complement U)$ (norm and associated distance d defined on E) imply $y + p^{-1}(0) \subset \Omega$: this will prove that V is closed in U, $V = U$.

Let $r > 0$ be such that $|\zeta| \leqslant r$ implies $x + \zeta(y - x) \in V$ and therefore $x + \alpha z + \zeta(y - x) \in \Omega$ for all $z \in p^{-1}(0)$, $\alpha \in \mathbb{C}$: with the notation δ used in Definition 5.1.1., $-\ln \delta(x + \alpha z, y - x)$ is a hypoharmonic function of $\alpha \in \mathbb{C}$ and remains $< -\ln r$; by the Liouville theorem for $u \in \mathscr{P}(\mathbb{C})$: $-\ln \delta(x + \alpha z, y - x) = -\ln \delta(x, y - x) < -\ln 2$ for all $\alpha \in \mathbb{C}$, in other words $|\zeta| \leqslant 2$, especially $\zeta = 1$, implies $x + z + \zeta(y - x) \in \Omega$ for all $z \in p^{-1}(0)$.

The canonical map $X \to X/p^{-1}(0)$ is linear and continuous; the image of Ω is open, and pseudoconvex in view of Proposition 5.1.3.a.   □

5.2 The Levi problem

The Levi problem deals with the equivalence of the 4 properties listed in Definition 5.2.1.: by Proposition 5.2.2, *in any space X*, each one of them entails the following one. The problem of their equivalence, *in the finite dimensional case*, was posed by E.E. Levi [Lev] and answered affirmatively by Oka, Bremermann [Brem], Norguet [Nor]; another method of proof is due to Hörmander [Hö]. An intermediate step in the solution of the problem was the famous Cartan-Thullen theorem [Ca Th]: any holomorphically convex domain is a domain of existence.

Definition 5.2.1. *Let Ω be a domain, i.e. a connected open set, in the l.c. space X. a) Ω is the domain of existence of a function $h \in \mathscr{A}(\Omega, \mathbb{C})$ (resp. a domain of holomorphy) if there do not exist a domain $\Omega_1 \not\subset \Omega \not\subset \complement \Omega_1$ and a connected component Ω_0 of $\Omega \cap \Omega_1$ such that $h_{|\Omega_0}$ [resp.: $f_{|\Omega_0}$ for every $f \in \mathscr{A}(\Omega, \mathbb{C})$] has an extension $\in \mathscr{A}(\Omega_1, \mathbb{C})$. b) Ω is holomorphically convex (resp.: pseudoconvex) if the convex hull $\hat{K}$ in Ω, with respect to $\mathscr{A}(\Omega, \mathbb{C})$ [resp.: to $\mathscr{P}(\Omega)$] of every compact subset K of Ω is such that $\hat{K} + V \subset \Omega$ for some neighbourhood V of the origin in X.*

Proposition 5.2.2. *Among the 4 properties: (1) Ω a domain of existence; (2) Ω a domain of holomorphy; (3) Ω holomorphically convex; (4) Ω pseudoconvex; each one entails the following one.*

Proof. We assume (2) and prove (3), the other two implications being obvious. Let K be a compact subset of Ω, V a balanced open neighbourhood of the origin in X such that $K + V \subset \Omega$ and $b \in \hat{K}$, the convex hull of K in Ω with respect to $\mathscr{A}(\Omega, \mathbb{C})$: for all $f \in \mathscr{A}(\Omega, \mathbb{C})$, the sequence of continuous polynomial maps

$$\varphi_n(b + x) = f(b) + \sum_{k=1}^{n} \frac{1}{k!} \hat{D}_b^k f(x)$$

(see Th. 2.3.5) converges to $f(b + x)$ for all x in $\omega(b)$, the largest balanced subset of $\Omega - b$; we claim that it converges for all $x \in V$ and that $\sup_{n \in \mathbb{N}^*} |\varphi_n|$ is locally bounded on $b + V$, from which will follow, by Theorem 3.1.5.c, that its sum is analytic on $b + V$, hence is an analytic extension to $b + V$ of every $f \in \mathscr{A}(\Omega, \mathbb{C})$ restricted to the connected component containing b of $\Omega \cap (b + V)$.

Let $x_0 \in V$ and W be another balanced neighbourhood of the origin in X such that $x_0 + W \subset V$ and

$$|f(a + e^{i\theta} x_0 + y)| \leqslant M \quad \text{for all } a \in K, \theta \in \mathbb{R}, y \in W;$$

then, by the generalized inequalities of Cauchy,

$$\frac{1}{k!} |\hat{D}_a^k f(x_0 + y)| \leqslant M \quad \text{for all } a \in K, k \in \mathbb{N}^*, y \in W,$$

$$\frac{1}{k!} |\hat{D}_b^k f(x_0 + y)| \leqslant M \quad \text{for all } k \in \mathbb{N}^*, y \in W$$

by Theorem 3.1.5.b, hence

$$\frac{1}{k!} |\hat{D}_b^k f| \leqslant \theta^k M \text{ on } \theta(x_0 + W) \quad \text{for all } \theta \in \,]0, 1[,$$

and finally $\sup_{n \in \mathbb{N}^*} |\varphi_n| \leqslant |f(b)| + M \dfrac{\theta}{1 - \theta}$ on $b + \theta(x_0 + W)$. $\square$

By this proof, (2) does not only imply (3), but also an important link between a compact subset K of Ω and its convex hull $\hat{K}$ in Ω with respect to $\mathscr{A}(\Omega, \mathbb{C})$: any balanced open neighbourhood V of the origin in X which satisfies $K + V \subset \Omega$ also satisfies $\hat{K} + V \subset \Omega$; the same property was obtained in 5.1.1, under the assumption (4), for the convex hull with respect to $\mathscr{P}(\Omega)$.

Also note that, since $e^{x'} \in \mathscr{A}(\Omega, \mathbb{C})$ for all $x' \in X'$, both hulls $\hat{K}$ are included in the closed convex hull of K in X (in the ordinary sense), and this last hull is precompact. In fact, given a closed convex neighbourhood V of the origin in X, there are $a_1, \ldots, a_m \in K$ such that $K \subset \bigcup_{j=1}^{m} (a_j + V)$, and the closed convex hull of K in X is included in $L + V$, where $L = \left\{ \sum_{j=1}^{m} \alpha_j a_j \colon \alpha_j \geqslant 0, \sum \alpha_j = 1 \right\}$ is compact.

The Levi problem was settled in the finite dimensional case by the equivalence of properties (1) through (4), but it is open again since complex analysis has developed in the infinite dimensional case. In the latter case, even the equivalence of (1) and (2) may fail:

Proposition 5.2.3. *Any convex open set in X is a domain of holomorphy (which improves Prop. 5.1.2.b), but maybe not a domain of existence, even if X is a Banach space.*

Proof. Let Ω be convex, Ω_0 and Ω_1 as in Definition 5.2.1.a, b a boundary point of Ω_0 in Ω_1; $b \notin \Omega$ implies the existence of $x' \in X'$ such that $\mathscr{R}e\, x' < \mathscr{R}e\, x'(b)$ on Ω, then $\dfrac{1}{x'(b) - x'}$ restricted to Ω_0 has no extension $\in \mathscr{A}(\Omega_1, \mathbb{C})$.

Now let A be an uncountable infinite set and $X = c_0(A)$ the Banach space, with the sup norm, of all maps $x \colon A \to \mathbb{C}$ such that $\{\alpha \in A \colon |x(\alpha)| > \varepsilon\}$ is finite for all $\varepsilon > 0$. By Example 3.1.11, for every function f analytic on the open unit ball of $c_0(A)$, there is a countable subset D of A such that, for x, y in the ball, $x(\alpha) = y(\alpha)$ for all $\alpha \in D$ implies $f(x) = f(y)$; then f has an analytic extension to the open set $\left\{ x \in c_0(A) \colon \sup_{\alpha \in D} |x(\alpha)| < 1 \right\}$. $\square$

In the same Banach space $c_0(A)$, Josefson [Jo]$_1$ has constructed a pseudoconvex domain which is not even a domain of holomorphy. In view of such counter-examples, the assumptions of countability in the following theorems are not surprising, and probably cannot be lifted, whether one makes classical arguments to serve again (Th. 5.2.4) or comes down to the finite dimensional case by an approximation process (Th. 5.2.7).

Theorem 5.2.4. *Let X be the inductive limit of metrizable l.c. spaces X_n, $n \in \mathbb{N}^*$, with the property: in each X_n, the origin has a neighbourhood V_n which is relatively compact in X_{n+1}.*

a) *Any holomorphically convex domain Ω in X is a domain of holomorphy.*
b) *If Ω is holomorphically convex and its boundary is also the boundary of $\bar{\Omega}$, then Ω is a domain of existence.*

Proof. a) Let d_n be the distance in X_n and K_n a balanced compact subset of X_{n+1} containing V_n:

$$K_{m,n} = \left\{ x \in mK_n : d_{n+1}(x, X_{n+1} \setminus \Omega) \geqslant \frac{1}{m} \right\}, \; m \in \mathbb{N}^*,$$

defines an increasing sequence of compact subsets of Ω such that each point $\in \Omega \cap X_n$ has a neighbourhood in $\Omega \cap X_n$ contained in one of them. Therefore a sequence in $\mathscr{A}(\Omega, \mathbb{C})$ which converges uniformly on each $K_{m,n}$ has a limit $\in \mathscr{A}(\Omega, \mathbb{C})$ by Proposition 1.5.1.b.: $\mathscr{A}(\Omega, \mathbb{C})$ is a Fréchet space with the semi-norms $f \mapsto \sup_{K_{m,n}} |f|$.

Now let Ω_0, Ω_1 be as in Definition 5.2.1.a and b a boundary point of Ω_0 in Ω_1 which is the limit of a sequence $(y_k) \subset \Omega_0$: such a point b is obtained, from a continuous map $y : [0,1] \to \Omega_1$ with $y(0) \in \Omega_0$, $y(1) \in \Omega_1 \setminus \Omega_0$, by setting $b = y(t_0)$, $t_0 = \inf\{t \in [0,1] : y(t) \notin \Omega_0\}$. If $f \in \mathscr{A}(\Omega, \mathbb{C})$ satisfies $\sup_k |f(y_k)| = +\infty$, $f_{|\Omega_0}$ cannot have any extension $\in \mathscr{A}(\Omega_1, \mathbb{C})$; if no such f exists, then $f \mapsto \sup_k |f(y_k)|$ is a lower semicontinuous seminorm on $\mathscr{A}(\Omega, \mathbb{C})$ for the above defined topology, hence a continuous one: there are a constant C and a compact subset K of Ω (the union of finitely many $K_{m,n}$) such that $\sup_k |f(y_k)| \leqslant C \sup_K |f|$ for all $f \in \mathscr{A}(\Omega, \mathbb{C})$, hence also $\sup_k |f(y_k)| \leqslant C^{1/n} \sup_K |f|$ for all $n \in \mathbb{N}^*$ and finally $\sup_k |f(y_k)| \leqslant \sup_K |f|$ for all $f \in \mathscr{A}(\Omega, \mathbb{C})$; this means (y_k) in the convex hull of K, a contradiction.

b) First note that each mK_n, as a compact subset of the metrizable space X_{n+1}, is separable in X_{n+1}; then X is separable for the inductive limit topology. In fact, given $a \in X$ and A a neighbourhood of a in X, there are $m, n \in \mathbb{N}^*$ such that $a \in mK_n$ and $A \cap (mK_n)$ is a neighbourhood of a in mK_n; let D be a countable set dense in X.

If Ω is not the domain of existence of the function $h \in \mathscr{A}(\Omega, \mathbb{C})$ we can find: 1°) Ω_0, Ω_1 as in Definition 5.2.1.a such that $h_{|\Omega_0}$ has an extension $\tilde{h} \in \mathscr{A}(\Omega_1, \mathbb{C})$; 2°) $a \in D \cap \Omega_0$, $p \in \mathrm{csn}(X)$ and $\alpha > 0$ such that $a + p^{-1}([0, 2\alpha[) \subset \Omega_1$ but $a + p^{-1}([0, \alpha]) \not\subset \Omega$ (this on account of the assumption made on the boundary of Ω); 3°) $b \in D$ such that $p(b - a) < 2\alpha$ and $\dfrac{a+b}{2} \in \Omega_1 \setminus \bar{\Omega}$, then $M \in \mathbb{N}^*$ such that $|\tilde{h}[a + e^{i\theta}(b - a)]| \leqslant M$ for all $\theta \in \mathbb{R}$ and therefore $\dfrac{1}{k!} |\hat{D}_a^k h(b - a)| \leqslant M$ for all $k \in \mathbb{N}$; 4°) an integer n such that $a, b \in X_n$ and another integer m such that $K_{m,n}$ is a neighbourhood of a in $\Omega \cap X_n$, hence $\dfrac{1}{k!} |\hat{D}_a^k h(b - a)| \leqslant \dfrac{1}{r^k} \sup_{K_{m,n}} |h|$ if $|\zeta| \leqslant r$ implies $a + \zeta(b - a) \in K_{m,n}$.

Given a, b, M: the set

(∗) $\left\{ h \in \mathscr{A}(\Omega, \mathbb{C}) : \dfrac{1}{k!} |\hat{D}_a^k h(b-a)| \leqslant M \text{ for all } k \in \mathbb{N}^* \right\}$

is closed in $\mathscr{A}(\Omega, \mathbb{C})$ for the topology chosen in a); for every h in this set, $\zeta \mapsto h[a + \zeta(b-a)]$ has an analytic extension from $U(a, b)$, the connected component containing 0 of $\{\zeta \in \Delta(0, 1): a + \zeta(b-a) \in \Omega\}$ to the open disc $\Delta(0, 1)$. Such an extension cannot exist for all $h \in \mathscr{A}(\Omega, \mathbb{C})$: in fact, if a sequence $(\zeta_k) \subset U(a, b)$ tends to a boundary point of $U(a, b)$ in $\Delta(0, 1)$, by the proof of a) there is an $h \in \mathscr{A}(\Omega, \mathbb{C})$ which is unbounded on the sequence $(a + \zeta_k(b-a))$. So the h for which an analytic extension from $U(a, b)$ to $\Delta(0, 1)$ exists make up a proper linear subspace of $\mathscr{A}(\Omega, \mathbb{C})$ and a proper linear subspace has an empty interior for any vector space topology.

For the topology we have chosen, and for all a, b as above, $M \in \mathbb{N}^*$: (∗) is a closed set, without any interior point, in the Baire space $\mathscr{A}(\Omega, \mathbb{C})$; $\{h \in \mathscr{A}(\Omega, \mathbb{C}): \Omega$ is not domain of existence of $h\}$ is a meagre set.   □

A complete answer to the Levi problem in a space X is the fact that any pseudoconvex domain is a domain of existence. This was proved: by Gruman [Gr] when X is a Hilbert space; by Gruman-Kiselman [Gr Ki] when X is a Banach space with a Schauder basis; more generally, when X is a metrizable l.c. space with an equi-Schauder basis, by Dineen-Noverraz-Schottenloher [Di No Sch], whose arguments we shall use in the proof of Theorem 5.2.7.

Definition 5.2.5. *A sequence* $(e_n)_{n \in \mathbb{N}^*}$ *in a l.c. space X is:*
a) *a Schauder basis of X if every $x \in X$ is the sum of a unique series* $\sum_{n=1}^{\infty} x_n e_n$, $x_n \in \mathbb{C}$ *for all $n \in \mathbb{N}^*$;* b) *an equi-Schauder basis if moreover the linear maps* $u_n: \sum_{n=1}^{\infty} x_n e_n \mapsto \sum_{m=1}^{n} x_m e_m$ *are equicontinuous.*

In the latter case: for all $p \in \mathrm{csn}(X)$ we also have $\hat{p} = \sup_{n \in \mathbb{N}^*} p \circ u_n \in \mathrm{csn}(X)$; since $\hat{p} \geqslant p$, the topology of X can be defined by a family (an increasing sequence if X is metrizable) of seminorms $\hat{p}$ such that $\hat{p} \circ u_n \leqslant \hat{p}$ for all $n \in \mathbb{N}^*$. In fact $u_m \circ u_n = u_{\inf(m, n)}$ for all $(m, n) \in (\mathbb{N}^*)^2$.

The existence of a Schauder basis of X implies X separable: a countable dense subset can be found in the dense linear subspace $E_{\infty} = \bigcup_{n \in \mathbb{N}^*} E_n$, where E_n is the n-dimensional linear subspace of X spanned by $e_1, \ldots, e_n$; note that $u_m|_{E_n} = \mathrm{id}$ for all $m \geqslant n$. This notation will be used in Proposition 5.2.6 and Theorem 5.2.7.

Proposition 5.2.6. *Any Schauder basis of a Fréchet space is equi-Schauder.*

Proof. Let Π be a countable family of seminorms p defining the topology of X and $\hat{\Pi}$ the countable family of seminorms $\hat{p} = \sup_{n \in \mathbb{N}^*} p \circ u_n$, $p \in \Pi$: since $\hat{p} \geqslant p$, the identity map $(X, \hat{\Pi}) \to (X, \Pi)$ is continuous; if we show that $(X, \hat{\Pi})$ is complete too, each $\hat{p} \in \hat{\Pi}$ shall be continuous for the topology of (X, Π) by the Banach open mapping theorem ([Bo]$_2$, Chap. I, §3, Th. 1).

Let $(x^k)_{k \in \mathbb{N}^*}$ be a Cauchy sequence in $(X, \hat{\Pi})$, $x^k = \sum_{n=1}^{\infty} x_n^k e_n$. For every $p \in \Pi$ and $\varepsilon > 0$, there is a $k_0(p, \varepsilon)$ such that k, $k' \geqslant k_0(p, \varepsilon)$ implies $\hat{p}(x^k - x^{k'}) \leqslant \varepsilon$ or $p\left[\sum_{m=1}^{n} (x_m^k - x_m^{k'})e_m \right] \leqslant \varepsilon$ for all n, which proves: for $n = 1$ and $p(e_1) > 0$, that $(x_1^k)_{k \in \mathbb{N}^*}$ is a Cauchy sequence, $x_1^k \to x_1$ as $k \to \infty$; for $n = 2$ and $p(e_2) > 0$, that $x_2^k \to x_2$ as $k \to \infty$, and so on.

Now $k \geqslant k_0(p, \varepsilon)$ implies $p\left[\sum_{m=1}^{n} (x_m^k - x_m)e_m \right] \leqslant \varepsilon$ for all n or $\hat{p}(x^k - x) \leqslant \varepsilon$ provided that the series $\sum_{n=1}^{\infty} x_n e_n$ converges to x. In order to prove this, we use $p\left[\sum_{m=n}^{n'} (x_m^{k_0} - x_m)e_m \right] \leqslant 2\varepsilon$ for all n, for all $n' \geqslant n$, with $k_0 = k_0(p, \varepsilon)$; for $n \geqslant n_0(p, \varepsilon)$ we have $p\left(\sum_{m=n}^{n'} x_m^{k_0} e_m \right) \leqslant \varepsilon$, hence $p\left(\sum_{m=n}^{n'} x_m e_m \right) \leqslant 3\varepsilon$. $\square$

Theorem 5.2.7. *Let X be a metrizable l.c. space with an equi-Schauder basis: any pseudoconvex domain Ω in X such that E_∞ contains a dense subset of the boundary of Ω is a domain of existence.*

Proof schedule. (A) The case in which the topology of X is defined by a sequence of norms. (B) The case in which it is defined by one norm: in this case, the assumption "E_∞ contains a dense subset of the boundary of Ω" may be dropped, which is dubious in the general case. (C) X is a metrizable l.c. space.

(A) Let the topology of X be defined by an increasing sequence of norms p_n, $n \in \mathbb{N}^*$, such that $p_n \circ u_m \leqslant p_n$ for all $m \in \mathbb{N}^*$, and also by the distance d given by the formula in 1.1(A); on the other hand, let $D \subset E_\infty \cap \Omega$ be a countable dense subset of Ω; we choose an indexation $D = \{a_j : j \in \mathbb{N}^*\}$ in which every point $a \in D$ appears an infinity of times (see e.g. the proof of Prop. 4.1.9).

$u_n(\Omega)$ is an open subset of E_n containing $E_n \cap \Omega$. The proof rests on an increasing sequence $(\omega_n)_{n \in \mathbb{N}^*}$ of pseudoconvex open sets whose union is Ω, such that $u_n(\omega_n) = E_n \cap \Omega$: such a sequence was constructed by Schottenloher [Sch] as follows. With the notation δ used in Definition 5.1.1: since $\delta(x, 0) = +\infty$ for all $x \in \Omega$, we have

$$\gamma_n(x) = \inf_{m \geqslant n} \delta[x, u_m(x) - x] > 0 \quad \text{for all } x \in \Omega, \, n \in \mathbb{N}^*,$$

γ_n is a lower semicontinuous function on Ω and $-\ln \gamma_n$, as an upper semicontinuous supremum $< +\infty$ of p.h.h. functions, is p.h.h. on Ω; since $\gamma_n(x)$ increases to $+\infty$ for all $x \in \Omega$, Ω is the union of the increasing sequence of pseudoconvex (Prop. 5.1.3.c) open sets

$$\omega_n = \{x \in \Omega : \gamma_n(x) > 1\}.$$

$\omega_n \supset E_n \cap \Omega$, hence $u_n(\omega_n) \supset E_n \cap \Omega$, but $E_n \cap \Omega \supset u_n(\omega_n)$ because $x \in \omega_n$ implies $\delta[x, u_n(x) - x] > 1$, $x + \zeta[u_n(x) - x] \in \Omega$ for all $\zeta \in \bar{\Delta}(0, 1)$, especially for $\zeta = 1$.

For $(n_0, n_1, M) \in (\mathbb{N}^*)^3$, $n_0 < n_1$, $E_{n_1} \cap \omega_{n_0} \supset E_{n_0} \cap \Omega$ and $K(n_0, n_1, M) =$
$\left\{ x \in E_{n_1} \cap \omega_{n_0} : p_1(x) \leqslant M, \, p_{n_1}(x - y) \geqslant \dfrac{1}{M} \text{ for all } y \in E_{n_1} \setminus \omega_{n_0} \right\}$ is a compact subset of $E_{n_1} \cap \omega_{n_0}$ with the following properties.

(1) Any function analytic on a neighbourhood of $K(n_0, n_1, M)$ in E_{n_1} can be uniformly approximated on $K(n_0, n_1, M)$ by functions $\in \mathscr{A}(E_{n_1} \cap \Omega, \mathbb{C})$. In fact, if δ_{n_0} is the analogue of δ for ω_{n_0}:

$$\inf\{p_{n_1}(x - y) : y \in E_{n_1} \setminus \omega_{n_0}\} = \inf\{\delta_{n_0}(x, z) : z \in E_{n_1}, \, p_{n_1}(z) = 1\}$$

for all $x \in E_{n_1} \cap \omega_{n_0}$, therefore $-\ln \inf[p_{n_1}(x - y) : y \in E_{n_1} \setminus \omega_{n_0}]$ is a continuous p.s.h. function of $x \in E_{n_1} \cap \omega_{n_0}$ (or the constant $-\infty$ if $E_{n_1} \subset \omega_{n_0}$); $K(n_0, n_1, M)$ is its own convex hull in $E_{n_1} \cap \omega_{n_0}$ with respect to $\mathscr{P}(E_{n_1} \cap \omega_{n_0})$, hence also with respect to $\mathscr{A}(E_{n_1} \cap \omega_{n_0}, \mathbb{C})$ since $E_{n_1} \cap \omega_{n_0}$ is pseudoconvex ([Hö], Th. 4.3.4). Moreover the convex hull in $E_{n_1} \cap \Omega$ of any compact set $J \subset E_{n_1} \cap \omega_{n_0}$ is contained in $E_{n_1} \cap \omega_{n_0}$ because $\inf_J \gamma_{n_0} > 1$; then ([Hö], Th. 4.3.3) $K(n_0, n_1, M)$ is also its own convex hull in $E_{n_1} \cap \Omega$, hence our claim ([Hö], Th. 4.3.2).

(2) The image of $K(n_0, n_1, M)$ under u_{n_0} is a compact subset of $E_{n_0} \cap \Omega$; given $f \in \mathscr{A}(E_{n_0} \cap \Omega, \mathbb{C})$ and $\varepsilon > 0$, there is $g \in \mathscr{A}(E_{n_1} \cap \Omega, \mathbb{C})$ such that $|g - f \circ u_{n_0}| \leqslant \dfrac{\varepsilon}{2}$ on $K(n_0, n_1, M)$; given $b \in (E_{n_1} \cap \Omega) \setminus K(n_0, n_1, M)$, there also exists $h \in \mathscr{A}(E_{n_1} \cap \Omega, \mathbb{C})$ such that $|h - f \circ u_{n_0}| \leqslant \varepsilon$ on $K(n_0, n_1, M)$ and $|h(b)|$ is arbitrarily large.

(3) For every compact set $L \subset \Omega$: $K(n_0, n_1, M) \supset u_{n_1}(L)$ for sufficiently large n_0, n_1, M. In fact, let $L \subset \omega_{v_0}$, $n \geqslant v_1$ imply

$$d[x, u_n(x)] \leqslant \frac{1}{2} d(L, \mathbb{C}\omega_{v_0}) = \eta \quad \text{for all } x \in L$$

and $\sum_{n > v_2} \frac{1}{n^2} \leqslant \frac{\eta}{2}$; then $n_0 \geqslant v_0$, $n_1 \geqslant v_1$, v_2, $x \in u_{n_1}(L)$, $y \in \mathbb{C}\omega_{n_0}$ imply $\frac{\pi^2}{6}$

$\frac{p_{n_1}(x - y)}{1 + p_{n_1}(x - y)} > \frac{\eta}{2}$; on the other hand $\sup_{u_n(L)} p_1 \leqslant \sup_L p_1$ for all $n \in \mathbb{N}^*$.

Everything is now ready for the construction of h having Ω as its domain of existence. Starting from an integer n_0 such that $E_{n_0} \cap \Omega \neq \emptyset$ and an $f_0 \in \mathscr{A}(E_{n_0} \cap \Omega, \mathbb{C})$, we successively choose:

$- n_1 > n_0$ such that $E_{n_1} \ni a_1$ and $\beta(a_1)$ as the largest open d-ball with centre a_1 in $E_{n_1} \cap \Omega$;
$- M_1$ such that $K_1 = K(n_0, n_1, M_1) \neq \emptyset$;
$- b_1 \in \beta(a_1) \backslash K_1$, which is possible since $\beta(a_1)$ is not relatively compact in $E_{n_1} \cap \Omega$;
$- f_1 \in \mathscr{A}(E_{n_1} \cap \Omega, \mathbb{C})$ such that $|f_1 - f_0 \circ u_{n_0}| \leqslant 1/2$ on K_1 and $|f_1(b_1)| \geqslant 1$;
$- n_2 > n_1$ such that $E_{n_2} \ni a_2$ and $\beta(a_2)$ as the largest open d-ball with centre a_2 in $E_{n_2} \cap \Omega$;
$- M_2 > M_1$ such that $K_2 = K(n_1, n_2, M_2) \ni b_1$;
$- b_2 \in \beta(a_2) \backslash K_2$;
$- f_2 \in \mathscr{A}(E_{n_2} \cap \Omega, \mathbb{C})$ such that $|f_2 - f_1 \circ u_{n_1}| \leqslant \frac{1}{2^2}$ on K_2 and $|f_2(b_2)| \geqslant 2$;
and so on with $K_m \ni b_1, b_2, \ldots, b_{m-1}$.

The sequence of functions $f_m \circ u_{n_m} \in \mathscr{A}(\omega_{n_m}, \mathbb{C})$, $m \in \mathbb{N}$, converges uniformly on every compact set $L \subset \Omega$. In fact, for sufficiently large m: $u_{n_m}(L) \subset K_m = K(n_{m-1}, n_m, M_m)$ and $|f_m - f_{m-1} \circ u_{n_{m-1}}| \leqslant \frac{1}{2^m}$ on K_m implies $|f_m \circ u_{n_m} - f_{m-1} \circ u_{n_{m-1}}| \leqslant \frac{1}{2^m}$ on L; then $h = \lim(f_m \circ u_{n_m}) \in \mathscr{A}(\Omega, \mathbb{C})$ since X is metrizable (Prop. 3.1.2.a). Moreover $b_j \in K_m$ for all $m > j$ and $u_{n_m}(b_j) = b_j$ for all $m \geqslant j$ imply $h(b_j) = f_j(b_j) + \sum_{m > j} (f_m - f_{m-1} \circ u_{n_{m-1}})(b_j)$, $|h(b_j)| \geqslant j - \frac{1}{2^j}$.

The last inequality proves that Ω is the domain of existence of h. In fact, let Ω_0, Ω_1 be as in Definition 5.2.1 and c a boundary point of Ω_0 in Ω_1; if $h|_{\Omega_0}$ has an extension $\in \mathscr{A}(\Omega_1, \mathbb{C})$, then c has an open neighbourhood Ω_2 in Ω_1 such that h is bounded on $\Omega_0 \cap \Omega_2$; if $a_j \in D \cap \Omega_0 \cap \Omega_2$ is chosen nearer to $\mathbb{C}\Omega_0$ than to $\mathbb{C}\Omega_2$ and j is large enough, then $\beta(a_j) \subset \Omega_0 \cap \Omega_2$ by the construction in the beginning of the proof, although $\sup_{\beta(a_j)} |h| \to \infty$ with j.

(B) In this particular case, the proof is much simpler, essentially because $\inf(\|x - y\|: y \notin \omega_{n_0}) > 0$ for all $x \in \omega_{n_0}$, therefore $-\ln \inf\{\|x - y\|: y \notin \omega_{n_0}\}$ is a continuous p.s.h. function of $x \in \omega_{n_0}$; if we set

$$A(n_0, M) = \left\{ x \in \omega_{n_0}: \|x\| \leqslant M, \|x - y\| \geqslant \frac{1}{M} \text{ for all } y \notin \omega_{n_0} \right\},$$

then $E_n \cap A(n_0, M)$ is for all $n > n_0$ a compact subset of $E_n \cap \omega_{n_0}$ with properties (1), (2), (3) in (A). Since the integer n_1 no longer needs to be preassigned, the $\beta(a_j)$ can be replaced by the following open sets β_j: given a countable fundamental family B of open sets in X, each of which is connected, $(\beta_j)_{j \in \mathbb{N}^*}$ is the family of connected components of the $\Omega \cap \beta$, $\beta \in B$, β intersecting the boundary of Ω; again the indexation is such that every such connected component occurs an infinity of times in the sequence.

The construction of h now runs this way. Starting from n_0 and f_0, we successively choose: M_1 such that $A(n_0, M_1) \neq \emptyset$, $b_1 \in E_\infty \cap [\beta_1 \setminus A(n_0, M_1)]$, $n_1 > n_0$ such that $b_1 \in E_{n_1} \cap \omega_{n_1}$, $f_1 \in \mathscr{A}(E_{n_1} \cap \Omega, \mathbb{C})$ such that $|f_1 - f_0 \circ u_{n_0}| \leqslant 1/2$ on $K_1 = E_{n_1} \cap A(n_0, M_1)$ and $|f_1(b_1)| \geqslant 1$, $M_2 > M_1$ such that $A(n_1, M_2) \ni b_1$, and so on. With the notation $\Omega_0, \Omega_1, \Omega_2$ as in (A), and $\Omega_2 \in B$: h cannot be bounded on $\Omega_0 \cap \Omega_2$ because $\Omega_0 \cap \Omega_2$ contains some β_j.

(C) Now let the p_n be seminorms; by discarding the first elements of the increasing sequence (p_n), we may assume that $\Omega \supset a + p_0^{-1}([0, \alpha])$ for some $a \in \Omega$ and $\alpha > 0$; then (Prop. 5.1.6) $\Omega = \Omega + p_0^{-1}(0)$, and Ω is the inverse image, under the canonical map $\psi: X \to X/p_0^{-1}(0)$, of a pseudoconvex domain $\tilde{\Omega}$ in $\tilde{X} = X/p_0^{-1}(0)$ endowed with the quotient topology, which is defined by the increasing sequence of norms

$$\tilde{p}_n(\tilde{x}) = \inf\{p_n(x): \psi(x) = \tilde{x}\}.$$

Given an equi-Schauder basis $(e_n)_{n \in \mathbb{N}^*}$ of X, with $p_n \circ u_m \leqslant p_n$ for all $m \in \mathbb{N}^*$, the linear maps $\tilde{u}_m = \psi \circ u_m \circ \psi^{-1}$ are well defined because $u_m[p_0^{-1}(0)] \subset p_0^{-1}(0)$, and equicontinuous because

$$\tilde{p}_n \circ \tilde{u}_m(\tilde{x}) \leqslant p_n \circ u_m(x) \leqslant p_n(x) \quad \text{for all } x \in \psi^{-1}(\tilde{x});$$

finally the $\tilde{p}_n$, $\tilde{u}_n$, $\tilde{E}_n = \psi(E_n)$ have all the properties of the p_n, u_n, E_n used in (A). Then the construction made in (A) can be repeated with $\psi[\beta(a_j)]$, $\tilde{p}_n$, $\tilde{u}_n$, $\tilde{E}_n$ substituted for $\beta(a_j)$, p_n, u_n, E_n, and leads to a function $h \circ \psi \in \mathscr{A}(\Omega, \mathbb{C})$ such that $\sup_{\beta(a_j)} |h \circ \psi| \to \infty$ with j.

5.3 Boundedness of p.s.h. functions and entire maps

Let again $p \in \mathrm{csn}(X)$, $q \in \mathrm{csn}(Z)$. In 3.3.5, the (p, q)-radius of boundedness $R(f, p, q)$ around a point a of an entire map $f : X \to Z$ was defined and computed by means of the Cauchy-Hadamard formula (4). Taking into account the fact that $q \circ f \in \mathscr{P}(X)$, we now define the *p-radius of boundedness* $R(u, p)(x)$, around an arbitrary point $x \in X$, of a nonconstant (hence unbounded from above by the Liouville theorem) function $u \in \mathscr{P}(X)$:

$$R(u, p)(x) = \sup\{r \geqslant 0 : u \text{ bounded from above on } x + p^{-1}([0, r])\}.$$

Proposition 5.3.1. *Either $R(u, p)$ is infinite everywhere; or the set of points $\in X$ where $R(u, p) > 0$ is a pseudoconvex open set Ω (which may be $\emptyset$ or X), $R(u, p)$ is finite and uniformly continuous on Ω, $-\ln R(u, p) \in \mathscr{P}(\Omega)$.*

Proof. We write $R(x)$ since u, p are given; if $R(x_0) > 0$, then $p(x - x_0) < R(x_0)$ ensures $R(x) \geqslant R(x_0) - p(x - x_0) > 0$. From now on we consider the second case only: for all $n \in \mathbb{N}$, $u^{-1}([-\infty, n[)$ is a pseudoconvex open set (Prop. 5.1.3.c) $\neq X$, $y_n(x) = \inf\{p(x - y) : u(y) \geqslant n\}$ a uniformly continuous function, $\omega_n = \{x \in X : y_n(x) > 0\}$ a pseudoconvex open set and $-\ln y_n|_{\omega_n}$, a finite continuous function $\in \mathscr{P}(\omega_n)$ (Prop. 5.1.6.a). The sequences (y_n) and (ω_n) are increasing, $\Omega = \bigcup_{n \in \mathbb{N}} \omega_n$ and $R = \lim y_n$ on Ω. $\square$

If x belongs to Ω, so does every point $x' \in x + p^{-1}(0)$, and $R(x') = R(x)$. If $x, x' \in \Omega$ and $p(x' - x) > 0$, the question of equality in $|R(x') - R(x)| \leqslant p(x' - x)$ arises; strangely enough, the answer $[\mathrm{Ki}]_1$ has a geometric aspect:

Proposition 5.3.2. *Let $p \in \mathrm{csn}(X)$ be locally uniformly convex in the following sense: given $x \in X$, with $p(x) = 1$, and $\delta > 0$, there is an $\varepsilon > 0$ such that $p(x') = 1$, $p\left(\dfrac{x + x'}{2}\right) \geqslant 1 - \varepsilon$ imply $p(x' - x) \leqslant \delta$. Also let $R = R(u, p) > 0$ and finite everywhere; then $|R(x') - R(x)| < p(x' - x)$ whenever $p(x' - x) > 0$.*

Proof. First note that, if $p(x) = p(x') = 1$, then $p[(1 - t)x + tx']$ is a convex function of $t \in \mathbb{R}$ equal to 1 for $t = 0$ or $t = 1$; hence follows, for $0 < t < 1$:

$$\inf(t, 1 - t)[2 - p(x + x')] \leqslant 1 - p[(1 - t)x + tx']$$

$$\leqslant \sup(t, 1 - t)[2 - p(x + x')]$$

Given $t \in]0, 1[$, $(1 - t)x + tx'$ may be substituted for $\dfrac{x + x'}{2}$ in the definition of local uniform convexity.

Now let $R(a) - R(x) = p(a - x) > 0$: by substituting $u[a + R(a).x]$ for $u(x)$, we may take $a = 0$, $R(a) = 1$, set $p(x) = t \in]0, 1[$ or $R(x) = 1 - t$ and $x = ty$, $p(y) = 1$; for all $k \in \mathbb{N}^*$, since u is bounded from above on $p^{-1}\left(\left[0, 1 - \dfrac{1}{k}\right]\right)$ and $x + p^{-1}\left(\left[0, 1 - t - \dfrac{1}{k}\right]\right)$, but not so on $x + p^{-1}\left(\left[0, 1 - t + \dfrac{1}{k}\right]\right)$, we can find $z_k \in X$ such that $u(z_k) \geqslant k$, $p(z_k) \geqslant 1 - \dfrac{1}{k}$ and $|p(z_k - x) - (1 - t)| \leqslant \dfrac{1}{k}$, hence $p(z_k) \leqslant p(x) + (1 - t) + \dfrac{1}{k}$, $|p(z_k) - 1| \leqslant \dfrac{1}{k}$; let $\dfrac{1}{k} < 1 - t$.

Setting $y_k = \dfrac{z_k - x}{p(z_k - x)}$, we have $p(y_k) = p(y) = 1$ and $p[(1 - t)y_k + ty] \to 1$ as $k \to \infty$; since y and $t \in]0, 1[$ are given, this implies $p(y_k - y) \to 0$, hence $p(z_k - y) \to 0$; but $u(z_k) \to \infty$ and, by the assumption, u has a p-radius of boundedness > 0 around y. $\quad\square$

The result is fairly general since, by a theorem of Kadec [Kad], any separable Banach space can be endowed with a locally uniformly convex norm equivalent to the initial one. Moreover [Brez] the usual norms on the spaces l^p, $1 < p < \infty$, are uniformly convex (i.e. in the statement of Prop. 5.3.2, ε depends only on δ) but the usual norm on l^1 is not so, which accounts for Example 5.3.3(B) below.

Examples 5.3.3. (A) Let X be an infinite dimensional Banach space and $(x_k')_{k \in \mathbb{N}^*}$ a sequence in X' such that $\|x_k'\| = 1$ but weak* convergent to the constant 0 (Rem. 3.3.4(A)); the radius of boundedness of the entire function $f(x) = \sum\limits_{k=1}^{\infty} \langle x, x_k' \rangle^k$ is 1 around the origin (ibid) and $\geqslant 1$ everywhere, therefore 1 everywhere by the maximum principle for the p.s.h. function $-\ln R(f, \| \ \|, |\ |)$.

In fact, let $a \in X$: given $r \in]0, 1[$, if $|x_k'(a)| \leqslant \dfrac{1 - r}{2}$ for all $k > k_0$, then $|f(a + x)| \leqslant \sum\limits_{k \leqslant k_0} [|x_k'(a)| + r]^k + \sum\limits_{k > k_0} \left(\dfrac{1 + r}{2}\right)^k$ for $\|x\| \leqslant r$.

(B) $[\text{Ki}]_1$ Proposition 5.3.2 fails for the usual norm $\|x\| = \sum\limits_{k=0}^{\infty} |\xi_k|$ of $l^1(\mathbb{N})$.

In fact, the radius of boundedness of the entire function $f(x) = \sum\limits_{k=1}^{\infty} (e^{-\xi_0}\xi_k)^k$ around $a = (\alpha_0, \alpha_1, \ldots)$ is $1 + \mathcal{R}e\,\alpha_0$ when $\mathcal{R}e\,\alpha_0 \geqslant 0$, $\exp(\mathcal{R}e\,\alpha_0)$ when

$\mathscr{Re}\,\alpha_0 \leqslant 0$; hence follows $|R(x') - R(x)| = \|x' - x\|$ when $\xi_k = \xi'_k = 0$ for all $k \geqslant 1$, ξ_0 and $\xi'_0 \in \mathbb{R}_+$.

5.4 The growth of p.s.h. functions and entire maps

The most precise results are obtained for functions $\in \mathscr{P}(X)$ with a *slow growth*, hence the letter S in the following notation: for $\sigma \in \mathbb{R}^*_+$, $S_\sigma(X)$ is the class of functions $w \in \mathscr{P}(X)$ such that

$$\limsup_{\mathbb{C} \ni \zeta \to \infty} \frac{w(\zeta x)}{\ln|\zeta|} \leqslant \sigma \quad \text{for all } x \in X.$$

For $n \in \mathbb{N}^*$, the class $S_n(X)$ contains $\ln(q \circ \varphi)$ for all continuous polynomial maps $\varphi \colon X \to Z$ with degree n and all $q \in \mathrm{csn}(Z)$ (proof of Prop. 2.2.11.a); therefore Theorem 5.4.2 and Proposition 5.4.3 contain an addition to the *Banach-Steinhaus* Theorem 3.1.8.b.

Proposition 5.4.1. *Let* $w \in S_\sigma(X)$. a) *The* p.s.h. *functions* $X^2 \ni (x, y) \mapsto \dfrac{w(\zeta x + y)}{\ln|\zeta|}$, $\zeta \in \mathbb{C}$, $|\zeta| \geqslant e$, *make up a locally bounded from above family.* b) $S_\sigma(X)$ *also contains the translated functions* $x \mapsto w(x + y)$, $y \in X$.

Proof. Let $p \in \mathrm{csn}(X)$ be such that $p(x) \leqslant 1$ implies $w(x) \leqslant M$; then everywhere $w \leqslant M + \sigma . \ln^+ p$ by the argument used for Proposition 2.2.11.a. From

$$w(\zeta x + y) \leqslant M + \sigma . \ln|\zeta| + \sigma . \ln^+ p\left(x + \frac{y}{\zeta}\right), |\zeta| \geqslant 1,$$

follow: a) given $(x_0, y_0) \in X^2$, with $c = \sup_{|\zeta| \geqslant e} p\left(x_0 + \frac{y_0}{\zeta}\right)$,

$$\frac{w(\zeta x + y)}{\ln|\zeta|} \leqslant M + \sigma + \sigma . \ln(c + 1)$$

$$\text{if } p(x - x_0) \leqslant \frac{2}{3}, p(y - y_0) \leqslant \frac{2}{3}, |\zeta| \geqslant e;$$

b) given $(x, y) \in X^2$, $\limsup_{\zeta \to \infty} \dfrac{w(\zeta x + y)}{\ln|\zeta|} \leqslant \sigma$. $\quad\square$

Theorem 5.4.2. *Let X be a Banach space and $\mathscr{W}$ a family of continuous functions $\in S_\sigma(X)$. Either there is a constant C such that $w(x) \leqslant C + \sigma\,.$* $\ln^+ \|x\|$ *for all $w \in \mathscr{W}$, $x \in X$, or the set $N = \left\{x \in X:\ \sup\limits_{w \in \mathscr{W}} w(x) < \infty\right\}$ is meagre, unipolar, and $N \cap \Omega$ strictly pluripolar in Ω for every bounded open set Ω.*

Proof. X being a normed space, the first statement in the proof of Proposition 5.4.1 now reads: if $\|x\| \leqslant r$ implies $w(x) \leqslant M$, then $w(x) \leqslant M + \sigma\,.$ $\ln^+ \dfrac{\|x\|}{r}$ for all $x \in X$, hence $\sup\limits_{\|x\| \geqslant 1} [w(x) - \sigma\,.\ln\|x\|] < \infty$ for all $w \in S_\sigma(X)$.

Assume that no constant C exists; let the sequence $(w_k)_{k \in \mathbb{N}^*} \subset \mathscr{W}$ be such that $C_k = \sup\limits_{\|x\| \geqslant 1} [w_k(x) - \sigma\,.\ln\|x\|]$ is positive and tends to infinity with k; then each function $\dfrac{w_k}{C_k}$ is continuous and p.s.h. on X,

$$\frac{w_k(x)}{C_k} \leqslant 1 + \frac{\sigma}{C_k} \ln^+ \|x\| \quad \text{for all } x \in X.$$

The sequence $\left(\dfrac{w_k}{C_k}\right)_{k \in \mathbb{N}^*}$ is locally bounded from above and $u = \limsup\limits_{k \to \infty} \dfrac{w_k}{C_k} \leqslant 1$ everywhere, u^* as well, therefore $u^* = \text{const.} \leqslant 1$ by the Liouville theorem for subharmonic functions on $\mathbb{C}$; we claim that actually u^* is the constant 1.

Let $u_k = \sup\left(\dfrac{w_k}{C_k}, \dfrac{w_{k+1}}{C_{k+1}}, \ldots\right)$: if $u^*(0) < 1$, we can find (Coroll. 4.3.9.c) $k_0 \in \mathbb{N}^*$ such that $u_{k_0}^*(0) < 1$, then $r > 0$ such that $u_{k_0}(x) \leqslant \lambda < 1$ for $\|x\| \leqslant r$, hence

$$w_k(x) \leqslant \lambda C_k + \sigma\,.\ln^+ \frac{\|x\|}{r} \quad \text{for all } k \geqslant k_0, x \in X,$$

in contradiction for large k with the definition of C_k. Since u^* is the constant 1, the set $\{x \in X: u(x) < 1\}$ is meagre (Ex. 4.1.5.h') and unipolar (Th. 4.5.6.b).

Now let $\Omega = \{x \in X: \|x\| < 1\}$: $\{x \in \Omega: u(x) < 1\}$ strictly pluripolar in Ω will follow, by Proposition 4.5.10.b, from $\{x \in \Omega: u_{k_0}(x) < 1\}$ strictly pluripolar in Ω for all $k_0 \in \mathbb{N}^*$. Since the latter set is meagre in Ω, let $u_{k_0}(a) = 1$, $a \in \Omega$: for all $n \in \mathbb{N}^*$ there is a $k(n) \geqslant k_0$ such that $\dfrac{w_{k(n)}(a)}{C_{k(n)}} \geqslant 1 - \dfrac{1}{n^2}$, and $u_{k_0}(x) < 1$ implies $\sum\limits_{n=1}^{\infty}\left[\dfrac{w_{k(n)}(x)}{C_{k(n)}} - 1\right] = -\infty.$

But $x \in N$ implies $u(x) \leqslant 0$; by the results we have got, N is meagre (already known, by Th. 3.1.8, in a special case) and $N \cap \Omega$ strictly pluripolar in Ω; hence follows easily $N \cap (R\Omega)$ strictly pluripolar in $(R\Omega)$ for all $R > 0$: in fact, for $R > 1$, the relations

$$\sup_{\|x\| \geqslant 1} \ [w_k(Rx) - \sigma . \ln \|x\|]$$

$$= \sup_{\|x\| \geqslant R} \ [w_k(x) - \sigma . \ln \|x\|] + \sigma . \ln R$$

$$\geqslant \sup_{\|x\| = R} \ w_k(x) \geqslant \sup_{1 \leqslant \|x\| \leqslant R} \ [w_k(x) - \sigma . \ln \|x\|]$$

prove the left hand member $\geqslant C_k$, hence $\left(\dfrac{1}{R}N\right) \cap \Omega$ strictly pluripolar in Ω.

Finally we obtain N unipolar by applying our results to the functions $\mathbb{C} \supset \zeta \mapsto w(a + \zeta b)$, $w \in \mathscr{W}$, where $E = \{a + \zeta b \colon \zeta \in \mathbb{C}\}$ is any complex line in X. In fact, such a function either is the constant $-\infty$ or belongs to $S_\sigma(\mathbb{C})$ by Proposition 5.4.1.b; if $\{\zeta \in \mathbb{C} \colon a + \zeta b \in N\}$ is not polar, there is a constant C such that $w(a + \zeta b) \leqslant C + \sigma . \ln^+ |\zeta|$ for all $w \in \mathscr{W}$, $\zeta \in \mathbb{C}$, hence $N \cap E = E$. $\square$

If X is a l.c. space, even a Fréchet space, the statement regarding strict pluripolarity has to be dropped, and what is left requires only a shorter proof.

Proposition 5.4.3. *Let $\mathscr{W}$ be a family of continuous functions $\in S_\sigma(X)$. Either there are $p \in \mathrm{csn}(X)$ and a constant C such that $w \leqslant C + \sigma . \ln^+ p$ for all $w \in \mathscr{W}$, and then $\sup\limits_{w \in \mathscr{W}} w$ has an upper regularized $\in S_\sigma(X)$; or the set $N = \{x \in X \colon \sup\limits_{w \in \mathscr{W}} w(x) < \infty\}$ is meagre (and unipolar if X is Baire).*

Proof. If N is not meagre: since $F^{(m)} = \{x \in X \colon w(x) \leqslant m$ for all $w \in \mathscr{W}\}$ is closed for all $m \in \mathbb{N}^*$, there are $a \in X$, $p \in \mathrm{csn}(X)$ such that $p(x) \leqslant 1$ implies $w(a + x) \leqslant m$ for all $w \in \mathscr{W}$; then (Prop. 5.4.1), for all $x \in X$:

$$w(a + x) \leqslant m + \sigma . \ln^+ p(x),$$

$$w(x) < m + \sigma[1 + \ln^+ p(a)] + \sigma . \ln^+ p(x).$$

If no pair (p, C) exists, then N is meagre, $N \neq X$ when X is Baire, and then N is unipolar by the argument used in the last paragraph of the proof of Theorem 5.4.2; but it may occur that, for no open set Ω, $N \cap \Omega$ is pluripolar in Ω.

As a counter-example, let $X = \mathbb{C}^{\mathbb{N}^*}$, the space of complex sequences $x = (\xi_1, \xi_2, \ldots)$ with the product topology, defined by the seminorms $x \mapsto$

$|\xi_n|, n \in \mathbb{N}^*$; let $Z = \mathbb{C}$ and $\mathscr{W}$ the family of functions $w_k(x) = \ln|\xi_k|, k \in \mathbb{N}^*$.

No pair (p, C) exists, $N = \left\{ x \in X : \sup_{n \in \mathbb{N}^*} |\xi_n| < \infty \right\}$ is meagre, and unipolar

since $\left\{ \zeta \in \mathbb{C} : \sup_{n \in \mathbb{N}^*} |\alpha_n + \zeta\beta_n| < \infty \right\}$ is at most a singleton or the whole of $\mathbb{C}$.

But let Ω be open in X, $w \in \mathscr{P}(\Omega)$, $N \cap \Omega \subset w^{-1}(-\infty)$, $a = (\alpha_1, \alpha_2, \dots) \in \Omega$ and $\lambda > w(a)$: there are $n_0 \in \mathbb{N}^*$ and $\delta > 0$ such that $|\xi_n - \alpha_n| < \delta$ for all $n \leqslant n_0$ implies $x \in \Omega$ and $w(x) < \lambda$; if $b = (\beta_1, \beta_2, \dots)$ is such that $\beta_n = 0$ for all $n \leqslant n_0$, then $a + \zeta b \in \Omega$ and $w(a + \zeta b) < \lambda$ for all $\zeta \in \mathbb{C}$, hence $w(a + b) = w(a)$ by the Liouville theorem. But $a + b \in N$ if $\beta_n = -\alpha_n$ for all $n > n_0$, hence $w(a) = -\infty$. $\square$

Apart from the case of a slow growth, we compare $w(\zeta x)$ with a power of $|\zeta|$, $\ln w(\zeta x)$ with $\ln|\zeta|$; for instance, if $w = \ln|f|$, $f \in \mathscr{A}(X, \mathbb{C})$, the *order* in the classical sense of the entire function $\zeta \mapsto f(\zeta x)$ is

$$\rho(x) = \limsup_{\mathbb{C} \ni \zeta \to \infty} \frac{\ln\ln|f(\zeta x)|}{\ln|\zeta|}.$$

Definition and Theorem 5.4.4. *Given a nonconstant $w \in \mathscr{P}(X)$, let $\rho(x) = 0$ if $w(\zeta x) = w(0)$ for all $\zeta \in \mathbb{C}$, $\rho(x) = \limsup_{r \to \infty} \dfrac{\ln u(x, r)}{\ln r} \geqslant 0$ if $u(x, r) = \sup_{|\zeta| = r} w(\zeta x)$ tends to infinity with r.*

a) *The upper regularized ρ^* of the function ρ is a constant (hence also the upper bound of ρ) $\in [0, +\infty]$, defining the order of the p.s.h. function w.*
b) *The cone $\{x \in X = \rho(x) < \rho^*\}$ is either X or unipolar; if X is a Fréchet space, the latter case only can occur. Moreover this cone is meagre if w is continuous, in particular if $w = \ln(q \circ f)$, $f \in \mathscr{A}(X, Z)$, $q \in \mathrm{csn}(Z)$.*
c) *The translated functions $x \mapsto w(x + y)$, $y \in X$, have the same order ρ^*.*

A *cone* is any set A such that $\zeta A \subset A$ for all $\zeta \in \mathbb{C}$.

Proof. a) $u(x, r)$ satisfies the assumptions of Theorem 4.5.13.a, defining an inverse function $v(x, s)$, if $w(0) < 0$, which we assume without altering the growth of w. If $u(x, r) \to +\infty$ with r, from the relation $u[x, v(x, s)] = s$ follows $\rho(x) = \limsup_{s \to \infty} \dfrac{\ln s}{\ln v(x, s)}, -\dfrac{1}{\rho(x)} = \limsup_{s \to \infty} \dfrac{-\ln v(x, s)}{\ln s}$, which remains true if $u(x, r) = w(0)$ for all $r \in \mathbb{R}_+$.

Now every point $x_0 \in X$ has an open neighbourhood V such that $x \in V$ implies $u(x, 1) \leqslant s_0$ and therefore $v(x, s) \geqslant 1$ for all $s \geqslant s_0$: the regularized function $-\dfrac{1}{\rho^*}$ of $-\dfrac{1}{\rho}$ is p.h.h. (Th. 4.1.7.c) and $\leqslant 0$, hence is a constant $\in [-\infty, 0]$.

b) The set $\{x \in X : \rho(x) < \rho*\}$ is either X or unipolar by Theorem 4.5.6.b.
It is a cone because $u\left(\zeta x, \dfrac{r}{|\zeta|}\right) = u(x, r)$ for all $r \in \mathbb{R}_+$ $(\zeta \in \mathbb{C}*)$ implies
$\rho(\zeta x) = \rho(x)$, while $\rho(0) = 0$.

If w is continuous, $u(., r)$ is continuous for all $r \in \mathbb{R}_+$, and $v(., s)$ for all
$s \in \mathbb{R}_+$ by the argument used in the beginning of the proof of Theorem
4.5.13.a; then the cone is meagre (Ex. 4.1.5.h').

c) Let us show that

$$\limsup_{|\zeta| \to \infty} \frac{\ln w(\zeta a + b)}{\ln |\zeta|} \leqslant \rho*$$

if $\rho*$ is finite and a, b linearly independent. Since $v(x, s)$ is an increasing
function of s, for each integer $k \geqslant 2$ the function $x \mapsto \sup\limits_{s \geqslant k} \dfrac{-\ln v(x, s)}{\ln s}$ belongs
to $\mathscr{P}^1(X)$; if X is Fréchet, by Proposition 4.3.10, given $\varepsilon > 0$ and K compact
in X, there is an integer $k_0 \geqslant 2$ such that $x \in K$ implies

$$\frac{-\ln v(x, s)}{\ln s} \leqslant -\frac{1}{\rho* + \varepsilon} \quad \text{for all } s \geqslant k_0$$

or $\dfrac{\ln u(x, r)}{\ln r} \leqslant \rho* + \varepsilon$ whenever $u(x, r) \geqslant k_0$, finally

$$\ln w(\zeta x) \leqslant \sup[(\rho* + \varepsilon) \ln|\zeta|, \ln k_0] \quad \text{for all } \zeta \in \mathbb{C}.$$

Choosing a norm on the subspace E of X spanned by a, b makes it a
Fréchet space; so this estimate holds when x runs over a compact subset
of E, especially for $\|x\| = 1$, hence

$$\ln w(x) < \sup[(\rho* + \varepsilon) \ln\|x\|, \ln k_0] \quad \text{for all } x \in E,$$

in particular for $x = \zeta a + b$. $\square$

The last argument proves that, if X is a finite dimensional normed space:
$$\rho* = \limsup_{\|x\| \to \infty} \frac{\ln w(x)}{\ln \|x\|}.$$

Examples 5.4.5. (A) If X is not Baire, it may happen that $\rho(x) < \rho*$ for all
$x \in X$, even if w is continuous. Let U be an open set in $\mathbb{R}^m$, $(K_n)_{n \in \mathbb{N}*}$ an
exhaustion of U by compact sets $K_n \subset \mathring{K}_{n+1}$, μ_n a positive Radon measure
carried by the boundary of K_n, $\int d\mu_n = 1$; finally let $X = \mathscr{K}(U)$, the space
of continuous functions $U \to \mathbb{C}$ with compact support, be endowed with
the norm $\sup\limits_{U} |x|$. (This example is taken from [Lel]$_4$).

The series $\sum_{n=1}^{\infty} \dfrac{1}{n^2} \exp[\int x \, d\mu_n]^n$ converges uniformly on some neighbour-

hood of every point $x_0 \in X$ since $\sup_U |x - x_0| \leqslant 1$, $n_0 = \sup\{n \in \mathbb{N}^*:$

$\int x_0 \, d\mu_n \neq 0\}$ imply $|\int x \, d\mu_n| \leqslant 1$ for all $n > n_0$; then the sum $f(x)$ of the
series is an analytic function on X. The same integer n_0 is the order of the
entire function

$$\zeta \to f(\zeta x_0) = \sum_{n=1}^{n_0} \frac{1}{n^2} \exp[\zeta \int x_0 \, d\mu_n]^n,$$

and $\rho(x_0) < \infty$ for all $x_0 \in X$ for $w = \ln|f|$; but $\rho^*(x_0) = \infty$ because, given
$\varepsilon > 0$ and $n \in \mathbb{N}^*$, one can find $x \in X$ satisfying $\sup_U |x - x_0| \leqslant \varepsilon$ and
$\int x \, d\mu_n \neq 0$.

(B) If φ is a continuous n-homogeneous polynomial map $X \to Z$ and
$q \in \mathrm{csn}(Z)$, $w = q \circ \varphi$ either is the constant 0 or has the order n; if $\varphi =$

$\varphi(0) + \sum_{k=1}^{n} \varphi_k$, where each φ_k is a continuous k-homogeneous polynomial

map $X \to Z$, $w = q \circ \varphi$ again has the order n unless $q \circ \varphi_n \equiv 0$, and more-

over $\lim_{\zeta \to \infty} \dfrac{w(\zeta x)}{|\zeta|^n} = q \circ \varphi_n(x)$ for all $x \in X$. In the situation where, for some

$\sigma \in \mathbb{R}_+^*$, $t(x) = \limsup_{|\zeta| \to \infty} \dfrac{w(\zeta x)}{|\zeta|^\sigma} < \infty$ for all $x \in X$, the type function t

satisfies $t(\alpha x) = |\alpha|^\sigma t(x)$ for all $x \in X$, $\alpha \in \mathbb{C}^*$: such functions will be con-
sidered later on (Def. 5.4.7 and Th. 5.4.9).

Another application of the inverse function Theorem 4.5.13 regards the
growth of a function $w \in \mathscr{P}(X)$ on the affine subspaces $x + E$, where E is a
given linear subspace of a finite dimension m: if $(e_1, \ldots, e_m)$ is a basis of E

and $z = \sum_{j=1}^{m} \zeta_j e_j$ the generic point of E, E admits as a topological supple-

ment the intersection of the kernels of m continuous linear functionals
extending (according to the Hahn-Banach theorem) the functionals $z \mapsto \zeta_j$.
So we can substitute $X \times E$ for X; we choose a norm on E and denote by
Ω an open set in X.

Proposition 5.4.6. *Given a nonconstant* $w \in \mathscr{P}(\Omega \times E)$, *let* $c(x) = 0$ *if*

$w(x, z) = w(x, 0)$ *for all* $z \in E$, $c(x) = \limsup_{r \to \infty} \dfrac{\ln u(x, r)}{\ln r} \geqslant 0$ *if* $u(x, r) =$

$\sup_{\|z\| = r} w(x, z)$ *tends to infinity with* r. *We assume* $u(x, 1) \leqslant 0$ *for all* $x \in \Omega$.

a) *The upper regularized* c^* *of the function* c *is such that* $-\dfrac{1}{c^*}$ *is p.h.h. on* Ω.

b) *Let Ω be connected: if $c(a) = \infty$ for some $a \in \Omega$, the set $\{x \in \Omega: c(x) < \infty\}$ is unipolar in Ω.*

c) *Let moreover $u(x,r) \to \infty$ for at least one $x \in \Omega$ and A a subset of Ω which is not strictly pluripolar in Ω but satisfies $M(r) = \sup_{x \in A} u(x,r) < \infty$ for all $r \in \mathbb{R}_+$: then*

$$c(x) \leqslant \frac{1}{R_1^A(x)} \limsup_{r \to \infty} \frac{\ln M(r)}{\ln r} \quad \text{for all } x \in \Omega.$$

R_1^A was defined, and $R_1^A > 0$ proved, in Proposition 4.5.12.

Proof. a) See the proof of Theorem 5.4.4.a. b) Theorem 4.5.6.a applies to $-\frac{1}{c}$. c) See Theorem 4.5.13.b.   □

Definition 5.4.7. *A function $w \in \mathscr{P}(X)$ is σ-homogeneous, $\sigma \in \mathbb{R}_+^*$, if $w(\zeta x) = |\zeta|^{\sigma} w(x)$ for all $x \in X$, $\zeta \in \mathbb{C}^*$; the number σ is the degree of w.*

For instance: a) the seminorms $p \in \text{csn}(X)$ are 1-homogeneous functions $\in \mathscr{P}(X)$; the fact that $\ln p \in \mathscr{P}(X)$ unless $p \equiv 0$ will be generalized by Theorem 5.4.10;

b) if φ is a continuous n-homogeneous polynomial map $X \to Z$ and $q \in \text{csn}(Z)$, then $q \circ \varphi$ is n-homogeneous. Two properties of these functions $q \circ \varphi$, namely the *translation lemma* 2.2.11.b and the Banach-Steinhaus Theorem 3.1.8.a, can forthwith be generalized, with exactly the same proof.

Proposition 5.4.8. a) *Let $w \in \mathscr{P}(X)$ be σ-homogeneous and V a balanced set in X: $w \leqslant M$ on $a + V$ implies $w \leqslant M$ on V and $w \leqslant \lambda^{\sigma} M$ on λV for all $\lambda > 0$.*

b) *w homogeneous implies $w \geqslant 0$ everywhere and $w(0) = 0$.*

c) *Let $\mathscr{W}$ be a family of continuous homogeneous functions $w \in \mathscr{P}(X)$: if the set $N = \{x \in X: \sup_{w \in \mathscr{W}} w(x)]^{1/d^\circ w} < \infty\}$ is not meagre, there is $p \in \text{csn}(X)$ such that $w \leqslant p^{d^\circ w}$ for all $w \in \mathscr{W}$.*

Proof of b). For all $x \in X \setminus \{0\}$, $w(re^{i\theta}x)$ is an increasing function of r since it does not depend on $\theta \in \mathbb{R}$, hence $w(x) \geqslant 0$, and $w(0) = \lim_{r \to 0} w(rx) = 0$.   □

Theorem 5.4.9. *Let the nonconstant $w \in \mathscr{P}(X)$ be continuous and, for some number $\sigma \in \mathbb{R}_+^*$:*

$$t(x) = 0 \quad \text{if } w(\zeta x) = w(0) \quad \text{for all } \zeta \in \mathbb{C},$$

$$t(x) = \limsup_{|\zeta| \to \infty} \frac{w(\zeta x)}{|\zeta|^{\sigma}} \quad \text{if not;}$$

let the set $\{x \in X: t(x) < \infty\}$ *be nonmeagre. Then:* a) *the upper regularized* t^* *of the type function* t *is a* σ-*homogeneous function* $\in \mathscr{P}(X)$, *hence* $t(x) < \infty$ *for all* $x \in X$; b) *there are* $p \in \mathrm{csn}(X)$ *and a constant* C *such that* $w \leqslant C + p^\sigma$, *therefore: either* $t \equiv 0$ *and the order* ρ^* *of* w *is* $\leqslant \sigma$; *or* $\rho^* = \sigma$.

Similar estimates were already obtained: for the class $S_\sigma(X)$ (Prop. 5.4.3); for the functions $w = \ln(q \circ f)$, where $q \in \mathrm{csn}(Z)$ and $f \in A(X, Z)$ has the exponential type (Th. 3.3.9); the latter estimate was the most precise one.

Proof. a) Again we set $u(x, r) = \sup\limits_{|\zeta|=r} w(\zeta x)$ and assume $w(0) < 0$ in order to use Theorem 4.5.13.a:

$$t(x) = \lim_{s \to \infty} \sup \frac{s}{[v(x, s)]^\sigma}.$$

Since $v(\zeta x, s) = \frac{1}{|\zeta|} v(x, s)$ for all $s \in \mathbb{R}_+$ ($\zeta \in \mathbb{C}^*$), the continuous p.s.h. functions $\frac{s}{[v(., s)]^\sigma}$ are σ-homogeneous; since $v(x, .)$ is also continuous, $t(x) < \infty$ is equivalent to $\sup\limits_{s \in \mathbb{R}_+} \frac{s}{[v(x, s)]^\sigma}$ finite; by Proposition 5.4.8.c (with a constant degree σ) there is $p \in \mathrm{csn}(X)$ such that $\frac{s}{v^\sigma(., s)} \leqslant p^\sigma$ for all $s \in \mathbb{R}_+$ or $\frac{u(., r)}{r^\sigma} \leqslant p^\sigma$ for all $r \in \mathbb{R}_+^*$: both families are locally bounded from above, hence $t^* \in \mathscr{P}(X)$ and the estimate in b) when the assumption $w(0) < 0$ is dropped.

b) This estimate implies $\rho(x) \leqslant \sigma$ for all $x \in X$, $\rho^* \leqslant \sigma$; on the other hand, $t(x) > 0$ implies $\rho(x) \geqslant \sigma$. $\square$

Theorem 5.4.10. *If* $w \in \mathscr{P}(X) \backslash \{0\}$ *is* σ-*homogeneous,* $\sigma \in \mathbb{R}_+^*$, *then* $\ln w \in \mathscr{P}(X)$, *hence also* $w^\alpha = \exp(\alpha . \ln w)$ *for all* $\alpha > 0$ *and, more generally,* $w_1^{\alpha_1} \ldots w_m^{\alpha_m}$ *if* $w_j \in \mathscr{P}(X) \backslash \{0\}$ *is* σ_j-*homogeneous and* $\alpha_j > 0$ *for all* $j \in \{1, \ldots, m\}$.

Proof. Let $(a, b) \in X^2$ and f be any function $\in \mathscr{A}(\mathbb{C}, \mathbb{C})$: $w[(a + \zeta b) \exp f(\zeta)]$ is a subharmonic function of $\zeta \in \mathbb{C}$ (Th. 4.2.1), equal by the assumption to $w(a + \zeta b) . \exp[\sigma . \mathscr{R}e \, f(\zeta)]$; since $\sigma . \mathscr{R}e \, f$ is any harmonic function on $\mathbb{C}$, we have to prove that an upper semicontinuous function $v: \mathbb{C} \to [-\infty, +\infty[$, such that $e^{v+h} \in \mathscr{P}(\mathbb{C})$ for all h harmonic on $\mathbb{C}$, either $\in \mathscr{P}(\mathbb{C})$ or is the constant $-\infty$.

Given any open disc $\Delta = \Delta(\alpha, \rho)$ in $\mathbb{C}$, there is a sequence (h_n) of harmonic functions on $\mathbb{C}$ such that $h_n(\alpha + \rho e^{i\theta})$ decreases to $v(\alpha + \rho e^{i\theta})$ for all $\theta \in \mathbb{R}$; then $\exp(v - h_n) \leqslant 1$ on the circumference of Δ entails $\exp(v - h_n)(\alpha) \leqslant 1$, $v(\alpha) \leqslant h_n(\alpha) = MV \, h_n(\alpha + \rho e^{i\theta})$, which decreases to $MV \, v(\alpha + \rho e^{i\theta})$. $\square$

Corollary 5.4.11. *A set $A \subset X$ is the set of zeros of a σ-homogeneous function $\in \mathscr{P}(X)$ if and only if A is a cone and $A = w^{-1}(-\infty)$ for some $w \in \mathscr{P}(X)$.*

Proof. If $A = w_0^{-1}(0)$ for some σ-homogeneous $w_0 \in \mathscr{P}(X)$, then $w = \ln w_0 \in \mathscr{P}(X)$. Conversely, if $A = w^{-1}(-\infty)$ is a cone, let again $v(x, s)$ be associated by Theorem 4.5.13.a to $u(x, r) = \sup_{|\zeta|=r} w(\zeta x)$: for all $s \in \mathbb{R}_+$, $\sigma \in \mathbb{R}_+^*$, $\dfrac{1}{v^\sigma(., s)}$ is a σ-homogeneous function $\in \mathscr{P}(X)$, which vanishes if and only if $u(x, r) = w(0) = -\infty$ for all $r \in \mathbb{R}_+$, i.e. if and only if $x \in A$.

Theorem 5.4.12. a) *Let $(\varphi_k)_{k \in \mathbb{N}^*}$ be a sequence of continuous k-homogeneous polynomial maps $X \to \mathbb{C}$: if the sequence $(|\varphi_k|^{1/k})_{k \in \mathbb{N}^*}$ is locally bounded from above, then $\left(\limsup_{k \to \infty} |\varphi_k|^{1/k}\right)^*$ is a 1-homogeneous function $\in \mathscr{P}(X)$.*

b) *Conversely, let X be a Banach space with a Schauder basis: given a 1-homogeneous function $w \in \mathscr{P}(X)$, there is a sequence of continuous k-homogeneous polynomial maps $\varphi_k: X \to \mathbb{C}$ such that the sequence $(|\varphi_k|^{1/k})_{k \in \mathbb{N}^*}$ is locally bounded from above and $w = \left(\limsup_{k \to \infty} |\varphi_k|^{1/k}\right)^*$.*

Proof of b). $\Omega = w^{-1}([0, 1[)$ is a balanced pseudoconvex domain, therefore (Th. 5.2.7.B) the domain of existence of an $h \in \mathscr{A}(\Omega, \mathbb{C})$; let $\varphi_k = \dfrac{1}{k!}(\hat{D}_0^k h)$ for all $k \in \mathbb{N}^*$. By Theorem 2.3.5 and Corollary 2.4.5, Ω is the largest balanced open set where $\sup_{k \in \mathbb{N}^*} |\varphi_k| < \infty$ or, equivalently, $\limsup_{k \to \infty} |\varphi_k|^{1/k} < 1$; then $w = \left(\limsup_{k \to \infty} |\varphi_k|^{1/k}\right)^*$, and by Theorem 3.1.8.a there is $p \in \mathrm{csn}(X)$ such that $|\varphi_k| \leqslant p^k$ for all $k \in \mathbb{N}^*$. $\square$

5.5 The density number for a p.s.h. function

First let u be subharmonic on an open set U in $\mathbb{C}$, $\alpha \in U$: for $0 < r < d(\alpha, \complement U)$, both

$$MV u(\alpha + re^{i\theta}) \text{ and } \sup_{\theta \in \mathbb{R}} u(\alpha + re^{i\theta}) = \sup_{\overline{\Delta}(\alpha, r)} u$$

are increasing convex functions of $\ln r$, tending to $u(\alpha)$ as $r \to 0$; we claim that, even if this common limit is $-\infty$, their difference tends to 0.

In fact, let $\mu = \dfrac{\Delta u}{2\pi}$ be the positive Riesz measure for u; given $R \in$ $]0, d(\alpha, \complement U)[$ there is a harmonic function h on $\Delta(\alpha, R)$ such that, for $0 < r < R$:

$$u(\alpha + re^{i\theta}) = h(\alpha + re^{i\theta}) + \int\limits_{0 < |\zeta - \alpha| < R} \ln|\alpha + re^{i\theta} - \zeta|\, d\mu(\zeta)$$

$$+ \mu(\{\alpha\})\ln r;$$

$$\sup_{\theta \in \mathbb{R}} \ln|\alpha + re^{i\theta} - \zeta| - MV\ln|\alpha + re^{i\theta} - \zeta|$$

$$= \ln(r + |\zeta - \alpha|) - \sup(\ln r, \ln|\zeta - \alpha|)$$

$$= \inf\left[\ln\left(1 + \frac{|\zeta - \alpha|}{r}\right), \ln\left(1 + \frac{r}{|\zeta - \alpha|}\right)\right] \leqslant \ln 2$$

entails

$$\sup_{\overline{\Delta}(\alpha, r)} u - MV u(\alpha + re^{i\theta}) \leqslant \sup_{\overline{\Delta}(\alpha, r)} h - MV h(\alpha + re^{i\theta})$$

$$+ \mu[\Delta(\alpha, R)\backslash\{\alpha\}]\ln 2 \leqslant \varepsilon$$

for a suitably chosen R and r sufficiently small, hence our claim.

On the other hand, the Green formula and the mollifying process described in 4.2(B) yield, for $0 < r_0 < r_1 < d(\alpha, \complement U)$:

$$MV u(\alpha + r_1 e^{i\theta}) - MV u(\alpha + r_0 e^{i\theta}) = \int\limits_{r_0}^{r_1} \mu[\Delta(\alpha, r)]\frac{dr}{r}$$

$$= \int\limits_{r_0}^{r_1} \mu[\overline{\Delta}(\alpha, r)]\frac{dr}{r},$$

from which follows that, for $0 < r < d(\alpha, \complement U)$, $\mu[\Delta(\alpha, r)]$ and $\mu[\overline{\Delta}(\alpha, r)]$ are the left and right hand side derivatives of the increasing convex function $\ln r \to MV u(\alpha + re^{i\theta})$; finally

$$\mu(\{\alpha\}) = \lim_{r \to 0} \frac{d\, MV u(\alpha + re^{i\theta})}{d\ln r} = \lim_{r \to 0} \frac{MV u(\alpha + re^{i\theta})}{\ln r}$$

$$= \lim_{r \to 0}\left[\frac{1}{\ln r}\sup_{\overline{\Delta}(\alpha, r)} u\right]$$

is the density of the measure μ, in other words the density number for the corresponding function u, at the point α: we denote it by $\nu_u(\alpha)$.

In particular, let $f \in \mathscr{A}(U, Z)$, Z s.c., and $f(\zeta) = \sum\limits_{k \in \mathbb{N}} c_k(\zeta - \alpha)^k$ be the Taylor expansion of f around α: if $q \in \mathrm{csn}(Z)$ and $q(c_k) > 0$ for some k, then

$\ln(q \circ f)$ is subharmonic on the connected component of U which contains α, and the density number for $\ln(q \circ f)$ at α is $\inf\{k \in \mathbb{N}: q(c_k) > 0\}$.

Following Lelong [Lel]$_7$, we now define a density number for a p.s.h. function on an open set Ω in a l.c. space X.

Definition and Theorem 5.5.1. *The density number $v_w(a) \in \mathbb{R}_+$ for a function $w \in \mathscr{P}(\Omega)$ at a point $a \in \Omega$ can be defined in the following equivalent ways.*

a) *For all $x \in X$,* $\lim\limits_{r \to 0}\left[\dfrac{1}{\ln r}\sup\limits_{|\zeta| \leqslant r} w(a + \zeta x)\right] = \lambda_w(x)$ *exists; the lower regularized of the function λ_w is a constant (hence also the lower bound of $\lambda_w) \in \mathbb{R}_+$, defining $v_w(a)$. This lower bound is a minimum if either X is a Fréchet space or X a Baire space and w continuous.*

b) $v_w(a) = \lim\limits_{r \to 0}\left(\dfrac{1}{\ln r}\sup\limits_{a+rW} w\right)$ *for any balanced neighbourhood W of the origin in $\Omega - a$ such that $\sup\limits_{a+W} w < \infty$, and then the inequality*

(1) $\sup\limits_{a+rW} w \leqslant \sup\limits_{a+W} w + v_w(a) \cdot \ln r \quad$ *for all $r \in \,]0, 1]$*

is an analogue of the Schwarz lemma.

c) *If $\mathscr{V} \subset \mathscr{P}(\Omega)$ is locally bounded from above and $w = \sup\limits_{v \in \mathscr{V}} v$, then $v_{w*}(a) = \inf\limits_{v \in \mathscr{V}} v_v(a)$.*

Proof. a) For sufficiently small $|\zeta|$, $w(a + \zeta x)$ is either the constant $-\infty$, and then $\lambda_w(x) = +\infty$, or a subharmonic function of ζ, and then $\lambda_w(x)$ is the density number $\in \mathbb{R}_+$ for this function at the point 0; the latter case occurs for some $x \in X$ since $w \in \mathscr{P}(\Omega)$.

On the other hand, $\sup\limits_{|\zeta| \leqslant r} w(a + \zeta x) = \sup\limits_{\theta \in \mathbb{R}} w(a + re^{i\theta}x)$ is a p.s.h. function of $x \in \dfrac{\omega(a)}{r}$ [$\omega(a)$ denoting the largest balanced subset of $\Omega - a$] for all $r > 0$; for $0 < r \leqslant r_0 < 1$ we have a locally bounded from above family of p.s.h. functions on $\dfrac{\omega(a)}{r_0}$ on account of the inequality

$$\frac{1}{\ln 1/r}\sup\limits_{\theta \in \mathbb{R}} w(a + re^{i\theta}x) \leqslant \frac{1}{\ln 1/r_0}\sup\limits_{\theta \in \mathbb{R}} w^+(a + r_0 e^{i\theta}x).$$

Then (Th. 4.1.7.c) the upper regularized of $-\lambda_w$ is p.s.h., but $\leqslant 0$ and $\not\equiv -\infty$.

If X is Fréchet, the set $\{x \in X: \lambda_w(x) > v_w(a)\}$ is unipolar (Th. 4.5.6.b); if w is continuous, it is meagre (Ex. 4.1.5.h').

b) Let $a + W \subset \Omega$ and $w \leqslant s$ on $a + W$. From the convexity of $\ln r \mapsto \sup\limits_{|\zeta| \leqslant r} w(a + \zeta x)$ follow for $0 < r \leqslant 1$:

(1) $\qquad \sup\limits_{|\zeta| \leqslant r} w(a + \zeta x) \leqslant s + \lambda_w(x) . \ln r \leqslant s + v_w(a) . \ln r \quad$ for all $x \in W$,

$$\liminf \left(\frac{1}{\ln r} \sup_{a + rW} w \right) \geqslant v_w(a).$$

But for all $x \in W$ we also have

$$\limsup_{r \to 0} \left(\frac{1}{\ln r} \sup_{a + rW} w \right) \leqslant \lim_{r \to 0} \left[\frac{1}{\ln r} \sup_{|\zeta| \leqslant r} w(a + \zeta x) \right] = \lambda_w(x),$$

and $\lambda_w(\zeta x) = \lambda_w(x)$ for all $\zeta \in \mathbb{C}^*$.

c) For all $v \in \mathscr{V}$: $v \leqslant w^*$ implies $v_v(a) \geqslant v_{w^*}(a)$ by the second definition of the density number, hence $v_{w^*}(a) \leqslant n = \inf\limits_{v \in \mathscr{V}} v_v(a)$. On the other hand, for all $v \in \mathscr{V}$ and $r \in \,]0,1]$, from (1) follows $\sup\limits_{a + rW} v \leqslant \sup\limits_{a + W} w^* + n \ln r$, hence $\sup\limits_{a + rW} \left(\sup\limits_{v \in \mathscr{V}} v \right) \leqslant \sup\limits_{a + W} w^* + n \ln r$, $\sup\limits_{a + rW} w^* \leqslant \sup\limits_{a + W} w^* + n \ln r$ if W is open, and finally $v_{w^*}(a) \geqslant n$ if $\sup\limits_{a + W} w^* < \infty$, by the second definition of the density number. $\quad \square$

Both definitions give the density number $+\infty$ for the constant $-\infty$. If $w(a) > -\infty$, $\lambda_w \equiv 0$; on the contrary, if $w(a) = -\infty$, $\lambda_w(0) = +\infty$ and the set $\{x \in X : \lambda_w(x) > v_w(a)\}$ is a cone.

For $u, v \in \mathscr{P}(\Omega)$ and $\alpha, \beta > 0$: $\lambda_{\alpha u + \beta v}(x) = \alpha . \lambda_u(x) + \beta . \lambda_v(x)$, for all $x \in X$, since we also have

$$\lambda_w(x) = \lim_{r \to 0} \frac{MV \, w(a + re^{i\theta} x)}{\ln r}$$

and therefore $v_{\alpha u + \beta v}(a) \geqslant \alpha . v_u(a) + \beta . v_v(a)$. Further information, more precisely, the equality instead of the last inequality, will be obtained through an adaptation of the Legendre transform due to Kiselman [Ki]₃, originating in the following theorem [Ki]₂.

Theorem 5.5.2. *Let T be an open interval, $W = \{z \in \mathbb{C} : \mathscr{R}e \, z \in T\}$ and the function $u : \Omega \times T \to [-\infty, +\infty[$ have the property that $[(x, z) \to u(x, \mathscr{R}e \, z)] \in \mathscr{P}(\Omega \times W)$, which implies that, for all $x \in \Omega$, $u(x, .)$ is either the constant $-\infty$ or a real convex function on T; then $v(x) = \inf\limits_{t \in T} u(x, t)$, $v(x, s) = \inf\limits_{t \in T} [u(x, t) - st]$, $s \in \mathbb{R}$, define p.h.h. functions $v, v(., s)$ on Ω.*

Proof. This function v is upper semicontinuous; if $u_n = \sup(u, -n)$ and $v_n(x) = \inf_{t \in T} u_n(x, t)$ for all $n \in \mathbb{N}$, the sequence (v_n) decreases to v; so we may assume $u \geqslant 0$. Now let χ be a convex function $T \to \mathbb{R}_+$ that is not constant on any subinterval and tends to $+\infty$ at either bound of T, for instance:

$$\chi(t) = \ln \frac{1}{(t - t_0)(t_1 - t)} \text{ if } T = \,]t_0, t_1[, (t_0, t_1) \in \mathbb{R}^2; \chi(t) = \frac{1}{1 - t} - t \text{ if } T =$$

$\,]-\infty, 1[$; and so on. Then $[z \to \chi(\mathscr{R}e\, z)] \in \mathscr{P}(W)$; if $u_n(x, t) = \dfrac{\chi(t)}{n} + u(x, t)$

and $v_n(x) = \inf_{t \in T} u_n(x, t)$ for all $n \in \mathbb{N}^*$, again the sequence (v_n) decreases to v; so we may assume that, for all $x \in \Omega$, $u(x, t) = v(x)$ occurs for a single value $t = w(x)$, which remains in a compact subset of T when x runs over a compact subset of Ω, in particular $m \leqslant w(a + e^{i\theta}b) \leqslant M$ for all $\theta \in \mathbb{R}$ if $\Omega \supset \{a + \zeta b \colon |\zeta| \leqslant 1\}$.

The bounded function $\theta \mapsto w(a + e^{i\theta}b)$ is measurable since $w(a + e^{i\theta}b) \leqslant t$ if and only if $u(a + e^{i\theta}b, t) < u\left(a + e^{i\theta}b, t + \dfrac{1}{n}\right)$ for all $n \in \mathbb{N}^*$; by the Lusin criterion for measurability, for all $\varepsilon > 0$ there is a continuous function f, $m \leqslant f \leqslant M$, on the boundary of $\Delta = \Delta(0, 1)$, such that $f(e^{i\theta}) = w(a + e^{i\theta}b)$ except on an open subset of $d\theta$-measure $\leqslant \varepsilon$; there is also an $h \in \mathscr{A}(\Delta, \mathbb{C})$ such that, for all $\theta \in \mathbb{R}$, $\mathscr{R}e\, h(\zeta) \to f(e^{i\theta})$ as $\Delta \ni \zeta \to e^{i\theta}$.

Then $[\zeta \mapsto u(a + \zeta b, \mathscr{R}e\, h(\zeta))] \in \mathscr{P}(\Delta)$ implies

$$v(a) \leqslant u(a, \mathscr{R}e\, h(0)) \leqslant MV u(a + re^{i\theta}b, \mathscr{R}e\, h(re^{i\theta})) \text{ for all } r \in \,]0, 1[,$$

$$\text{and} \quad \limsup_{r \to 1} u(a + re^{i\theta}b, \mathscr{R}e\, h(re^{i\theta})) \leqslant u(a + e^{i\theta}b, f(e^{i\theta}))$$

for all $\theta \in \mathbb{R}$ implies, by the Fatou lemma:

$$v(a) \leqslant MV u(a + e^{i\theta}b, f(e^{i\theta}))$$

$$\leqslant MV v(a + e^{i\theta}b) + \frac{\varepsilon}{2\pi} \sup_{\substack{\theta \in \mathbb{R} \\ m \leqslant t \leqslant M}} u(a + e^{i\theta}b, t). \qquad \square$$

Corollary 5.5.3. *Given* $a \in \Omega$, *let* $\omega(a)$ *be the largest balanced subset of* $\Omega - a$.

a) *For all* $w \in \mathscr{P}(\Omega)$, $s > v_w(a)$ *and* $R > 0$, *the set* $\left\{x \in \dfrac{\omega(a)}{R} \colon \lambda_w(x) > s\right\}$ *is pluripolar in* $\dfrac{\omega(a)}{R}$. b) *For all* $u, v \in \mathscr{P}(\Omega)$ *and* $\alpha, \beta > 0$: $v_{\alpha u + \beta v}(a) = \alpha \cdot v_u(a) + \beta \cdot v_v(a)$.

Proof. Since $\lambda_w(\zeta x) = \lambda_w(x)$ for all $\zeta \in \mathbb{C}^*$, all values assumed by λ_w are assumed on $\dfrac{\omega(a)}{R}$.

a) $u(x,t) = \sup\limits_{\theta \in \mathbb{R}} w(a + e^{t+i\theta}x)$ satisfies the assumption of the theorem on $\dfrac{\omega(a)}{R} \times \,]-\infty, \ln R[$; for all $x \in \dfrac{\omega(a)}{R}$, $u(x,\,.\,)$ is either the constant $-\infty$ or a real convex increasing function on an open interval containing $]-\infty, \ln R]$; in both cases, $v(x,s) = -\infty$ if $s < \lim\limits_{t \to -\infty} \dfrac{u(x,t)}{t} = \lambda_w(x)$, $v(x,s) > -\infty$ if $s > \lambda_w(x)$.

If $s > v_w(a)$, then $\lambda_w(x) < s$ occurs for some $x \in \dfrac{\omega(a)}{R}$, $v(.,s)$ is p.s.h. on $\dfrac{\omega(a)}{R}$ and equal to $-\infty$ on $\left\{ x \in \dfrac{\omega(a)}{R} : \lambda_w(x) > s \right\}$.

b) Let $v_{\alpha u + \beta v}(a) = \varepsilon + \alpha \,.\, v_u(a) + \beta \,.\, v_v(a)$, $\varepsilon > 0$: then for all $x \in \omega(a)$ we have

$$\lambda_{\alpha u + \beta v}(x) = \alpha \,.\, \lambda_u(x) + \beta \,.\, \lambda_v(x) \geqslant \varepsilon + \alpha \,.\, v_u(a) + \beta \,.\, v_v(a)$$

and therefore either $\lambda_u(x) \geqslant \dfrac{\varepsilon}{2\alpha} + v_u(a)$ or $\lambda_v(x) \geqslant \dfrac{\varepsilon}{2\beta} + v_v(a)$; but the open set $\omega(a)$ cannot be the union of two pluripolar subsets of $\omega(a)$. $\square$

Examples 5.5.4. (A) If $w \in \mathscr{P}(X) \backslash \{0\}$ is σ-homogeneous (Def. 5.4.7), $\lambda_{\ln w}(x)$ at the origin (Th. 5.4.10) is either σ or $+\infty$.
(B) Let $f \in \mathscr{A}(\Omega, Z)$, Z s.c., and $q \in \mathrm{csn}(Z)$ be such that $\ln(q \circ f) \in \mathscr{P}(\Omega)$: the density number for $\ln(q \circ f)$ at a point $a \in \Omega$ is an integer, namely 0 if $q \circ f(a) > 0$ and, if $q \circ f(a) = 0$, the smallest $k \in \mathbb{N}^*$ such that $q \circ (\hat{D}_a^k f) \not\equiv 0$.

Proof. This integer, say k_0, exists, for otherwise the continuity of q would imply $\ln(q \circ f) \equiv -\infty$ on $a + \omega(a)$. If $q \circ f(a) = 0$ and $\ln(q \circ f)(a + \zeta x)$ is a subharmonic function of ζ near 0, the density number for this function at 0 is an integer $\geqslant k_0$ for every $x \in X$, but $= k_0$ for some x; in this example, the meagre cone $\{x \in X : \lambda(x) > v(a)\}$ is pluripolar in X, as the set of zeros of the k_0-homogeneous p.s.h. function $q \circ (\hat{D}_a^{k_0} f)$. $\square$

Proposition 5.5.5. *Given $w \in \mathscr{P}(\Omega)$: the density number $v_w(x)$ is an upper semicontinuous function of $x \in \Omega$, and $\{x \in \Omega : v_w(x) > 0\}$ a pluripolar subset of Ω.*

Proof. Let W be a balanced convex neighbourhood of the origin such that $a + 2W \subset \Omega$ and $w \leqslant 0$ on $a + 2W$, which we assume without altering the density numbers for w. Given $\varepsilon \in]0,1[$, let $r \in]0,1[$ be such that $\sup\limits_{a+rW} w \geqslant [\varepsilon + v_w(a)]\ln r$, and $r' = r^{1-\varepsilon}$; then $b \in a + (r' - r)W$ implies

$$a + rW \subset b + r'W \quad \text{and} \quad b + W \subset a + 2W,$$

therefore, on account of (1):

$$v_w(b).\ln r' \geqslant \sup_{b+r'W} w \geqslant [\varepsilon + v_w(a)]\ln r, \ v_w(b) \leqslant \frac{\varepsilon + v_w(a)}{1 - \varepsilon}.$$

Finally $v_w(x) > 0$ implies $w(x) = -\infty$. $\quad\square$

Theorem 5.5.6. *Let Ω and Γ be open neighbourhoods of the origins in X and Z(s.c.) respectively, $v \in \mathscr{P}(\Gamma)$. a) If $f \in \mathscr{A}(\Omega, Z)$ vanishes at the origin of X, but not identically on a neighbourhood of the origin, then $v_{v \circ f}(0) \geqslant k_0$. $v_v(0)$, where k_0 is the smallest $k \in \mathbb{N}^*$ such that $\hat{D}_0^k f \not\equiv 0$. b) Given $p \in \mathrm{csn}(X)$, $p \not\equiv 0$, $\Pi \in \mathrm{csn}(Z)$ such that*

$$p^{-1}([0,1]) \subset \Omega, \ \Pi^{-1}([0,1]) \subset \Gamma, \ s = \sup_{\Pi^{-1}([0,1])} v < \infty$$

and $k_0 \in \mathbb{N}^*$, let

$$A(p, \Pi, k_0) = \left\{ f \in \mathscr{A}(\Omega, Z): f(0) = 0, \ \sup_{p^{-1}([0,1])} (\Pi \circ f) < \infty, \right.$$

$$\left. \hat{D}_0^k f \equiv 0 \text{ for all } k < k_0 \right\}.$$

Then $k_0 . v_v(0) = \inf\{v_{v \circ f}(0): f \in A(p, \Pi, k_0)\}$. c) The set $\{f \in A(p, \Pi, k_0): v_{v \circ f}(0) > k_0 . v_v(0)\}$ is unipolar in $A(p, \Pi, k_0)$ [a property which does not depend on a topology of $A(p, \Pi, k_0)$] if either Z is a Fréchet space or Z a Baire space and v continuous.
    By Theorem 4.3.1, $v \circ f$ is p.h.h. on $f^{-1}(\Gamma)$.

Proof. a) Choose $\Pi \in \mathrm{csn}(Z)$ as in b). Given $f \in \mathscr{A}(\Omega, Z)$ with $f(0) = 0$, there is a $p \in \mathrm{csn}(X)$ such that $p(x) \leqslant 1$ implies $x \in \Omega$, $\Pi \circ f(x) \leqslant M$; then, by the generalized inequalities of Cauchy, $p(x) \leqslant r < 1$ implies $\Pi \circ f(x) \leqslant \dfrac{Mr^{k_0}}{1 - r}$. Now choose $r_0 \in]0,1[$ such that

$$(*) \qquad \frac{Mr_0^{k_0}}{1 - r_0} \leqslant 1.$$

$v \circ f$ is p.h.h. and $\leqslant s$ on $p^{-1}([0, r_0[)$ and, for $r \leqslant r_0$:

$$\sup\{v \circ f(x): p(x) \leqslant r\} \leqslant \sup\left\{v(z): \Pi(z) \leqslant \frac{Mr^{k_0}}{1-r} = \rho\right\},$$

hence the result since $k_0 = \lim\limits_{r \to 0} \dfrac{\ln \rho}{\ln r}$.

b) Given $\varepsilon > 0$, let $b \in Z\backslash\{0\}$ be such that (with the notation of 5.5.1) $\lambda_v(b) \leqslant \varepsilon + v_v(0)$, $a \in X$ and $x' \in X'$ such that $p(a) = 1, |x'| \leqslant p, \langle a, x' \rangle = 1$: then $f(x) = \langle x, x' \rangle^{k_0} b$ defines an $f \in A(p, \Pi, k_0)$ for which

$$\sup_{\theta \in \mathbb{R}} v \circ f(re^{i\theta}a) = \sup_{\varphi \in \mathbb{R}} v(r^{k_0}e^{i\varphi}b) \geqslant k_0[\varepsilon + \lambda_v(b)] \ln r$$

for sufficiently small r, hence $\lambda_{v \circ f}(a) \leqslant k_0 . \lambda_v(b)$.

c) We may endow $A(p, \Pi, k_0)$ with the seminorms of pointwise convergence, namely $f \mapsto q \circ f(x)$, $x \in \Omega$, $q \in \text{csn}(Z)$, and the seminorm $\|f\| = \sup\limits_{p^{-1}([0,1])} (\Pi \circ f)$. Given $M > 0$, choose $r_0 \in \,]0, 1[$ satisfying $(*)$; for $0 < r \leqslant r_0$ and $p(x) \leqslant 1$, $w_{r,x}(f) = \dfrac{v \circ f(rx)}{\ln \dfrac{1}{r}}$ is a p.s.h. function of $f \in A(p, \Pi, k_0)$ with $\|f\| < M$, and then $w_{r,x}(f) \leqslant s^+/\ln\dfrac{1}{r_0}$; on the bounded from above family $\{w_{r,x}: 0 < r \leqslant r_0, p(x) \leqslant 1\}$, let $\mathcal{F}$ be the filter whose basis is made up of the sets $\left\{w_{r,x}: 0 < r \leqslant \dfrac{r_0}{n}, p(x) \leqslant 1\right\}$, $n \in \mathbb{N}^*$: by Theorem 4.1.7.c, the function $A(p, \Pi, k_0) \ni f \mapsto -v_{v \circ f}(0) = \limsup\limits_{\mathcal{F}} w_{r,x}(f)$ has a p.s.h. upper regularized, which is $\leqslant 0$, hence a constant, namely

$$-\inf\{v_{v \circ f}(0): f \in A(p, \Pi, k_0)\} = -k_0 . v_v(0).$$

Now (Th. 4.5.6.b) the set $\{f \in A(p, \Pi, k_0): v_{v \circ f}(0) > k_0 . v_v(0)\}$ is either $A(p, \Pi, k_0)$ or a unipolar subset of $A(p, \Pi, k_0)$. But, if Z is Fréchet or Z Baire and v continuous, by 5.5.1.a we can choose $b \in Z\backslash\{0\}$ so that $\lambda_v(b) = v_v(0)$, and then the proof of b) gives $a \in X$ such that $\lambda_{v \circ f}(a) = k_0 . v_v(0)$. $\quad\square$

Exercises

5.2.1. Let Ω_1, Ω_2 be open sets in l.c. spaces X_1, X_2 respectively. Prove that Ω_1, Ω_2 domains of holomorphy implies $\Omega_1 \times \Omega_2$ domain of holomorphy, and conversely, by means of Proposition 3.2.1, if X_1, X_2 are Fréchet.

5.2.2. a) Let $f \in \mathscr{A}(X, \mathbb{C})$, $f^{-1}(0) \neq \emptyset$ and X; show that, for all $y \in X$, $\ln \sup \{r > 0:$ $|\zeta| \leqslant r$ implies $f(x + \zeta y) \neq 0\}$ is a plurisuperharmonic function of $x \in X \setminus f^{-1}(0)$.
b) Let $X = \mathscr{A}(\mathbb{C}^m, \mathbb{C})$, endowed with the topology of compact convergence, $a \in \mathbb{C}^m$, $\Omega = \{x \in X : x(a) \neq 0\}$; using a) with $f(x, z) = x(z)$ for $(x, z) \in X \times \mathbb{C}^m$, denoting by d the distance on $\mathbb{C}^m$ associated to a norm, show that $\ln d[a, x^{-1}(0)]$ is a plurisuperharmonic function of $x \in \Omega$. What can be said about the set $\{x \in X: x^{-1}(0) = \emptyset\}$?

5.4.5. [Lel]$_4$ With the notation of Example 5.4.5(A): in the open unit ball Ω of $\mathscr{K}(U)$, is the union of the strictly pluripolar subsets $\{x \in \Omega : \int x \, d\mu_n = 0\}$ pluripolar?

5.4.6. Let Ω be connected and $f \in \mathscr{A}(\Omega \times \mathbb{C}, \mathbb{C})$. Show that, if $f(a, .)$ has an infinity of zeros for some $a \in \Omega$, the set of $x \in \Omega$ for which $f(x, .)$ has a finite number of zeros (or none) is unipolar in Ω. *Hint*: use Jensen's formula ([He]$_3$, Th. 15.7).

5.5.1. Let $X = c_{00}(\mathbb{N}^*)$ with the norm $\|x\| = \sup_{k \in \mathbb{N}^*} |\xi_k|$. Choosing (how ?) the numbers $c_k > 0$ so that $w(x) = \sup_{k \in \mathbb{N}^*} (c_k \ln |\xi_k|)$ defines $w \in \mathscr{P}(X)$, compute the density number $v_w(0)$; with the notation of 5.5.1.a, show that the cone $\{x \in X: \lambda_w(x) > v_w(0)\}$ may be the whole of X.

Chapter 6
Analytic maps from a given domain to another one

Summary

6.1.A.   Green function and Green open sets in the complex plane.

6.1.B.   The Lindelöf principle and its generalization to infinite dimensions.

6.2.   Intrinsic pseudodistances: C_Ω, K_Ω and their infinitesimal forms c_Ω, k_Ω.

Th. 6.2.2. C_Ω, c_Ω are not only suprema, but actually maxima.

Th. 6.2.4. K_Ω in a convex domain $[\text{Lem}]_1$.

Prop. 6.2.7. $c_\Omega(0;\,.)$, $k_\Omega(0;\,.)$ in a balanced domain.

Th. 6.2.9. K_Ω is the integrated form of $k_\Omega [\text{Ve}]_1$.

Def. 6.3.1. Complex geodesics $[\text{Ve}]_2$.

Th. 6.3.2.   Complex geodesics with the same range $[\text{Ve}]_3$.

Th. 6.3.3.   Existence of complex geodesics in some convex domains of a reflexive Banach space.

Def. 6.3.5. Complex extremal points [Th Wh].

Prop. 6.4.1.
  and 2.   The generalized Schwarz lemma [Pa].

Th. 6.4.4.   Inequalities of Harris $[\text{Har}]_1$.

Th. 6.4.6.   The fixed point theorem of Earle-Hamilton [Ea Ha].

Th. 6.4.8.   The fixed point theorem of Hayden-Suffridge [Ha Su].

Th.6.4.9.   A fixed point theorem of Mazet-Vigué.

Chapter 6
Analytic maps from a given domain to another one

Throughout this chapter, letters such as U, V will denote *connected* open sets in $\mathbb{C}$, but $\Delta = \Delta(0, 1)$ the open unit disc; Ω and Γ *connected* open sets in s.c.l.c. spaces X and Y respectively; $f \in \mathscr{A}(\Omega, \Gamma)$ will mean $f \in \mathscr{A}(\Omega, Y)$, $f(\Omega) \subset \Gamma$.

6.1 A generalization of the Lindelöf principle

(A) Let Π be the *Poincaré distance* in Δ:

$$\Pi(\alpha, \beta) = th^{-1} \left| \frac{\beta - \alpha}{1 - \bar{\alpha}\beta} \right| \quad \text{for all } (\alpha, \beta) \in \Delta^2$$

or $\coth \Pi = \exp G_\Delta$ by formula (3) in 4.2(D). By the Schwarz-Pick lemma, for every $h \in \mathscr{A}(\Delta, \Delta)$:

$$\Pi[h(\alpha), h(\beta)] \leqslant h(\alpha, \beta) \quad \text{or} \quad G_\Delta[h(\alpha), h(\beta)] \geqslant G_\Delta(\alpha, \beta)$$

for all $(\alpha, \beta) \in \Delta^2$, and equality for one pair (α, β) with $\alpha \neq \beta$ implies that $h \in \text{Aut}(\Delta)$, i.e. h is a 1-1 map of Δ onto Δ.

The *Green function* G_U was defined in 4.2(D) for an arbitrary open set U. U is a *Green open set* if U has the equivalent properites: $\mathbb{C} \setminus U$ is not polar; there exists a nonconstant bounded harmonic function on U (Prop. 4.5.4.b); there exists a nonconstant superharmonic function > 0 on U; $G_U(\alpha, \beta) < \infty$ for one pair, or all pairs, $(\alpha, \beta) \in U^2$ with $\alpha \neq \beta$. For instance, any bounded or simply connected (except $\mathbb{C}$ itself) open set is Green.

If U is Green: $G_U(., \beta)$ is a potential on U, which means that any subharmonic function $\leqslant G_U(., \beta)$ on U is $\leqslant 0$, and implies that $G_U(\zeta, \beta)$ has the same upper bound for $|\zeta - \beta| = \rho < \text{dist}(\beta, \complement U)$ and for $|\zeta - \beta| \geqslant \rho$, hence

(1) $\quad \limsup_{\substack{\alpha \to \alpha_0 \\ \beta \to \beta_0}} G_U(\alpha, \beta) < \infty$

if α_0 is on the boundary of U (including ∞ if U is unbounded) and $\beta_0 \in U$. But (1) still holds if α_0, β_0 are distinct boundary points of U: in fact, let $\beta_0 \neq \infty$; by the first paragraph in the proof of Proposition 4.5.4.a, there are at least two points $\in \mathbb{C} \backslash U$ such that any disc centred in one of them has a nonpolar intersection with $\mathbb{C} \backslash U$; then the union U' of U with a sufficiently small open disc centred in β_0 is still a Green open set, and $G_U(\alpha, \beta) \leqslant G_{U'}(\alpha, \beta)$, which is bounded as $\alpha \to \alpha_0, \beta \to \beta_0$, by the first result.

Let $U' \subset U$: if $U \backslash U'$ is polar, the harmonic function $U' \supset \zeta \mapsto G_{U'}(\zeta, \beta) + \ln |\zeta - \beta|$ has a harmonic extension to U, hence $G_U \equiv G_{U'}$ on U'^2. Conversely, $G_U(\alpha, \beta) = G_{U'}(\alpha, \beta)$ for one pair $(\alpha, \beta) \in U'^2$, with $\alpha \neq \beta$, implies $U \backslash U'$ polar since otherwise the balayage in U of $G_U(., \beta)$ over $U \backslash U'$ would give a harmonic function > 0 and $\leqslant G_U(., \beta)$ on U'.

(B) *The lindelöf principle. Let* $h \in \mathscr{A}(U, V)$.
a) $G_V[h(\alpha), h(\beta)] \geqslant G_U(\alpha, \beta)$ *for all* $(\alpha, \beta) \in U^2$.
b) $G_V[h(\alpha), h(\beta)] = G_U(\alpha, \beta) < \infty$ *for one pair* $(\alpha, \beta) \in U^2$ *implies that* h *is* 1-1, *and* $V = h(U)$ *unless there exists a nonempty polar set in* $\overline{\mathbb{C}} = \mathbb{C} \cup \{\infty\}$ *(note that* $\{\infty\}$ *is polar) whose union with* U *is open in* $\overline{\mathbb{C}}$.

For instance: such a polar set does exist if $U = \mathbb{C} \backslash \overline{\Delta}$, but does not if U is regular, i.e. U bounded and $\mathbb{C} \backslash U$ nowhere thin (see 4.6(A)) on the boundary of U.

Proof. a) let h be nonconstant and V a Green open set; then U too is Green since $G_V[h(.), h(\beta)]$ is a nonconstant superharmonic function > 0 on U; $G_U(., \beta) - G_V[h(.), h(\beta)]$ is subharmonic and $< G_U(., \beta)$, hence $\leqslant 0$.
b) The assumption entails at once that U, V are Green and $V \backslash h(U)$, polar; the positive superharmonic function $G_{h(U)}[h(.), h(\beta)] - G_U(., \beta)$ vanishes at α, therefore identically; then, for all $\zeta \in U$, $G_{h(U)}[h(\zeta), h(.)] - G_U(\zeta, .)$ vanishes at β, therefore identically. Thus we have

$$G_{h(U)}[h(\alpha), h(\beta)] = G_U(\alpha, \beta) \quad \text{for all } (\alpha, \beta) \in U^2$$

and h is a 1-1 map of U onto $h(U)$; since $\mathbb{C} \backslash U$ is not polar, there is a nonconstant bounded harmonic function h_0 on U and, by the argument used in the second paragraph of the proof of Theorem 4.5.3, h^{-1} has a meromorphic extension to V which is a 1-1 map of V onto an open subset U' of $\overline{\mathbb{C}}$; finally $U' \backslash U$ is polar like $V \backslash h(U)$. $\square$

Thanks to the principle, a Green open set V can be substituted for a bounded open set in some questions, for instance: on account of (1), given $\alpha \in U$ and a compact set $K \subset V$, the maps $\in \mathscr{A}(U, V)$ whose values at α belong to K are equicontinuous at α; if U is not Green, a map $\in \mathscr{A}(U, V)$ has to be a constant. Since a substitute for bounded open sets is welcome in general l.c. spaces, we consider the following generalization of the Lindelöf principle: for all $(a, b) \in \Omega^2$, let

(2) $G_\Omega(a, b) = \inf\{G_{\varphi(\Omega)}[\varphi(a), \varphi(b)]: \varphi \in \mathscr{A}(\Omega, \mathbb{C}), \varphi \neq \text{const.}\},$

where $\varphi(\Omega)$ is an open connected set in $\mathbb{C}$, even if $\varphi \in \mathscr{G}(\Omega, \mathbb{C})$ (Prop. 2.4.7.a). Then

(3) $G_\Gamma[f(a), f(b)] \geqslant G_\Omega(a, b)$ for all $(a, b) \in \Omega^2, f \in \mathscr{A}(\Omega, \Gamma)$

since, for every $\varphi \in \mathscr{A}(\Gamma, \mathbb{C})$ such that $\varphi \circ f \neq \text{const}$:

$$G_{\varphi(\Gamma)}[\varphi \circ f(a), \varphi \circ f(b)] \geqslant G_{\varphi \circ f(\Omega)}[\varphi \circ f(a), \varphi \circ f(b)] \geqslant G_\Omega(a, b).$$

Either there is no nonconstant $\varphi \in \mathscr{A}(\Omega, \mathbb{C})$ whose range $\varphi(\Omega)$ is Green, and $G_\Omega \equiv +\infty$; or, for every such φ, $\{(x, y) \in \Omega^2: \varphi(x) = \varphi(y)\}$ is a closed unipolar subset of Ω^2, and $P = \{(x, y) \in \Omega^2: G_\Omega(x, y) = +\infty\}$ as well. P contains the diagonal $\{(x, x): x \in \Omega\}$; P is the diagonal if and only if the nonconstant $\varphi \in \mathscr{A}(\Omega, \mathbb{C})$ whose range is Green separate the points of Ω. G_Ω is obviously symmetric.

Theorem 6.1.1. a) *Either* $G_\Omega \equiv +\infty$ *or* $G_\Omega|_{\Omega^2 \setminus P}$ *is continuous* $\Omega^2 \setminus P \to]0, +\infty[$.
b) *If* $G_\Omega \not\equiv +\infty$, *for all* $a \in \Omega$, $G_\Omega(a, .)$ *is continuous plurisuperharmonic* $\Omega \to]0, +\infty]$.
c) $G_\Omega(x, y) \to +\infty$ *as* x *and* y *tend to* $a \in \Omega$.
   Is G_Ω continuous $\Omega^2 \to]0, +\infty]$? it will be proved only for a simply connected domain Ω (Prop. 6.2.3).

Proof. We first remark that $x \in a + \dfrac{\omega(a)}{R}$, $R > 1$, implies $G_\Omega(a, x) \geqslant \ln R$ since, for every nonconstant $\varphi \in \mathscr{A}(\Omega, \mathbb{C})$, $\zeta \mapsto \varphi[a + \zeta(x - a)]$ is an analytic map $\Delta(0, R) \to \varphi(\Omega)$, hence $G_{\varphi(\Omega)}[\varphi(a), \varphi(x)] \geqslant G_{\Delta(0, R)}(0, 1) = \ln R$.

a) Let $G_\Omega(a, b) < \infty$; choose a nonconstant $\psi \in \mathscr{A}(\Omega, \mathbb{C})$ and $R > 1$ so that $G_{\psi(\Omega)}[\psi(a), \psi(b)] < \ln R$, then $p \in \text{csn}(X)$ so that

(i) $p^{-1}([0, 2R + 1[) \subset (\Omega - a) \cap (\Omega - b)$;
(ii) $x \in \omega = a + p^{-1}([0, 1[)$, $y \in \omega' = b + p^{-1}([0, 1[)$ imply $G_{\psi(\Omega)}[\psi(x), \psi(y)] \leqslant M < \ln R$, a fortiori $G_\Omega(x, y) \leqslant M$.

Given $(x_0, y_0) \in \omega \times \omega'$ and a nonconstant $\varphi \in \mathscr{A}(\Omega, \mathbb{C})$ whose range is Green: either there is $x \in \omega$ such that $\varphi(x) = \varphi(y_0)$, and $G_{\varphi(\Omega)}[\varphi(x_0), \varphi(y_0)]$ $= G_{\varphi(\Omega)}[\varphi(x_0), \varphi(x)] \geqslant \ln R$ since $x \in x_0 + \dfrac{\omega(x_0)}{R}$; or $G_{\varphi(\Omega)}[\varphi(.), \varphi(y_0)]$ is pluriharmonic > 0 on ω, and (Prop. 4.4.3) $G_{\varphi(\Omega)}[\varphi(x_0), \varphi(y_0)]$ lies between $G_{\varphi(\Omega)}[\varphi(a), \varphi(y_0)] \cdot \exp[\pm H_\omega(a, x_0)]$. Similarly: either there is $y \in \omega'$ such that $\varphi(a) = \varphi(y)$, and

$$G_{\varphi(\Omega)}[\varphi(a), \varphi(y_0)] = G_{\varphi(\Omega)}[\varphi(y), \varphi(y_0)] \geqslant \ln R;$$

or $G_{\varphi(\Omega)}[\varphi(a), \varphi(.)]$ is pluriharmonic > 0 on ω', and $G_{\varphi(\Omega)}[\varphi(a), \varphi(y_0)]$ lies between $G_{\varphi(\Omega)}[\varphi(a), \varphi(b)] \cdot \exp[\pm H_{\omega'}(b, y_0)]$.

Let $J(x_0, y_0) = H_\omega(a, x_0) + H_{\omega'}(b, y_0)$, which tends to 0 as $x_0 \to a$, $y_0 \to b$: on account of property (ii) of ψ, we have

$$G_\Omega(x_0, y_0) \geqslant G_\Omega(a, y_0) \cdot \exp[-H_\omega(a, x_0)]$$

$$\geqslant G_\Omega(a, b) \cdot \exp[-J(x_0, y_0)];$$

besides G_Ω is upper semi-continuous.

$G_\Omega > 0$ on $\Omega^2 \backslash P$ will follow from the results of b).

b) $G_\Omega(a, x) \to +\infty$ as $x \to a$ by the above preliminary remark. If $G_\Omega(a, b) = +\infty$, then $\varphi(a) = \varphi(b)$ for all nonconstant $\varphi \in \mathscr{A}(\Omega, \mathbb{C})$ whose range is Green, and $G_\Omega(a, x) = G_\Omega(b, x)$ tends to $+\infty$ as $x \to b$. $G_\Omega(a, .)$ is a continuous infimum of positive plurisuperharmonic functions on Ω, and > 0 since $G_\Omega(a, a) = +\infty$.

c) Choose $p \in \mathrm{csn}(X)$ so that $\Omega \supset a + p^{-1}([0, 1[)$: for all $R > 1$, $p(x - a)$ and $p(y - a) < \dfrac{1}{2R + 1}$ imply $y \in x + \dfrac{\omega(x)}{R}$, $G_\Omega(x, y) \geqslant \ln R$. □

Corollary 6.1.2. *If $G_\Omega \equiv +\infty \not\equiv G_\Gamma$, the range of a map $\in \mathscr{A}(\Omega, \Gamma)$ is a strictly pluripolar subset of Γ.*

This makes the following remarks useful.

a) If a, b are joined by the complex line E and lie in the same connected component U of $E \cap \Omega$, $G_\Omega(a, b) < \infty$ demands that U be a Green open set: otherwise $\varphi(U)$ could not be a Green open set.

b) If $\Omega \neq X$ is convex, then $G_\Omega \not\equiv +\infty$, for the Hahn-Banach separation theorem gives a real continuous $\mathbb{R}$-linear functional $\not\equiv 0$ which is bounded from above on Ω, and this functional is the real part of an $x' \in X'$.

c) That $\Omega' \subset \Omega$ entails $G_{\Omega'} \leqslant G_\Omega$ on Ω'^2 is included in (3) above. We now present two cases of equality; the second one does not occur for the Green functions of open sets in $\mathbb{C}$.

Theorem 6.1.3. *Let $\Omega' \subset \Omega$. a) If $\Omega \backslash \Omega'$ is contained in a closed unipolar subset P of Ω, then $G_{\Omega'} \equiv G_\Omega$ on Ω'^2.*

b) *Let $\Omega \backslash \Omega'$ be closed and bounded in X; assume moreover that, for all linear subspaces E of X with a sufficiently large finite dimension, $E \cap \Omega'$ is con-*

nected (an assumption satisfied, for instance, if $\Omega \setminus \Omega'$ is convex): then again $G_{\Omega'} \equiv G_\Omega$ on Ω'^2.
c) *Conversely, if $G_{\Omega'} \equiv G_\Omega$ on Ω'^2, then every nonconstant $\varphi \in \mathscr{A}(\Omega', \mathbb{C})$ such that $\varphi(\Omega')$ is Green has a continuous extension $\Omega \cap \bar{\Omega}' \to \bar{\mathbb{C}}$.*

Proof. a) Given $(a, b) \in \Omega'^2$, we have to prove

$$(*) \qquad G_{\varphi(\Omega')}[\varphi(a), \varphi(b)] \geqslant G_\Omega(a, b)$$

for every nonconstant $\varphi \in \mathscr{A}(\Omega', \mathbb{C})$ such that $\varphi(\Omega')$ is Green. Since $\mathbb{C} \setminus \varphi(\Omega')$ is not polar, there is $\alpha \in \mathbb{C} \setminus \varphi(\Omega')$ such that the intersection of $\mathbb{C} \setminus \varphi(\Omega')$ with any neighbourhood of α is not polar; then $\varphi' = \dfrac{1}{\varphi - \alpha} \in \mathscr{A}(\Omega', \mathbb{C})$ and we may substitute φ' for φ in the left hand member of $(*)$. Now the intersection of $\mathbb{C} \setminus \varphi'(\Omega')$ with any neighbourhood of ∞ is not polar: by Theorem 4.5.3, φ' has an extension $\psi \in \mathscr{A}(\Omega, \mathbb{C})$ and, by Proposition 4.5.5.b, $\psi(\Omega) \setminus \varphi'(\Omega')$ is polar; the left hand side of $(*)$ is finally $G_{\psi(\Omega)}[\psi(a), \psi(b)]$.
b) For any E with a sufficiently large finite dimension, $E \cap \Omega'$ is connected and $E \cap (\Omega \setminus \Omega')$ compact; then [Kau], for any $\varphi \in \mathscr{A}(\Omega', \mathbb{C})$, $\varphi_{|E \cap \Omega'}$ has a unique extension $\in \mathscr{A}(E \cap \Omega, \mathbb{C})$; on account of this uniqueness, φ has an extension $\psi \in \mathscr{G}(\Omega, \mathbb{C})$ with the same range as φ since $\dfrac{1}{\varphi - \alpha}$ has an extension for all $\alpha \in \mathbb{C} \setminus \varphi(\Omega')$; if this common range is Green, $\psi \in \mathscr{A}(\Omega, \mathbb{C})$ (Th. 2.4.9.a) and again the left hand side of $(*)$ is $G_{\psi(\Omega)}[\psi(a), \psi(b)]$.
c) Given a boundary point a of Ω' in Ω and $M > 0$, by Theorem 6.1.1.c there is a neighbourhood W of a in X such that $(x, y) \in (W \cap \Omega')^2$ implies $G_\Omega(x, y) \geqslant M$ and therefore $G_{\varphi(\Omega')}[\varphi(x), \varphi(y)] \geqslant M$. By property (1) of the Green function in (A) above, any limit point, in the compact space $\bar{\mathbb{C}}^2$, of $(\varphi(x), \varphi(y))$ as $\Omega'^2 \ni (x, y) \to (a, a)$, must be a point (α, α) with $\alpha \in \bar{\mathbb{C}}$. $\square$

6.2 Intrinsic pseudodistances

The *Caratheodory pseudodistance*, already introduced in Proposition 4.4.4, is the *smallest* pseudodistance on Ω such that, when Ω is endowed with this pseudodistance and Δ with the Poincaré distance Π, all maps $\in \mathscr{A}(\Omega, \Delta)$ are contractions (hence $C_\Delta = \Pi$ by the Schwarz-Pick lemma):

$$(1) \qquad C_\Omega(a, b) = \sup\{\Pi[\varphi(a), \varphi(b)]: \varphi \in \mathscr{A}(\Omega, \Delta)\}$$

$$= \sup\{th^{-1}|\varphi(b)|: \varphi \in \mathscr{A}(\Omega, \Delta), \varphi(a) = 0\}$$

since the composition of φ with an element of $\mathrm{Aut}(\Delta)$ does not alter $\Pi[\varphi(a), \varphi(b)]$.

This supremum is always finite (Prop. 4.4.4.a) and a continuous pseudo-distance on Ω (Prop. 4.4.4.c); C_Ω and C_Γ make all maps $\in \mathscr{A}(\Omega, \Gamma)$ contractions (Prop. 4.4.4.b), in particular $\Omega' \subset \Omega$ entails $C_{\Omega'} \geqslant C_\Omega$ on Ω'^2. $th\, C_\Omega(a, .) \in \mathscr{P}(\Omega)$ for all $a \in \Omega$, $\ln th\, C_\Omega(a, .)$ as well unless $C_\Omega \equiv 0$.

Instead of (1), one may write

$$(2) \qquad \ln \coth C_\Omega(a, b) = \inf\{G_\Delta[\varphi(a), \varphi(b)]: \varphi \in \mathscr{A}(\Omega, \Delta)\} \geqslant G_\Omega(a, b)$$

and therefore $G_\Omega \equiv +\infty$ implies $C_\Omega \equiv 0$.

Given $a \in \Omega$, $c \in X$, $\varphi \in \mathscr{A}(\Omega, \Delta)$:

$$\frac{1}{\zeta} \frac{\varphi(a + \zeta c) - \varphi(a)}{1 - \overline{\varphi(a)}\varphi(a + \zeta c)} \to \frac{(\hat{D}_a^1 \varphi)(c)}{1 - |\varphi(a)|^2} \quad \text{as } \mathbb{C} \ni \zeta \to 0,$$

which leads to the *derived* form of the Caratheodory pseudodistance:

$$(3) \qquad c_\Omega(a; c) = \sup\left\{\frac{|(\hat{D}_a^1 \varphi)(c)|}{1 - |\varphi(a)|^2}: \varphi \in \mathscr{A}(\Omega, \Delta)\right\}$$

$$= \sup\{|(\hat{D}_a^1 \varphi)(c)|: \varphi \in \mathscr{A}(\Omega, \Delta), \varphi(a) = 0\}$$

as in (1). This sup is finite and $c_\Omega(a; .)$ a continuous seminorm on X, for $c \in \dfrac{\omega(a)}{R}$, $R > 0$, entails $c_\Omega(a; c) < \dfrac{1}{R}$ by formula (3) in Theorem 2.3.5; Proposition 6.2.1 below will provide further information.

c_Ω and c_Γ make any map $f \in \mathscr{A}(\Omega, \Gamma)$ a contraction in the sense that

$$(4) \qquad c_\Gamma[f(a); (\hat{D}_a^1 f)(c)] \leqslant c_\Omega(a; c) \quad \text{for all } a \in \Omega, c \in X$$

which proceeds from the formula in Theorem 3.1.10 for $\varphi \circ f$; in particular $\Omega' \subset \Omega$ implies $c_{\Omega'} \geqslant c_\Omega$ on $\Omega' \times X$.

Proposition 6.2.1. c_Ω *is continuous on* $\Omega \times X$; *either* $c_\Omega \equiv 0$ *or* $\ln c_\Omega \in \mathscr{P}(\Omega \times X)$.

Proof: Let $\Omega \supset a_0 + p^{-1}([0, 1[)$, $p \in \mathrm{csn}(X)$, $p(a - a_0) < \dfrac{1}{2}$ and $p(c) < R$: for every $\varphi \in \mathscr{A}(\Omega, \Delta)$, $\varphi \not\equiv 0$, $\varphi(a_0) = 0$, formula (2) in Theorem 2.3.5. gives

$$(*) \qquad (\hat{D}_a^1 \varphi)(c) = 2R . MV\left[e^{-i\theta}\varphi\left(a + e^{i\theta}\frac{c}{2R}\right)\right];$$

the right hand member is an analytic function (Th. 3.1.7.b), hence

$\ln |(\hat{D}_a^1 \varphi)(c)|$ a p.h.h. function of (a, c); $\ln \dfrac{1}{1 - |\varphi(a)|^2}$ is also a p.s.h. function

of a since $]-\infty, 1[\ni u \mapsto \ln \dfrac{1}{1 - u}$ is an increasing convex function.

Moreover our φ are equicontinuous; the right hand members of (*) are

equicontinuous functions of (a, c), and the $\dfrac{1}{1 - |\varphi(a)|^2}$ equicontinuous

functions of a sufficiently near a_0: either $c_\Omega \equiv 0$ or $\ln c_\Omega \in \mathscr{P}(\Omega \times X)$ as a continuous supremum of p.s.h. functions.   □

Theorem 6.2.2. a) *The suprema in* (1) *and* (3) *are actually maxima.*
b) *Given* $(a, b) \in \Omega^2$ [*resp.:* $(a, c) \in \Omega \times X$] *and* $\varepsilon > 0$, *there is a finite dimensional affine subspace* E *of* X, *containing* a, b [*resp.:* $a + \zeta c$ *for all* $\zeta \in \mathbb{C}$] *such that* $C_\Omega(a, b) \leq C_{E \cap \Omega}(a, b) \leq \varepsilon + C_\Omega(a, b)$ [*resp.:* $c_\Omega(a; c) \leq c_{E \cap \Omega}(a; c) \leq \varepsilon + c_\Omega(a; c)$].

Proof. a) let the sequence $(\varphi_n) \subset \mathscr{A}(\Omega, \Delta)$ be such that $\varphi_n(a) = 0$ for all n and

$$|\varphi_n(b)| \to th\, C_\Omega(a, b) \quad [\text{resp.:}\ |(\hat{D}_a^1 \varphi_n)(c)| \to c_\Omega(a; c)].$$

By Theorem 3.1.7.a, there is a $\psi \in \mathscr{A}(\Omega, \mathbb{C})$ which, on each compact set $K \subset \Omega$, is the uniform limit of a subsequence (φ_{n_K}) depending on K; from this follow $\psi(a) = 0$, $|\psi| < 1$ by the local maximum modulus principle (Prop. 2.3.7.b) and

$$|\psi(b)| = th\, C_\Omega(a, b) \quad [\text{resp.:}\ |(\hat{D}_a^1 \psi)(c)| = c_\Omega(a; c)].$$

b) Let $\mathscr{E}$ be the family of all finite dimensional affine subspaces of X containing a, b [resp.: $a + \zeta c$ for all $\zeta \in \mathbb{C}$]: $E, E' \in \mathscr{E}$, $E' \subset E$ imply

$$C_\Omega(a, b) \leq C_{E \cap \Omega}(a, b) \leq C_{E' \cap \Omega}(a, b)$$

$$[\text{resp.:}\ c_\Omega(a; c) \leq c_{E \cap \Omega}(a; c) \leq c_{E' \cap \Omega}(a; c)].$$

For every $E \in \mathscr{E}$ we can, by a) above, choose $\varphi_E \in \mathscr{A}(E \cap \Omega, \Delta)$ so that $\varphi_E(a) = 0$ and

$$|\varphi_E(b)| = th\, C_{E \cap \Omega}(a, b) \quad [\text{resp.:}\ |(\hat{D}_a^1 \varphi_E)(c)| = c_{E \cap \Omega}(a; c)].$$

Now let Λ [resp.: λ] be the infimum of $C_{E \cap \Omega}(a, b)$ [resp.: $c_{E \cap \Omega}(a; c)$] for $E \in \mathscr{E}$, and Φ an ultrafilter on $\mathscr{E}$ containing $\mathscr{E}_F = \{E \in \mathscr{E} : E \supset F\}$ for all $F \in \mathscr{E}$: by Theorem 2.1.5.b, with $U = F \cap \Omega$, $\mathscr{B} = \{\varphi_{E|F \cap \Omega} : E \supset F\}$, there is a $\psi_F \in \mathscr{A}(F \cap \Omega, \mathbb{C})$ with the property that, for each compact set $K \subset F \cap \Omega$

and $\varepsilon > 0$, the set $\left\{ E \in \mathscr{E}_F : \sup_K |\psi_F - \varphi_E| \leq \varepsilon \right\}$ belongs to Φ.

This ψ_F is unique; on account of this uniqueness, the ψ_F are the restrictions to the $F \cap \Omega$ of a $\psi \in \mathscr{G}(\Omega, \mathbb{C})$ which satisfies $\psi(a) = 0$, $|\psi| < 1$ as in a) and

$$|\psi(b)| = th \, \Lambda \, [\text{resp.:} \, |(\hat{D}_a^1 \psi)(c)| = \lambda]$$

since the set of $E \in \mathscr{E}$ such that $C_{E \cap \Omega}(a, b) \leqslant \Lambda + \varepsilon$ [resp.: $c_{E \cap \Omega}(a; c) \leqslant \lambda + \varepsilon$] also belongs to Φ. $\quad\square$

Theorem 6.2.2.b, due to Dineen-Timoney-Vigué [Di Ti Vi], has important consequences for convex domains: see below, after the definitions of K_Ω and k_Ω by formulas (7) and (9).

Proposition 6.2.3. *Let Ω be simply connected, that is to say: given two continuous paths x_0, x_1: $[0, 1] \to \Omega$ with $x_0(0) = x_1(0) = a_0$, $x_0(1) = x_1(1) = a_1$, there is a continuous map y: $[0, 1]^2 \to \Omega$ such that $y(t, 0) = x_0(t)$ and $y(t, 1) = x_1(t)$ for all $t \in [0, 1]$, $y(0, u) = a_0$ and $y(1, u) = a_1$ for all $u \in [0, 1]$. Then the inequality in (2) is an equality: $G_\Omega = \ln \coth C_\Omega$.*

Proof. We claim that every $\varphi \in \mathscr{A}(\Omega, \mathbb{C})$ such that $\varphi(\Omega)$ is Green factorizes into $\varphi = h \circ \psi$, where $\psi \in \mathscr{A}(\Omega, \Delta)$ and $h \in \mathscr{A}(\Delta, \varphi(\Omega))$; by the Lindelöf principle, this will imply, for all $(a, b) \in \Omega^2$:

$$G_{\varphi(\Omega)}[\varphi(a), \varphi(b)] \geqslant G_\Delta[\psi(a), \psi(b)] \geqslant \ln \coth C_\Omega(a, b).$$

We first choose h as a conformal mapping of Δ onto $\varphi(\Omega)$: $h(\Delta) = \varphi(\Omega)$, $h' \neq 0$ on Δ and, for every simply connected open set $\sigma \subset \varphi(\Omega)$, the restriction of h to each connected component of $h^{-1}(\sigma)$ is a 1-1 map onto σ; given $a \in \Omega$, we choose $\alpha \in \Delta$ such that $\varphi(a) = h(\alpha)$. Then, for every continuous path x: $[0, 1] \to \Omega$ starting from a, i.e. $x(0) = a$, there is a unique continuous path ξ: $[0, 1] \to \Delta$ starting from α, such that $h \circ \xi = \varphi \circ x$; since Ω is simply connected, the end point $\beta = \xi(1)$ depends only on $b = x(1)$, say $\beta = \psi(b)$; ψ satisfies $\varphi = h \circ \psi$ and is analytic since $\psi = h^{-1} \circ \varphi$ on each $\varphi^{-1}(\sigma)$. $\quad\square$

If Ω is not simply connected, the inequality in (2) is in general strict (Ex. 6.2.6(B) below and Part b) of the Lindelöf principle 6.1(B)).

The *Kobayashi pseudodistance* is the *largest* pseudodistance on Ω such that, when Ω is endowed with this pseudodistance and Δ with the Poincaré distance Π, all maps $\in \mathscr{A}(\Delta, \Omega)$ are contractions (then it is Π on Δ by the Schwarz-Pick lemma). The exact analogue of C_Ω would be

(5) $\quad d_\Omega(a, b) = \inf\{\Pi(\alpha, \beta): \exists g \in \mathscr{A}(\Delta, \Omega), g(\alpha) = a, g(\beta) = b\};$

here again, by composition with an element of $\mathrm{Aut}(\Delta)$, one may prescribe α and the Poincaré geodesic starting from α on which β lies.

$d_\Omega(a,b)$ is finite if and only if maps g as in (5) exist: this is a certainty if $b \in a + \omega(a)$; in fact, if $R > 1$ is the radius of the open disc $\{\zeta \in \mathbb{C}: \zeta(b - a) \in \omega(a)\}$, we may take $g(\zeta) = a + \zeta R(b - a)$, hence $d_\Omega(a,b) \leqslant$ $\Pi\left(0, \dfrac{1}{R}\right) = th^{-1}\dfrac{1}{R}$; $d_\Omega(a,b) \to 0$ as $b \to a$. Also note that (2) and the Lindelöf principle yield

(6) $\ln \coth C_\Omega \geqslant G_\Omega \geqslant \ln \coth d_\Omega$, $C_\Omega \leqslant d_\Omega$ on Ω^2.

The following theorem due to Lempert [Lem]$_1$ shows that, if Ω is convex, d_Ω is a pseudodistance and therefore solves the problem of the largest pseudodistance on Ω making all maps $\in \mathscr{A}(\Delta, \Omega)$ contractions.

Theorem 6.2.4. *If Ω is convex (or the image of a convex open set by a 1-1 bianalytic map), then d_Ω is a continuous pseudodistance $\Omega^2 \to \mathbb{R}_+$.*

Proof. We only have to check the triangle inequality $d_\Omega(a,c) \leqslant d_\Omega(a,b) + d_\Omega(b,c)$: d_Ω finite everywhere will follow by the argument of connectedness already used for H_Ω and C_Ω (Prop. 4.4.3.a and 4.4.4.a).

Let $f, g \in \mathscr{A}(\Delta, \Omega)$, $a = f(0)$, $b = f(\beta) = g(\beta)$, $c = g(\gamma)$, $0 < \beta < \gamma < 1$: choosing a sequence $(r_n) \subset \,]\sqrt{\gamma}, 1[$ tending to 1, and setting

$$\lambda_n(\zeta) = \frac{\zeta - \gamma}{\zeta - \beta} \frac{\zeta - (r_n^2/\gamma)}{\zeta - (r_n^2/\beta)}, \qquad \zeta \in \Delta,$$

we get a meromorphic function λ_n with a single pole β in Δ, and

$$\lambda_n(r_n e^{i\theta}) = \frac{\beta}{\gamma} \frac{r_n^2 - 2\gamma r_n \cos\theta + \gamma^2}{r_n^2 - 2\beta r_n \cos\theta + \beta^2} \in \,]0, 1[\quad \text{for all } \theta \in \mathbb{R};$$

since $(f - g)(\beta) = 0$, $h_n = \lambda_n f + (1 - \lambda_n)g \in \mathscr{A}(\Delta, X)$.

Now let Ω contain the origin; the gauge q of Ω is convex and continuous (1.5.A), hence p.s.h. (4.1.5.f); $q \circ h_n \in \mathscr{P}(\Delta)$ and $q \circ h_n \leqslant \lambda_n(q \circ f) + (1 - \lambda_n)(q \circ g) < 1$ on the circumference of $\Delta(0, r_n)$, hence $q \circ h_n < 1$ on $\overline{\Delta}(0, r_n)$. Replacing r_n by 1, we obtain in the same way a meromorphic function λ and $h = \lambda f + (1 - \lambda)g \in \mathscr{A}(\Delta, X)$, $h(0) = f(0) = a$, $h(\gamma) = g(\gamma) = c$; since $h = \lim h_n$ pointwise, $q \circ h \leqslant 1$ on Ω, actually $q \circ h < 1$ by the maximum principle (Prop. 4.1.2.b), $h \in \mathscr{A}(\Delta, \Omega)$.

Finally $d_\Omega(a,c) \leqslant \Pi(0, \gamma) = \Pi(0, \beta) + \Pi(\beta, \gamma)$; for all $\varepsilon > 0$, f and β can be chosen so that $\Pi(0, \beta) \leqslant \varepsilon + d_\Omega(a, b)$, then g and γ so that $\Pi(\beta, \gamma) \leqslant \varepsilon + d_\Omega(b, c)$. $\square$

If Ω is not convex, even if it is star-shaped, in general d_Ω no longer satisfies the triangle inequality: the following counter-example is also due to Lempert [Lem]$_1$. In $\mathbb{C}^2$, let

$$\Omega_n = \left\{ (\xi, \eta) \in \Delta^2 : |\xi\eta| < \frac{1}{n} \right\}, \ n \in \mathbb{N}^*, \text{ and } a = \left(\frac{1}{3}, 0 \right), \ b = \left(0, \frac{1}{3} \right),$$

hence $d_{\Omega_n}(0, a) \leqslant th^{-1} \frac{1}{3}$, $d_{\Omega_n}(0, b) \leqslant th^{-1} \frac{1}{3}$ for all n. The triangle inequality would imply $d_{\Omega_n}(a, b) \leqslant th^{-1} \frac{3}{5}$ and the existence, for each $n \in \mathbb{N}^*$, of $g_n = (\varphi_n, \psi_n) \in \mathscr{A}(\Delta, \Omega_n)$, $\beta_n \in \bar{\Delta} \left(0, \frac{4}{5} \right)$, such that $|\varphi_n \psi_n| < \frac{1}{n}$, $g_n(0) = a$, $g_n(\beta_n) = b$, hence $\varphi_n(0) = \psi_n(\beta_n) = \frac{1}{3}$, then no subsequence of the sequence $(\varphi_n \psi_n)$ could converge to 0 uniformly on compact sets.

Therefore, in the general case, for the construction of the Kobayashi pseudodistance K_Ω one has to set

(7) $$K_\Omega(a, b) = \inf \left\{ \sum_{j=1}^{m} d_\Omega(a_{j-1}, a_j) : a_0 = a, a_1, \ldots, a_{m-1} \in \Omega, a_m = b \right\}$$

which indeed defines a continuous pseudodistance $\Omega^2 \to \mathbb{R}_+$; K_Ω and K_Γ make all maps $\in \mathscr{A}(\Omega, \Gamma)$ contractions, in particular $\Omega' \subset \Omega$ entails $K_{\Omega'} \geqslant K_\Omega$ on Ω'^2. Since C_Ω is a pseudodistance, from (6) follows

(8) $$C_\Omega \leqslant K_\Omega \leqslant d_\Omega \text{ on } \Omega^2.$$

If Ω is convex, the second inequality is an equality by Theorem 6.2.4, and the first one too as a consequence of Theorem 6.2.2.b. In fact, for a convex domain Ω in a finite dimensional space $\mathbb{C}^m$, Lempert has proved, by an elaborate argument [Lem]$_2$ which will not be presented here, $C_\Omega = K_\Omega$ on Ω^2; and, for a convex domain Ω in a l.c. space X, obviously $K_\Omega \leqslant K_{E \cap \Omega}$ on $(E \cap \Omega)^2$ if E is any finite dimensional affine subspace of X. The equality $C_\Omega = d_\Omega$ will be fundamental for the construction of complex geodesics in a convex domain (Th. 6.3.3.c).

The Poincaré distance $\Pi(\alpha, \beta)$ is also the infimum of $\int_\alpha^\beta \dfrac{|d\zeta|}{1 - |\zeta|^2}$ along all continuous paths, with piecewise continuous derivative, joining α, β in Δ; a similar relation can be expected, and will be obtained below (Th. 6.2.9.b), between K_Ω and k_Ω defined as follows on $\Omega \times X$:

(9) $k_\Omega(a; c)$

$$= \inf \left\{ \frac{|\tau|}{1 - |\alpha|^2} : \alpha \in \Delta, \tau \in \mathbb{C}, \exists g \in \mathscr{A}(\Delta, \Omega), g(\alpha) = a, \tau . g'(\alpha) = c \right\}$$

$$= \inf \{ |\tau| : \tau \in \mathbb{C}, \exists g \in \mathscr{A}(\Delta, \Omega), g(0) = a, \tau . g'(0) = c \}$$

since, for $h \in \mathrm{Aut}(\Delta)$ with $\alpha = h(\beta)$, $(g \circ h)'(\beta) = g'(\alpha)h'(\beta)$ and $h'(\beta)$ is any complex number with modulus $= \dfrac{1 - |\alpha|^2}{1 - |\beta|^2}$; for the same reason, one may replace $\tau \in \mathbb{C}$ by $\tau \in \mathbb{R}_+$ in (9), and $k_\Omega(a; \zeta c) = |\zeta| k_\Omega(a; c)$ for all $\zeta \in \mathbb{C}$.

Maps g as in (9) certainly exist: if R is the radius of the open disc $\{\zeta \in \mathbb{C}: \zeta c \in \omega(a)\}$, one may take $g(\zeta) = a + \dfrac{\zeta}{\tau} c$ if $|\tau| \geqslant \dfrac{1}{R}$, hence $k_\Omega(a; c) \leqslant \dfrac{1}{R}$. If a g in (9) satisfies $g(0) = a$, by Theorem 3.1.10: for any $\varphi \in \mathscr{A}(\Omega, \Delta)$ satisfying $\varphi(a) = 0$ we have $(\varphi \circ g)'(0) = (\hat{D}_a^1 \varphi)(g'(0))$, hence

(10) $c_\Omega \leqslant k_\Omega$ on $\Omega \times X$

by the Schwarz lemma, and the same relation with any $f \in \mathscr{A}(\Omega, \Gamma)$ instead of φ, hence

(11) $k_\Gamma[f(a); (\hat{D}_a^1 f)(c)] \leqslant k_\Omega(a; c)$ for all $a \in \Omega$, $c \in X$

to be compared with (4). *If Ω is convex, the equality in (10) is another consequence of Theorem 6.2.2.b; if not, see Proposition 6.2.7 below.*

Proposition 6.2.5. *Given V in $\mathbb{C}$, Ω in X, let there exist $g \in \mathscr{A}(V, \Omega)$ and $\varphi \in \mathscr{A}(\Omega, V)$ such that $\varphi \circ g = \mathrm{id}_V$; then every one of the above defined quantities for Ω is related as follows to the same quantity for V: for all $(\alpha, \beta) \in V^2$*

(12) $G_\Omega[g(\alpha), g(\beta)] = G_V(\alpha, \beta)$;

(13) $C_\Omega[g(\alpha), g(\beta)] = C_V(\alpha, \beta)$, $K_\Omega[g(\alpha), g(\beta)] = K_V(\alpha, \beta)$;

(14) $c_\Omega[g(\alpha); g'(\alpha)] = c_V(\alpha; 1)$, $k_\Omega[g(\alpha); g'(\alpha)] = k_V(\alpha; 1)$.

Proof. The inequalities $\geqslant$ in (12), $\leqslant$ in (13) and (14), do not depend on the existence of φ; the opposite inequalities are obtained by setting $a = g(\alpha)$, $b = g(\beta)$, hence $\alpha = \varphi(a)$, $\beta = \varphi(b)$, $(\hat{D}_a^1 \varphi)(g'(\alpha)) = 1$. $\square$

Examples 6.2.6. (A) Let Ω be balanced and convex: its gauge $q \in \mathrm{csn}(X)$ (1.5.A); given $b \in \Omega$ with $0 < q(b) < 1$, there is a $\varphi \in X'$ such that $|\varphi| \leqslant q$, $\varphi(b) = q(b)$, and Proposition 6.2.5 with $V = \Delta$, $g(\zeta) = \dfrac{\zeta}{q(b)} b$, gives

(15) $C_\Omega(0, b) = K_\Omega(0, b) = th^{-1} q(b)$, $c_\Omega(0; c) = k_\Omega(0; c) = q(c)$

since $c_\Delta(0; 1) = k_\Delta(0; 1) = 1$ by the Schwarz lemma.

If $q(a) = \rho < 1$, then $(1 - \rho)\Omega \subset \Omega - a \subset (1 + \rho)\Omega$, and the invariance of both pseudodistances under translations, give

(16) $\qquad th^{-1}\dfrac{q(b-a)}{1+q(a)} \leqslant C_\Omega(a,b) = K_\Omega(a,b) \leqslant th^{-1}\dfrac{q(b-a)}{1-q(a)}.$

(B) Given U in $\mathbb{C}$ and $x'_0 \in X'$, let $\Omega = x'^{-1}_0(U)$; if $a, b \in \Omega$ are such that $\alpha = x'_0(a) \neq x'_0(b) = \beta$ (the only case of interest), Proposition 6.2.5 with $V = U$, $g(\zeta) = \dfrac{\beta-\zeta}{\beta-\alpha}a + \dfrac{\zeta-\alpha}{\beta-\alpha}b$, shows that G_Ω, C_Ω, K_Ω have the same values on (a,b) as G_U, C_U, K_U on (α, β), c_Ω and k_Ω have the same values on (a,c) as c_U and k_U on $(x'_0(a), x'_0(c))$.

(C) Now let Ω be convex and contain the origin, Q (a Minkowski functional, see 1.5(A)) and q (a seminorm) be the gauges of Ω and its largest balanced subset ω respectively. Given $c \in X$ with $q(c) > 0$, we can choose $\theta \in \mathbb{R}$ so that $\dfrac{e^{i\theta}}{q(c)}c$ lies on the boundary of Ω, i.e. $Q(e^{i\theta}c) = q(c)$, then, by the Hahn-Banach theorem ([Du Sch], Chap. II, §3) a continuous $\mathbb{R}$-linear functional $u: X \to \mathbb{R}$ such that $u \leqslant Q \leqslant q$ on X, but $u(e^{i\theta}c) = q(c)$, hence $u(ie^{i\theta}c) = 0$ since q is a seminorm. The corresponding $x'_0 \in X'$ defined by $x'_0(x) = u(x) - iu(ix)$ for all $x \in X$ satisfies $|x'_0| \leqslant q$ on X, $\mathscr{R}e\, x'_0 < 1$ on Ω and $x'_0(e^{i\theta}c) = q(c) = |x'_0(c)|$; setting $U = \{\zeta \in \mathbb{C}: \mathscr{R}e\, \zeta < 1\}$ we have $\omega \subset \Omega \subset x'^{-1}_0(U)$, hence by (A) and (B): $q(c) = c_\omega(0;c) \geqslant c_\Omega(0;c) \geqslant c_{x'^{-1}_0(U)}(0;c)$
$= c_U(0; x'_0(c)) = \dfrac{1}{2}q(c).$

Given $(a,b) \in \Omega^2$ with $q(a-b) > 0$, the same argument with $c = a - b$ gives an inequality due to Harris [Har]₁:

$$th\, C_\Omega(a,b) \geqslant th\, C_{x'^{-1}_0(U)}(a,b) = th\, C_U[x'_0(a), x'_0(b)]$$
$$= \left|\frac{x'_0(a) - x'_0(b)}{2 - x'_0(a) - \overline{x'_0(b)}}\right|$$

implies, since $2 - \alpha - \bar{\beta} = 2(1 - \mathscr{R}e\,\alpha) + \overline{\alpha - \beta}$ for all $(\alpha, \beta) \in \mathbb{C}^2$:

(17) $\qquad th\, C_\Omega(a,b) \geqslant \dfrac{q(a-b)}{2[1 - \mathscr{R}e\, x'_0(a)] + q(a-b)}$
$$\geqslant \dfrac{q(a-b)}{2[1 + q(a)] + q(a-b)}.$$

Finally, if $Q(b) > 0$, we can choose $u \leqslant Q$ on X with $u(b) = Q(b)$, and obtain

(18) $\qquad th\, C_\Omega(0,b) \geqslant \left|\dfrac{x'_0(b)}{2 - x'_0(b)}\right| \geqslant \dfrac{Q(b)}{2 - Q(b)}$

which proves that $C_\Omega(a, b) \to +\infty$, for a given $a \in \Omega$, as b tends to a boundary point of Ω.

(D) If Ω is balanced, pseudoconvex but not convex, two accurate computations, the second one due to Barth [Ba], show that the inequality in (10) is in general strict:

Proposition 6.2.7. *Let Ω balanced.*

a) $c_\Omega(0; .)$ *is the gauge of the convex hull* $\hat{\Omega}$ *of* Ω.
b) *If* Ω *is pseudoconvex,* $k_\Omega(0; .)$ *is the gauge of* Ω.

Proof. Let $\hat{q}$ (a seminorm) and q be the gauges of $\hat{\Omega}$ and Ω.
a) We already know that $c_\Omega(0; c) \leqslant k_\Omega(0; c) \leqslant q(c)$ and $c_\Omega(0; c) \geqslant c_{\hat{\Omega}}(0; c) = \hat{q}(c)$ for all $c \in X$. Now let $\varepsilon > 0$ and $1 - \varepsilon \leqslant \hat{q}(c) < 1$: since $c \in \hat{\Omega}$, there are m points $c_j \in \Omega$ and m numbers $\alpha_j > 0$ such that $\sum_{j=1}^{m} \alpha_j = 1$ and $c = \sum_{j=1}^{m} \alpha_j c_j$;
then $\hat{q}(c) \leqslant c_\Omega(0; c) \leqslant \sum_{j=1}^{m} \alpha_j c_\Omega(0; c_j) \leqslant \sum_{j=1}^{m} \alpha_j q(c_j) < 1 \leqslant \dfrac{\hat{q}(c)}{1 - \varepsilon}$. From this follows $c_\Omega(0; c) = \hat{q}(c)$ for all $c \in X$, since both members are continuous seminorms, if $\hat{q} \not\equiv 0$.

If $\hat{q} \equiv 0$ or $\hat{\Omega} = X$: $c_\Omega(0; c) \not\equiv 0$ would imply the existence of $\varphi \in \mathscr{A}(\Omega, \Delta)$ such that $\varphi(0) = 0$, $\hat{D}_0^1 \varphi \not\equiv 0$; but, by formula (2) in Theorem 2.3.5, $|\varphi| < 1$ on Ω implies $|\hat{D}_0^1 \varphi(x)| < 1$ for all $x \in \Omega$, hence also for all $x \in \hat{\Omega}$.
b) Ω pseudoconvex entails $q \in \mathscr{P}(X)$ (and conversely by Prop. 5.1.4) since, with the notation of Definition 5.1.1.:

$$q(y) = \frac{1}{\delta(0, y)} = \exp[-\ln \delta(0, y)] \quad \text{for all } y \in X.$$

If $g \in \mathscr{A}(\Delta, \Omega)$ satisfies $g(0) = 0$, $\tau \cdot g'(0) = c$: $g_1(\zeta) = \dfrac{g(\zeta)}{\zeta}$ defines a map $g_1 \in \mathscr{A}(\Delta, X)$, $q \circ g_1 \in \mathscr{P}(\Delta)$, $q \circ g_1(0) \leqslant 1$ by the maximum principle, i.e. $|\tau| \geqslant q(c)$. $\square$

Corollary 6.2.8. *Let Ω be balanced and pseudoconvex in X, f a 1-1 bianalytic map of Ω onto Γ.*

a) *If Γ is convex, Ω too.*
b) *If Γ also is balanced pseudoconvex and $f(0) = 0$, then $\Gamma = (\hat{D}_0^1 f)(\Omega)$.*

Proof. a) Let $b = f(0)$: by (11), $k_\Omega(0; c) = k_\Gamma[b; (\hat{D}_0^1 f)(c)]$ for all $c \in X$; since Γ is convex, $k_\Gamma(b; .) = c_\Gamma(b; .)$ is a seminorm, then $k_\Omega(0; .)$, the gauge of Ω, and Ω are convex.

b) $c \in \Omega$ (resp.: $c' \in \Gamma$) if and only if $k_\Omega(0; c) < 1$ [resp. $k_\Gamma(0; c') < 1$].   □

In the next theorem, part b) means that K_Ω is the *integrated form* of k_Ω; the proof below was essentially given by Royden [Ro] in a finite dimensional space; that it goes over to l.c. spaces was pointed out by Vesentini [Ve]$_1$.

Theorem 6.2.9. a) k_Ω *is upper semicontinuous on* $\Omega \times X$.

b) *For* $(a, b) \in \Omega^2$, *let* $K_\Omega^i(a, b)$ *be the infimum of the integrals* $\int_0^1 k_\Omega(x(t); x'(t)) \, dt$ *along all continuous paths* $x: [0, 1] \to \Omega$, *with piecewise continuous derivative, joining* a *and* b: $x(0) = a$, $x(1) = b$. *Then* $K_\Omega = K_\Omega^i$ *on* Ω^2.

Since $k_\Omega(a; .)$ is 1-homogeneous, the integral does not depend on the parametrization of the path x, and K_Ω^i is symmetric.

Proof. a) Let $k_\Omega(a_0; c_0) < \lambda$, $g_0 \in \mathscr{A}(\Delta, \Omega)$, $g_0(0) = a_0$, $\tau \cdot g_0'(0) = c_0$, $|\tau| < \lambda$; choosing $r \in {]0, 1[}$ such that $\left| \dfrac{\tau}{r} \right| < \lambda$ and substituting, if necessary, $\zeta \mapsto g_0(r\zeta)$ for g_0 and $\dfrac{\tau}{r}$ for τ, we may assume that g_0 is also a continuous map $\bar{\Delta} \to \Omega$. Now, for $(a - a_0)$ and $(c - c_0)$ sufficiently near the origin, $g(\zeta) = g_0(\zeta) + (a - a_0) + \dfrac{\zeta}{\tau}(c - c_0)$ defines $g \in \mathscr{A}(\Delta, \Omega)$ with $g(0) = a, \tau \cdot g'(0) = c$.

b$_1$) We first claim that $K_\Omega^i[g(\alpha), g(\beta)] \leqslant \Pi(\alpha, \beta)$ for all $(\alpha, \beta) \in \Delta^2$ and $g \in \mathscr{A}(\Delta, \Omega)$, which will ensure $K_\Omega^i \leqslant K_\Omega$ since obviously K_Ω^i is a pseudo-distance on Ω. In fact $k_\Delta(\alpha; \gamma) = \dfrac{|\gamma|}{1 - |\alpha|^2}$ for all $(\alpha, \gamma) \in \Delta \times \mathbb{C}$ by the Pick lemma and

$$\Pi(\alpha, \beta) = \int_\alpha^\beta \frac{|d\xi|}{1 - |\xi|^2} = \int_0^1 k_\Delta(\xi(t); \xi'(t)) \, dt$$

along the Poincaré geodesic ξ joining α, β; setting $x = g \circ \xi$ we have $k_\Omega(x(t); x'(t)) \leqslant k_\Delta(\xi(t); \xi'(t))$ by (11).

b$_2$) Now let x be a $\mathscr{C}^1$ path $[0, 1] \to \Omega$: we claim that, given $t \in [0, 1]$ and $\varepsilon > 0$, the estimate

$$K_\Omega(x(t); x(u)) \leqslant [\varepsilon + k_\Omega(x(t); x'(t))] |u - t|$$

holds for u sufficiently near t. The first item in the argument is $p[x(u) - x(t) - (u - t)x'(t)] = o(|u - t|)$ as $u \to t$ for all $p \in \mathrm{csn}(X)$, which we write for brevity $x(u) - x(t) = (u - t)x'(t) + o(|u - t|)$. If $x'(t) = 0$, we choose $p \in \mathrm{csn}(X)$ so that Ω contains $\Omega' = x(t) + p^{-1}([0, 1[)$; then, for u sufficiently near t, $x(u) \in \Omega'$ and (Ex. 6.2.6(A))

$$K_\Omega(x(t), x(u)) \leqslant K_{\Omega'}(x(t), x(u)) = th^{-1} p(x(u) - x(t)) = o(|u - t|).$$

If $x'(t) \neq 0$, we choose: $x(t)$ as the origin, set for brevity $x(u) = a$, $x'(t) = c$; choose a $g \in \mathcal{A}(\Delta, \Omega)$ which is also a continuous map $\bar{\Delta} \to \Omega$ and satisfies $g(0) = 0$, $\tau \cdot g'(0) = c$, with $0 < |\tau| < \varepsilon + k_\Omega(0; c)$; then $\ell \in X'$ such that $\ell(c) = \tau$, which entails $(\ell \circ g)'(0) = 1$ and the existence of $\rho \in {]0, 1]}$ such that the restriction of $\ell \circ g$ to $\delta = \Delta(0, \rho)$ is a 1-1 map of δ onto another open neighbourhood δ' of 0 in $\mathbb{C}$; let h be the reciprocal map of δ' onto δ and finally ω an open balanced convex neighbourhood of the origin in $\ker \ell$, such that $g(\Delta) + \omega \subset \Omega$.

Then $(\zeta, z) \mapsto g(\zeta) + z$ is an analytic map $\Delta \times \omega \to g(\Delta) + \omega$ and a 1-1 map of $\delta \times \omega$ onto an open neighbourhood W of the origin in Ω, namely

$$W = \{x \in \Omega: \ell(x) \in \delta', x - g \circ h \circ \ell(x) \in \omega\},$$

whose reciprocal map is $x \mapsto (h \circ \ell(x), x - g \circ h \circ \ell(x))$. If u is sufficiently near t, $a = x(u)$ lies in W, where it is the image of $b \in \delta \times \omega$, $K_\Omega(0, a) \leqslant K_{g(\Delta)+\omega}(0, a) \leqslant K_{\Delta \times \omega}(0, b)$ and $th\, K_{\Delta \times \omega}(0, b)$ is (Ex. 6.2.6(A)) the gauge of $\Delta \times \omega$ at b. In order to estimate this gauge, we write $a = (u - t)c + o(|u - t|)$, hence $b = ((u - t)\tau + o(|u - t|), o(|u - t|))$, $\dfrac{th\, K_{\Delta \times \omega}(0, b)}{|u - t|} \to |\tau|$ as $u \to t$.

b_3) Given a continuous path $x: [0, 1] \to \Omega$ with piecewise continuous derivative, $\{(x(t), x'_g(t)): 0 < t \leqslant 1\} \cup \{x(t), x'_d(t)): 0 \leqslant t < 1\}$ is a compact set in $\Omega \times X$, on which k_Ω is bounded. By the Fatou lemma, for all $\varepsilon > 0$ there is $\eta > 0$ such that, whenever the interval $[0, 1]$ is partitioned by $t_0 = 0, u_0, t_1, u_1, \ldots, t_{m-1}, u_{m-1}, t_m = 1$ into intervals with lengths $\leqslant \eta$, the function equal, on each partial interval, to the upper bound of $k_\Omega(x(t); x'(t))$ on the closure of this interval, has an integral $I \leqslant \varepsilon + \int_0^1 k_\Omega(x(t); x'(t))\, dt$.

But, by part b_2), the partition can be chosen so that $I + \varepsilon \geqslant \sum_{j=1}^{m} K_\Omega(x(t_j), x(u_{j-1})) + \sum_{j=0}^{m-1} K_\Omega(x(t_j), x(u_j)) \geqslant K_\Omega(x(0), x(1))$. $\square$

Corollary 6.2.10. *The integrated form C_Ω^i of c_Ω satisfies $C_\Omega \leqslant C_\Omega^i \leqslant K_\Omega$ on Ω^2.*

Proof. We only have to prove the first inequality, which may be strict if Ω is not convex $[\text{Vi}]_3$. Let x be a continuous path $[0, 1] \to \Omega$ with piecewise continuous derivative: for $\varphi \in \mathcal{A}(\Omega, \Delta)$

$$\Pi(\varphi \circ x(0), \varphi \circ x(1)) \leqslant \int_0^1 \frac{|(\varphi \circ x)'(t)|}{1 - |\varphi \circ x(t)|^2}\, dt$$

and $\dfrac{|(\varphi \circ x)'(t)|}{1 - |\varphi \circ x(t)|^2} = c_\Delta[(\varphi \circ x)(t); (\varphi \circ x)'(t)] \leqslant c_\Omega(x(t); x'(t))$ for all $t \in [0, 1]$ by (4). $\quad\square$

6.3 Complex geodesics and complex extremal points

The first notion, due to Vesentini [Ve]$_2$, was investigated by him [Ve]$_{3,4}$ and Vigué [Vi]$_4$; the second one was introduced by Thorp-Whitley [Th Wh].

Definition and Proposition 6.3.1. a) *A map* $g \in \mathscr{A}(\Delta, \Omega)$ *is a complex geodesic in* Ω *if it has the equivalent properties*

(1) $\qquad C_\Omega[g(\alpha), g(\beta)] = \Pi(\alpha, \beta)$ *for one pair* $(\alpha, \beta) \in \Delta^2$, $\alpha \neq \beta$, *or any*

$\qquad$ *pair* $(\alpha, \beta) \in \Delta^2$;

(2) $\qquad c_\Omega[g(\alpha); g'(\alpha)] = \dfrac{1}{1 - |\alpha|^2}$ *for one or any* $\alpha \in \Delta$.

b) *Such a map* g *also has the properties*

(3) $\qquad K_\Omega[g(\alpha), g(\beta)] = \Pi(\alpha, \beta) = d_\Omega[g(\alpha), g(\beta)]$ *for all* $(\alpha, \beta) \in \Delta^2$,

(4) $\qquad k_\Omega[g(\alpha); g'(\alpha)] = \dfrac{1}{1 - |\alpha|^2}$ *for all* $\alpha \in \Delta$.

From (1) and (3) follows $C_\Omega = C_\Omega^i = K_\Omega$ on $[g(\Delta)]^2$.

Proof. Let (1) hold for one pair $(\alpha_0, \beta_0) \in \Delta^2$, $\alpha_0 \neq \beta_0$, or (2) for $\alpha_0 \in \Delta$. By Theorem 6.2.2.a, there is a $\varphi \in \mathscr{A}(\Omega, \Delta)$ such that

$$C_\Omega[g(\alpha_0), g(\beta_0)] = \Pi[\varphi \circ g(\alpha_0), \varphi \circ g(\beta_0)] \quad \text{or}$$

$$c_\Omega[g(\alpha_0); g'(\alpha_0)] = \frac{|(\varphi \circ g)'(\alpha_0)|}{1 - |\varphi \circ g(\alpha_0)|^2};$$

then $\varphi \circ g \in \mathrm{Aut}(\Delta)$, $\psi = (\varphi \circ g)^{-1} \circ \varphi$ is another map $\in \mathscr{A}(\Omega, \Delta)$ such that $\psi \circ g = id_\Delta$, hence (1) through (4) by Proposition 6.2.5. $\quad\square$

The next theorem [Ve]$_3$ shows that a complex geodesic is determined by its range, up to composition with an element of $\mathrm{Aut}(\Delta)$, with a few consequences.

Theorem 6.3.2. a) *If 2 complex geodesics g_1, g_2 in Ω have the same range, i.e. $g_1(\Delta) = g_2(\Delta)$, there is $h \in \mathrm{Aut}(\Delta)$ such that $g_1 = g_2 \circ h$.*
b) *If moreover $g_1(\alpha) = g_2(\alpha)$, $g_1(\beta) = g_2(\beta)$ for some pair $(\alpha, \beta) \in \Delta^2$, $\alpha \neq \beta$, or $g_1(\alpha) = g_2(\alpha)$, $g_1'(\alpha) = g_2'(\alpha)$ for some $\alpha \in \Delta$, then $g_1 \equiv g_2$.*
c) *Let g be a complex geodesic in Ω and $f \in \mathcal{A}(\Omega, \Omega)$ satisfy $f[g(\Delta)] = g(\Delta)$: if 2 distinct points $\in g(\Delta)$ are fixed points of f, or one point $g(\alpha)$ a fixed point of f and $g'(\alpha)$ a fixed vector of the linear map $\hat{D}_{g(\alpha)}^1 f$, then $g(\zeta)$ is a fixed point of f for all $\zeta \in \Delta$.*

Proof. a) Since g_1, g_2 are 1-1 maps of Δ onto their common range, also isometries with respect to Π and C_Ω, there is a 1-1 map h of Δ onto Δ, also an isometry with respect to Π, hence a homeomorphism, such that $g_1 = g_2 \circ h$. If $x' \in X'$ is such that $x' \circ g_2 \neq \mathrm{const}$, the relation $(x' \circ g_1) = (x' \circ g_2) \circ h$ proves that h is holomorphic on the complement of the set of zeroes of $(x' \circ g_2)'$, a discrete set.
b) In the first case, h has 2 distinct fixed points; in the second case, $h(\alpha) = \alpha$ and $h'(\alpha) = 1$.
c) $f \circ g$ is a complex geodesic in Ω, with the same range as g. $\square$

We now discuss the existence of a complex geodesic in Ω joining distinct given points $a, b \in \Omega$: it cannot be a general property in a nonconvex Ω since there the inequalities $C_\Omega \leqslant C_\Omega^i$, $K_\Omega \leqslant d_\Omega$ may be strict. Moreover, the existence of a complex geodesic joining a, b implies $C_\Omega(a, b) > 0$, hence $G_\Omega(a, b) < \infty$, i.e. the existence of a $\varphi \in \mathcal{A}(\Omega, \mathbb{C})$ such that $\varphi(\Omega)$ is a Green set and $\varphi(a) \neq \varphi(b)$; then $\{\zeta \in \mathbb{C}: a + \zeta(b - a) \in \Omega\}$, if connected, is Green as well, and some assumption ensuring this is necessary. Finally, the proof given below depends on the fact that, given $a \in \Omega$, $\mathcal{B} = \{g \in \mathcal{A}(\Delta, \Omega): g(0) = a\}$ is a bounded set in $\mathcal{A}(\Delta, X)$ (see 2.1(E)); but this implies $\{x \in \Omega: d_\Omega(a, x) < \lambda\}$ bounded for all $\lambda > 0$, and this is an open set in a convex Ω since d_Ω is continuous (Th. 6.2.4).

This discussion accounts for the assumptions of the next theorem; the assumption made on Ω is obviously satisfied by any bounded Ω, but also by some unbounded Ω (Ex. 6.3.4). As mentioned above (Ex. 6.2.6(C)), the inequality in a) is due to Harris [Har]$_2$.

Theorem 6.3.3. *Let X be a Banach space, Ω a convex open set and $a_0 \in \Omega$ be such that $\omega(a_0)$ (the largest balanced subset of $\Omega - a_0$) is bounded. Then:*
a) *For all $(a, b) \in \Omega^2$,*

$$\mathrm{th}\, C_\Omega(a, b) \geqslant \frac{\|a - b\|}{\|a - b\| + 2R_0(1 + \|a - a_0\|/r_0)},$$

where r_0 (resp.: R_0) is the radius of the largest (resp. smallest) open ball

centred at the origin contained in (resp.: containing) $\omega(a_0)$; therefore C_Ω is a distance defining on Ω the initial topology, and $\omega(a)$ is bounded for all $a \in \Omega$;
b) For all $a \in \Omega$, $c_\Omega(a, .)$ is a norm equivalent to the initial one.
c) Let, moreover, X be reflexive; then, given $(a, b) \in \Omega^2$, $a \neq b$ [resp.: $(a, c) \in \Omega \times X$, $c \neq 0$], there is a complex geodesic $g \in \mathscr{A}(\Delta, \Omega)$ such that $a, b \in g(\Delta)$ [resp.: $g(0) = a, \tau . g'(0) = c, \tau > 0$].

Proof. a) The inequality ensues from (17) since the gauge q_0 of $\omega(a_0)$ satisfies $\dfrac{\|x\|}{R_0} \leqslant q_0(x) \leqslant \dfrac{\|x\|}{r_0}$ for all $x \in X$, and in turn it entails $\omega(a)$ contained in the open ball with centre at the origin, radius $R = 2R_0(1 + \|a - a_0\|/r_0)$. In fact, if $tx \in \omega(a)$, $t > 0$, and q denotes the gauge of $\omega(a)$:

$$\frac{\|tx\|}{R + \|tx\|} \leqslant th\, C_\Omega(a, a + tx) \leqslant th\, C_{\omega(a)}(0, tx) = q(tx),$$

hence $\|x\| \leqslant Rq(x)$ by letting $t \to 0$.
b) Example 6.2.6(C) includes also the inequality $c_\Omega(a; c) \geqslant \dfrac{1}{2} q(c) \geqslant \dfrac{\|c\|}{2R}$.

c) We shall use the equalities $C_\Omega = d_\Omega$, $c_\Omega = k_\Omega$, which hold in a convex domain Ω. For all $n \in \mathbb{N}^*$, let $g_n \in \mathscr{A}(\Delta, \Omega)$, $g_n(0) = a$ and: in the first case,

$$g_n(\beta_n) = b, th\, C_\Omega(a, b) \leqslant \beta_n \leqslant \frac{1}{n} + th\, C_\Omega(a, b); \text{ in the second case, } \tau_n . g'_n(0) =$$

$c, c_\Omega(a; c) \leqslant \tau_n \leqslant \dfrac{1}{n} + c_\Omega(a; c)$. For $|\zeta| \leqslant \rho < 1$, we have

$$C_\Omega[a; g_n(\zeta)] = d_\Omega[g_n(0), g_n(\zeta)] \leqslant \Pi(0, \zeta) \leqslant th^{-1}\rho,$$

hence, by the inequality in a), $\|g_n(\zeta) - a\| \leqslant$ some number depending only on ρ: the sequence (g_n) is bounded in $\mathscr{A}(\Delta, X)$ and, by Theorem 2.1.5.b, there is a $g \in \mathscr{A}(\Delta, X)$ with the property: for all $x' \in X'$, $x' \circ (g - g_{n_k}) \to 0$ uniformly on compact sets, for a subsequence (g_{n_k}) depending on x'.
   Then $g(0) = a$; by the Hahn-Banach separation theorem [Du Sch], $g(\Delta) \subset \bar{\Omega}$, i.e. $Q \circ g \leqslant 1$ on Δ if Q is the gauge of Ω, but actually $Q \circ g < 1$ since $Q \circ g \in \mathscr{P}(\Delta)$; moreover: in the first case, $\beta_n \to \beta = th\, C_\Omega(a, b)$, $a = g(0)$, $b = g(\beta)$; in the second case, $\tau_n \to \tau = c_\Omega(a; c)$, $\tau . g'(0) = c$, $c_\Omega[g(0)$; $g'(0)] = 1$. Either result proves that g is a complex geodesic. $\square$

Example 6.3.4. Let X be a Hilbert space and $(x_i)_{i \in I}$ the set of coordinates of $x \in X$ in an orthonormal basis of X; let $\mathscr{R}e\, x$ (resp.: $\mathscr{I}m\, x$) denote the point of X whose coordinates are the $\mathscr{R}e\, x_i$ (resp.: $\mathscr{I}m\, x_i$). Since $\|\mathscr{R}e\, y\| \leqslant \|\mathscr{R}e\, x\| + \|y - x\|$, $\Omega = \{x \in X: \|\mathscr{R}e\, x\| < 1\}$ is an open unbounded convex set in X, but $c \in \omega(a)$ implies $\|\mathscr{R}e\,c\| < 1$, $\|\mathscr{I}m\,c\| < 1$, $\|c\| < \sqrt{2}$: all sets $\omega(a)$, $a \in X$, are contained in a same ball.

Now let $\mathscr{R}e^+\,x$ be the point of X whose coordinates are the $\mathscr{R}e^+\,x_i = \sup(0, \mathscr{R}e\,x_i)$. Since $\|\mathscr{R}e^+(x+y)\| \leqslant \|\mathscr{R}e^+\,x + \mathscr{R}e^+\,y\|$, hence $\|\mathscr{R}e^+\,y\| \leqslant \|\mathscr{R}e^+\,x\| + \|y - x\|$, again $\Omega' = \{x \in X : \|\mathscr{R}e^+\,x\| < 1\}$ is an open unbounded convex set in X, each set $\omega'(a)$ is bounded because $c \in \omega'(a)$ implies $\|\mathscr{R}e^+(\pm c)\| < 1 + \|\mathscr{R}e^+(-a)\|$, $\|\mathscr{R}e^+(\pm ic)\| < 1 + \|\mathscr{R}e^+(-a)\|$, but the $\omega'(a)$ are no longer bounded together because $\mathscr{R}e^+\,a = \mathscr{I}m\,a = 0$ implies $a \in \omega'(a)$.

Definition 6.3.5. *Let C be a closed convex set in X. A boundary point a of C is*: a) *a real extremal point of C if $a \pm b \in C$ implies $b = 0$*; b) *a complex extremal point of C if $a + \bar{\Delta}b \subset C$ implies $b = 0$ or, equivalently, $\{a + b, a + jb, a + j^2 b\} \subset C$ implies $b = 0$.*

A real extremal point is, obviously, a complex one; a counter-example to the converse statement is the closed unit ball of $\ell^1(\mathbb{N}*)$. In fact, with the notation $\|x\| = \sum_{k \in \mathbb{N}*} |\xi_k|$ for the generic point: $\|a + b\| = \|a\| + \|b\|$ if and only if β_k / α_k is real > 0 whenever $\alpha_k \beta_k \neq 0$, therefore $\|a + \zeta b\| = \|a - \zeta b\| = \|a\| = 1$ for all $\zeta \in \bar{\Delta}$ demands $b = 0$; every boundary point a of the ball is complex extremal, but a is not real extremal if $\alpha_k \neq 0$ for at least 2 indeces k.

The next theorem is due to Thorp-Whitley [Th Wh] and Vesentini [Ve]₃.

Theorem 6.3.6. *Let U be a connected open set in $\mathbb{C}$, $\alpha \in U$ and $g \in \mathscr{A}(U, X)$ nonconstant: there is a $b \in X \backslash \{0\}$ such that the closed convex hull of $g(U)$ contains $g(\alpha) + \Delta b$; therefore $g(U)$ cannot contain any complex extremal point of its closed convex hull.*

Proof. Buckholtz [Bu] remarked that, for all $n \in \mathbb{N}*$, all roots of the polynomial $P_n(\zeta) = \sum_{k=0}^{n} \frac{(n\zeta)^k}{k!}$ lie in $\mathbb{C} \backslash \bar{\Delta}(0, \gamma)$, where $\gamma \in {]}0,1{[}$ is defined by $1 + \gamma + \ln \gamma = 0$: in fact $|\zeta| \leqslant \gamma$ entails $|\zeta e^{1-\zeta}| \leqslant 1$ and

$$|1 - e^{-n\zeta} P_n(\zeta)| = \left| (\zeta e^{1-\zeta})^n \sum_{k>n} \frac{n^k e^{-n}}{k!} \zeta^{k-n} \right| < 1 - e^{-n} P_n(1) < 1.$$

From this ensues that the roots $\beta_{\nu,n}$ of the polynomial

$$\zeta^n P_n\left(-\frac{1}{\zeta} \right) \equiv \zeta^n - n\zeta^{n-1} + \frac{n^2}{2!} \zeta^{n-2} + \cdots + \frac{(-n)^n}{n!}$$

lie in $\Delta\left(0, \dfrac{1}{\gamma}\right)$ and by the Newton identities they satisfy

$$\sum_{v=1}^{n} \beta_{v,n} = n, \quad \sum_{v=1}^{n} \beta_{v,n}^{k} = 0 \quad \text{for } k = 2,, \ldots, n.$$

Now let $g(\alpha) + \sum_{k\in\mathbb{N}^*} c_k(\zeta - \alpha)^k$ be the Taylor expansion of g around α and $U \supset \bar{\Delta}(\alpha, \rho)$: for all $p \in \mathrm{csn}(X)$, $p(c_k) \leqslant \dfrac{M(p)}{\rho^k}$ for all $k \in \mathbb{N}^*$ (Th. 2.1.2.b), hence, for $|\zeta - \alpha| < \gamma\rho$,

$$g(\alpha) + c_1(\zeta - \alpha) = \lim_{n\to\infty} \frac{1}{n} \sum_{v=1}^{n} g[\alpha + \beta_{v,n}(\zeta - \alpha)],$$

$g(\alpha) + c_1(\zeta - \alpha)$ belongs to the closed convex hull of $g(U)$, and $b = \gamma\rho c_1$ has the required property if $c_1 \neq 0$. If $c_1 = 0$ but $c_m \neq 0$, just replace $g(\zeta)$ by $\dfrac{1}{m}\sum_{\mu=1}^{m} g[\alpha + \omega_\mu(\zeta - \alpha)^{1/m}]$, where the ω_μ are the $\sqrt[m]{1}$. $\square$

Corollary 6.3.7. *Let Ω be balanced and convex, q its gauge, $b \in \Omega$ with $q(b) > 0$.*

a) $g(\zeta) = \dfrac{\zeta}{q(b)}b$ *for all $\zeta \in \Delta$ defines a complex geodesic in Ω, and the only one joining 0 and b, up to composition with an element of $\mathrm{Aut}(\Delta)$, if and only if $b/q(b)$ is a complex extremal point of $\bar{\Omega}$.*
b) *Under the same assumption on b: if 0 and b are fixed points of $f \in \mathcal{A}(\Omega, \Omega)$, $g(\zeta)$ is also a fixed point of f for all $\zeta \in \Delta$.*

Proof. a) g is a complex geodesic in Ω since $c_\Omega(g(0); g'(0)) = 1$ (Ex. 6.2.6(A)). Now let $g_1 \in \mathcal{A}(\Delta, \Omega)$ be another complex geodesic in Ω with $g_1(\alpha) = 0$, $g_1(\beta) = b$: since $\Pi(\alpha, \beta) = C_\Omega(0, b) = \Pi(0, q(b))$, by composition with an element of $\mathrm{Aut}(\Delta)$ we obtain $\alpha = 0$, $\beta = q(b)$, and then $q \circ g_1(\zeta) = \mathrm{th}\, C_\Omega(0, g_1(\zeta)) = |\zeta|$ for all $\zeta \in \Delta$, $g_2(\zeta) = g_1(\zeta)/\zeta$ defines a map $g_2 \in \mathcal{A}(\Delta, X)$; $q \circ g_2 = 1$, $\dfrac{b}{q(b)} \in g_2(\Delta) \subset \bar{\Omega}$, hence $g_2 = \mathrm{const.} = \dfrac{b}{q(b)}$.
Now let $c \in X\setminus\{0\}$ be such that $\dfrac{b}{q(b)} + \bar{\Delta}c \subset \bar{\Omega}$: for any function $\varphi \in \mathcal{A}(\Delta, \Delta)$ such that $|\varphi(\zeta)| \leqslant |\zeta|$ for all $\zeta \in \Delta$ and $\varphi[q(b)] = 0$, e.g. $\varphi(\zeta) = \dfrac{\zeta}{2}[\zeta - q(b)]$, $g + c\varphi$ is also a complex geodesic joining 0 and b in Ω.
b) Put $g_1 = f \circ g$ in the proof of a). $\square$

6.4 Automorphisms and fixed points

In this section, we shall denote by $\text{Aut}(\Omega)$ the group of 1-1 bianalytic maps h of Ω onto Ω (h and h^{-1} are analytic), and by $\text{Fix}(h)$ the set of fixed points of a map $h: \Omega \to \Omega$.

The classical *Schwarz lemma* was generalized (e.g. [He]$_2$, §6 and 7) to a map $\in \mathscr{A}(\Omega, \Gamma)$, where Ω, Γ are balanced convex open sets; Patrizio [Pa] remarked that the sole assumption Γ pseudoconvex can be substituted for Ω, Γ convex.

Proposition 6.4.1. *Let Ω and Γ be balanced open sets in X and Y, p and q their gauges, Γ pseudoconvex or, equivalently, $q \in \mathscr{P}(Y)$; finally $f \in \mathscr{A}(\Omega, \Gamma)$, $f(0) = 0$. Then:*

a) $q \circ f \leqslant p$ *on* Ω *and* $q \circ (\hat{D}_0^1 f) \leqslant p$ *on* X, *i.e.* $(\hat{D}_0^1 f)(\Omega) \subset \Gamma$;
b) *given* $x \in \Omega \backslash \{0\}$: $q \circ f(x) = p(x)$, $q \circ (\hat{D}_0^1 f)(x) = p(x)$, *are equivalent properties implying* $q \circ f = p$ *on* $(\mathbb{C}x) \cap \Omega$.

Proof. For the equivalence between Γ pseudoconvex and $q \in \mathscr{P}(Y)$, see the proof of Proposition 6.2.7.b; for the proof of a) and b), consider

$$f_x(\zeta) = \frac{f(\zeta x)}{\zeta}, \qquad 0 < |\zeta| < \frac{1}{p(x)}, \qquad f_x(0) = (\hat{D}_0^1 f)(x):$$

f_x is analytic, $q \circ f_x$ subharmonic, on $\Delta\left[0, \dfrac{1}{p(x)}\right]$, and $q \circ f_x \leqslant p(x)$. $\square$

If Γ is convex and $\overline{\Gamma}$ has only complex extremal boundary points (Def. 6.3.5), further information is available.

Proposition 6.4.2. *Let Ω and $\Gamma \neq Y$ be balanced open sets in X and Y, p and q their gauges, Γ convex, and $\overline{\Gamma}$ have only complex extremal boundary points; finally let $f \in \mathscr{A}(\Omega, \Gamma)$, $f(0) = 0$. Then:*
a) *given* $x \in \Omega \backslash \{0\}$: $q \circ f(x) = p(x)$ *or* $q \circ (\hat{D}_0^1 f)(x) = p(x)$ *implies* $f = \hat{D}_0^1 f$ *on* $(\mathbb{C}x) \cap \Omega$; b) *if* Ω *is pseudoconvex:* f *a 1-1 bianalytic map of* Ω *onto* Γ *implies* f *linear, hence* Ω *actually convex.*

Proof. a) Since all boundary points of $\overline{\Gamma}$ are complex extremal, q vanishes only at the origin; then, under either assumption, $p(x) = 0$ implies $f = \hat{D}_0^1 f = 0$ on $(\mathbb{C}x) \cap \Omega$. Now let $p(x) > 0$; then, since $c_\Gamma(0; c) = q(c)$,

$$\Delta \ni \zeta \mapsto \frac{\zeta}{q \circ f(x)} f(x) \quad \text{and} \quad \Delta \ni \zeta \mapsto f\left(\frac{\zeta}{p(x)} x\right)$$

are 2 complex geodesics joining the origin and $f(x)$, which correspond to the same values of $\zeta : 0$ and $p(x)$; these geodesics coincide by Corollary 6.3.7. b) By Proposition 6.4.1.a: $q \circ f \leqslant p$ on Ω and $p \circ f^{-1} \leqslant q$ on Γ. $\square$

Corollary 6.4.3. *Let $\Omega \neq X$ be balanced, convex, $\bar{\Omega}$ have only complex extremal boundary points, and $h \in \mathscr{A}(\Omega, \Omega)$: if $h(0) = 0$, then $Fix(h)$ is the intersection of Ω with a closed linear subspace of X.*

Proof. By Proposition 6.4.2.a, h and $\hat{D}_0^1 h$ have the same fixed points in Ω. $\square$

The Corollary may fail if Ω is only balanced and convex: a counter-example in the bidisc Δ^2 is the map $(\xi, \eta) \to \left(\xi, \dfrac{\xi^2 + \eta}{2} \right)$.

In a famous paper [Ca], H. Cartan proved the following properties of a *bounded* domain Ω, containing the origin, in a finite dimensional space X, and a map $h \in \mathscr{A}(\Omega, \Omega)$ with $h(0) = 0$: a) if $h'(0) = id_X$, then $h = id_\Omega$; b) if the Jacobian of h at the origin has modulus 1, then $h \in Aut(\Omega)$; c) if moreover Ω is a circular domain, i.e. $e^{i\theta}\Omega = \Omega$ for all $\theta \in \mathbb{R}$, then any $h \in Aut(\Omega)$ such that $h(0) = 0$ is linear.

Our aim now is an extension of those properties to a balanced convex domain Ω in s.c.l.c. space X, although in such a space (apart from normed spaces) bounded domains do not exist. A first result was Proposition 6.4.2.b, obtained under the additional assumption that all boundary points of $\bar{\Omega}$ are complex extremal; in Corollary 6.4.5, the weaker assumption that the gauge q of Ω vanishes only at the origin (i.e. is a norm) will be a better substitute for the boundedness of Ω since, in the finite dimensional case, Ω is bounded if and only if q is a norm.

For the modulus 1 of the Jacobian of h at the origin, we shall substitute the assumption that $(\hat{D}_0^1 h) \in Aut(X)$ and is an isometry for $c_\Omega(0; .)$, i.e. $q \circ (\hat{D}_0^1 h) = q$ on X (Ex. 6.2.6(A)). The main tool will be a fine result of Harris [Har]$_1$ which, in connection with property a) above, links the difference $h - (\hat{D}_0^1 h)$ to the difference $(\hat{D}_0^1 h) - id_X$.

Theorem 6.4.4. *Let Ω be a balanced convex open set with gauge q, and $h(x) = h(0) + \sum\limits_{k \in \mathbb{N}^*} \varphi_k(x)$, $x \in \Omega$, the homogeneous polynomial expansion of a map $h \in \mathscr{A}(\Omega, X)$.*

a) *If there is a number $\delta > 0$ such that $q[x + \zeta h(x)] \leqslant 1 + \delta$ for all $x \in \Omega$, $\zeta \in \Delta$, then for all $x \in X$*

$$q \circ h(0) \leqslant \delta, \, q \circ \varphi_1(x) \leqslant \delta eq(x),$$

$$q \circ \varphi_k(x) \leqslant \delta k^{k/k-1} q^k(x) \text{ for } k \geqslant 2.$$

b) *Let moreover* $h \in \mathscr{A}(\Omega, \Omega)$ *and* $h(0) = 0$: *then for all* $x \in \Omega$

$$q[h(x) - \varphi_1(x)] \leqslant \delta\chi \circ q(x), \qquad \delta = \sup_{x \in \Omega} q[\varphi_1(x) - x],$$

where χ *is a universal (i.e. independent of* Ω *and* h) *convex increasing function on* $[0, 1[$.

Proof. a) Let $\alpha \in \,]0, 1[$ be such that $\dfrac{1 - \alpha}{\alpha} < \delta$ (which will allow $\alpha \to 1$ at the end of the proof) and $\rho = \dfrac{1 - \alpha}{\alpha\delta}$: since

$$q[x + \zeta\rho h(x)] \leqslant (1 - \rho)q(x) + \rho q[x + \zeta h(x)] < \frac{1}{\alpha}$$

for all $x \in \Omega, \zeta \in \Delta$,

$$f_1(x, \zeta) = \alpha[x + \zeta\rho h(x)]$$

defines a map $f_1 \in \mathscr{A}(\Omega \times \Delta, \Omega)$, which we iterate with respect to the first variable by setting

$$f_n(x, \zeta) = f_1[f_{n-1}(x, \zeta), \zeta] \quad \text{for } n \geqslant 2.$$

Then, by induction on $n \in \mathbb{N}^*$: $f_n \in \mathscr{A}(\Omega \times \Delta, \Omega)$, $f_n(x, 0) = \alpha^n x$ and

$$\frac{\partial f_n}{\partial \zeta}(x, 0) = \alpha\rho[\alpha^{n-1}h(x) + \alpha^{n-2}h(\alpha x) + \cdots + h(\alpha^{n-1}x)]$$

$$= \alpha\rho\frac{1 - \alpha^n}{1 - \alpha}h(0) + n\alpha^n\rho\varphi_1(x) + \alpha^n\rho\sum_{k=2}^{\infty}\frac{1 - \alpha^{n(k-1)}}{1 - \alpha^{k-1}}\varphi_k(x).$$

The first inequality of Cauchy (Th. 2.1.2.b) now gives $q \circ \dfrac{\partial f_n}{\partial \zeta}(x, 0) \leqslant 1$ for all $x \in \Omega$, and the generalized ones (Th. 2.3.5.a)

(0) $q \circ h(0) \leqslant \dfrac{\delta}{1 - \alpha^n}$

(1) $q \circ \varphi_1(x) \leqslant \dfrac{q(x)}{n\alpha^n\rho} \leqslant \dfrac{\delta q(x)}{n\alpha^n(1 - \alpha)}$,

(2) $q \circ \varphi_k(x) \leqslant \dfrac{(1 - \alpha^{k-1})q^k(x)}{\alpha^n\rho[1 - \alpha^{n(k-1)}]} \leqslant \dfrac{\delta(k - 1)q^k(x)}{\alpha^n[1 - \alpha^{n(k-1)}]}.$

We conclude by letting $n \to \infty$ with α fixed in (0), $\alpha = e^{-1/n}$ in (1), $\alpha = 1 - \dfrac{\ln k}{(k-1)n}$ in (2).

b) Let x' be any element of the adjoint space X' such that $|x'| \leqslant q$: for all $x \in \Omega$, $\lambda_x(\zeta) = \dfrac{1}{\zeta} \langle h(\zeta x), x' \rangle$ defines $\lambda_x \in \mathscr{A}(\Delta, \mathbb{C})$, with $|\lambda_x| \leqslant 1$; for any such function, say $\lambda(\zeta) = \sum_{m \in \mathbb{N}} \alpha_m \zeta^m$, from the Pick lemma derive the inequalities $|\alpha_1| \leqslant 1 - |\alpha_0|^2$ directly, and $|\alpha_m| \leqslant 1 - |\alpha_0|^2$ for $m \geqslant 2$ as well, if one replaces $\lambda(\zeta)$ by $\dfrac{1}{m} \sum_{\mu=1}^{m} \lambda(\omega_\mu \zeta^{1/m})$, where the ω_μ are the $\sqrt[m]{1}$; a fortiori

$$|\alpha_0| + \frac{1}{2}|\alpha_m| \leqslant 1 \quad \text{for all } m \in \mathbb{N}^*.$$

For λ_x, the inequality gives, for all $m \geqslant 2$,

$$\left| \left\langle \varphi_1(x) + \frac{1}{2}\zeta\varphi_m(x), x' \right\rangle \right| \leqslant 1 \quad \text{for all } x \in \Omega, \zeta \in \Delta,$$

provided that $|x'| \leqslant q$ on X, hence also

$$q\left[\varphi_1(x) + \frac{1}{2}\zeta\varphi_m(x) \right] \leqslant 1, \qquad q\left[x + \frac{1}{2}\zeta\varphi_m(x) \right] \leqslant 1 + \delta.$$

Then the last inequality in a) gives the result with $\chi(u) = 2 \sum_{m=2}^{\infty} m^{m/m-1} u^m$.
$\square$

Corollary 6.4.5. *Let Ω be a balanced convex open set whose gauge q is a norm, $h \in \mathscr{A}(\Omega, \Omega)$, $h(0) = 0$.*

a) *If $\hat{D}_0^1 h = id_X$, then $h = id_\Omega$.*
b) *If $\hat{D}_0^1 h$ is an isometry for $c_\Omega(0; .)$, i.e. $q \circ h = q$ on Ω or $q \circ (\hat{D}_0^1 h) = q$ on X, and moreover $(\hat{D}_0^1 h) \in \mathrm{Aut}(X)$, then $h \in \mathrm{Aut}(\Omega)$ and is linear.*
c) *Besides any $h \in \mathrm{Aut}(\Omega)$ such that $h(0) = 0$ is linear.*

Proof. b) $q \circ h = q$ and $q \circ (\hat{D}_0^1 h) = q$ are equivalent properties (Prop. 6.4.1.b); $\hat{D}_0^1 h|_\Omega \in \mathrm{Aut}(\Omega)$ and $(\hat{D}_0^1 h)^{-1} \circ h = id_\Omega$ by a).
c) $h \in \mathrm{Aut}(\Omega)$ implies $q \circ h = q$ on Ω (Prop. 6.4.1.a) and $(\hat{D}_0^1 h) \in \mathrm{Aut}(X)$.
$\square$

It may be noted that, since q is a norm, $q \circ (\hat{D}_0^1 h) = q$ implies $\hat{D}_0^1 h$ injective; then, if the s.c. space X is metrizable, $\hat{D}_0^1 h$ surjective implies $(\hat{D}_0^1 h) \in \mathrm{Aut}(X)$ by the Banach open mapping theorem ([B0]$_2$, Chap. I, §3, Th. 1). But the assumption of convexity is essential: in [Vi]$_2$, Vigué constructs a Banach space X, a bounded balanced open set Ω in X and a map $h \in \mathscr{A}(\Omega, \Omega)$, $h(0) = 0$, such that $\hat{D}_0^1 h$ is a surjective isometry for $c_\Omega(0; .)$, but h is not injective.

In the 1-dimensional case: if a nonconstant $h \in \mathscr{A}(U, U)$ is such that $V = h(U)$ is relatively compact in U, $h(\bar{V})$ is a compact subset of V and one easily checks (see 6.1(A)) the existence of a number $k > 0$ such that $G_U(\alpha, \beta) \geqslant k + G_V(\alpha, \beta)$ for all $\alpha, \beta \in h(\bar{V})$; then, by the Lindelöf principle, $G_V[h(\alpha), h(\beta)] \geqslant k + G_V(\alpha, \beta)$, and the sequence of iterates of h converges to a constant, which is the unique fixed point of h. The *fixed point theorem of Earle-Hamilton* [Ea Ha] is a close generalization of this property.

Theorem 6.4.6. *Let $h \in \mathscr{A}(\Omega, \Omega)$ be such that $h(\Omega)$ is bounded and $h(\Omega) + W \subset \Omega$ for some neighbourhood W of the origin in X: then h has a unique fixed point.*

Proof. The set $h(\Omega) - h(\Omega)$, since bounded, is absorbed by W: there is a number $\delta > 0$ such that $\delta[h(\Omega) - h(\Omega)] \subset W$ and, for all $a \in \Omega$,

$$f_a(x) = h(x) + \delta[h(x) - h(a)]$$

defines a map $f_a \in \mathscr{A}(\Omega, \Omega)$ such that $f_a(a) = h(a)$, $\hat{D}_a^1 f_a = (1 + \delta)\hat{D}_a^1 h$, hence for all $(a, c) \in \Omega \times X$

$$(*) \qquad (1 + \delta)c_\Omega[h(a); (\hat{D}_a^1 h)(c)] = c_\Omega[f_a(a); (\hat{D}_a^1 f_a)(c)] \leqslant c_\Omega(a; c)$$

by formula (4) in §6.2.

Now consider the iterated maps $h_n \in \mathscr{A}(\Omega, \Omega)$ defined by $h_0 = id_\Omega$, $h_n = h \circ h_{n-1}$ for all $n \in \mathbb{N}^*$: by the formula in Theorem 3.1.10,

$$\hat{D}_a^1 h_n = (\hat{D}_{h_{n-1}(a)}^1 h) \circ (\hat{D}_a^1 h_{n-1})$$

and therefore the iteration of $(*)$ gives, for each $n \in \mathbb{N}$,

$$(1 + \delta)^n c_\Omega[h_n(a); (\hat{D}_a^1 h_n)(c)] \leqslant c_\Omega(a; c) \quad \text{for all } (a, c) \in \Omega \times X,$$

$$(1 + \delta)^n c_\Omega[x_n(t); x_n'(t)] \leqslant c_\Omega[x(t); x'(t)] \quad \text{for all } t \in [0, 1]$$

if x is any continuous path $[0, 1] \to \Omega$ with piecewise continuous derivative, and $x_n = h_n \circ x$; by integration $(1 + \delta)^n C_\Omega^i[h_n(a), h_n(b)] \leqslant C_\Omega^i(a, b)$ if the path x starts from a and ends in b.

On the other hand, for all $p \in csn(X)$ we have $M(p) = \sup_{h(\Omega)} p$ finite; for all $a \in \Omega$, $n \in \mathbb{N}^*$, h_n maps Ω into

$$h_n(a) + p^{-1}([0, 2M(p)]), \quad \text{hence (Ex. 6.2.6(A))}$$

$$th^{-1} p\left[\frac{h_n(a) - h_n(b)}{2M(p)}\right] \leqslant C_\Omega[h_n(a), h_n(b)] \leqslant \frac{C_\Omega^i(a, b)}{(1 + \delta)^n}.$$

Taking $b = h(a)$, we see that $(h_n(a))_{n \in \mathbb{N}}$ is a Cauchy sequence in X, whose limit is a fixed point of h; the last inequality also proves the uniqueness of this point. $\square$

In most applications of the Theorem, of course, Ω is bounded, hence X a Banach space.

Corollary 6.4.7. *Let Ω be a bounded convex open set in a reflexive Banach space X: if the map $h \in \mathscr{A}(\Omega, \Omega)$ has a weakly continuous extension to $\bar{\Omega}$, h (or its extension) has at least one fixed point $\in \bar{\Omega}$.*

Proof. Let Ω contain the origin. For all $r \in]0, 1[$, $rh \in \mathscr{A}(\Omega, \Omega)$ and $rh(\Omega) + (1 - r)\Omega \subset \Omega$; by the Theorem, rh has a unique fixed point $a(r) \in \Omega$. Since Ω is weakly relatively compact (Rem. 1.4.3), $a(r)$ has at least one limit point b, for the weakened topology, as $r \to 1$, and $b \in \bar{\Omega}$ since $\bar{\Omega}$ is also closed for this topology (Prop. 1.1.1.b). $\square$

A counter-example due to Hayden-Suffridge shows that the assumption of reflexivity cannot be dropped. In the Banach space $c_0(\mathbb{N})$ of complex sequences tending to 0, with the sup norm, the map $(\xi_0, \xi_1, \xi_2, \dots) \to (1, \xi_0, \xi_1, \dots)$ has no fixed point, although it is analytic and weakly continuous on the whole space (as the sum of a constant and a continuous linear map) and maps into itself any open ball centred at the origin with a radius > 1.

On the contrary, the Corollary is well suited to the Hilbert space; here we write the inner product $(. | .)$ in such a way that it is linear with respect to the second factor.

Theorem 6.4.8. *Let Ω be the open unit ball in the Hilbert space X.*

a) *The group $\mathrm{Aut}(\Omega)$ acts transitively on Ω: more precisely, given a, b, a', $b' \in \Omega$ with $C_\Omega(a, b) = C_\Omega(a', b')$, there is an $h \in \mathrm{Aut}(\Omega)$ such that $a' = h(a)$, $b' = h(b)$.*
b) *Every element $h \in \mathrm{Aut}(\Omega)$ extends to some open ball containing $\bar{\Omega}$ into a weakly continuous map, therefore h (or its extension) has at least one fixed point $\in \bar{\Omega}$.*
c) *For all $f \in \mathscr{A}(\Omega, \Omega)$, $\mathrm{Fix}(f)$ is either $\emptyset$ or the intersection of Ω with a closed affine subspace of X.*

Proof. a) For $a \in \Omega$, let $\alpha = \sqrt{1 - \|a\|^2}$,

$$l_a(x) = \frac{(a|x)}{1 + \alpha} a + \alpha x, \qquad h_a(x) = l_a\left[\frac{x - a}{1 - (a|x)}\right];$$

simple computations give

$$\|l_a(x - a)\|^2 + \alpha^2(1 - \|x\|^2) = |1 - (a|x)|^2,$$

hence $h_a \in \mathscr{A}(\Omega, \Omega)$, and $h_{-a} \circ h_a = id_\Omega$, hence $h_a \in \mathrm{Aut}(\Omega)$ for all $a \in \Omega$. From Example 6.2.6(A) follows, since $h_a(a) = 0$:

$$th\, C_\Omega(a, b) = \|h_a(b)\| = \|h_b(a)\|;$$

then the assumption on a, b, a', b' can be written $\|h_a(b)\| = \|h_{a'}(b')\|$ and implies the existence of a unitary operator $u: X \to X$ such that $h_{a'}(b') = u \circ h_a(b)$: $h = h_{-a'} \circ u \circ h_a$ has the required property.
b) Let $h \in \mathrm{Aut}(\Omega)$: if $h(0) = 0$, then $\|h(x)\| = \|x\|$ for all $x \in \Omega$ (Prop. 6.4.1.a), h is linear (Corol. 6.4.5), hence is unitary; consequently, if $h(a) = 0$, then $h = u \circ h_a$ for some unitary u, h extends to the open ball $\left\{ x \in X: \|x\| < \dfrac{1}{\|a\|} \right\}$ and by its expression is weakly continuous.
c) If $\mathrm{Fix}(f) \neq \emptyset$, let $a \in \mathrm{Fix}(f)$: then $0 \in \mathrm{Fix}(g)$ if $g = h_a \circ f \circ h_{-a}$, therefore (Coroll. 6.4.3) $\mathrm{Fix}(g)$ is the intersection of Ω with a closed linear subspace E of X, and $\mathrm{Fix}(f) = h_{-a}(E \cap \Omega)$ is the intersection of Ω with an affine subspace of X since $x \mapsto \dfrac{x + a}{1 + (a|x)}$ maps every complex line into another one. $\square$

Part b) above is the fixed point theorem of Hayden-Suffridge [HaSu]; c) was proved by Renaud [Re]. The essential fact that every element of $\mathrm{Aut}(\Omega)$ extends to a neighbourhood of $\bar{\Omega}$, was generalized (by a quite different method) to the open unit ball B of any Banach space X, by Kaup-Upmeier [Kau Up], who in the same paper also showed that the orbit $\{h(0): h \in \mathrm{Aut}(B)\}$ is the intersection of B with a closed linear subspace of X. This property of the orbit was extended by Vigué [Vi]$_1$ to any balanced bounded open set in a Banach space.

Theorem 6.4.9. *Let Ω be a bounded convex open neighbourhood of the origin in a reflexive Banach space X, $h \in \mathscr{A}(\Omega, \Omega)$, $h(0) = 0$. There is another map $f \in \mathscr{A}(\Omega, \Omega)$, $f(0) = 0$, with the properties: a) $f \circ h = h \circ f = f$; b) $f(\Omega) = \mathrm{Fix}(f) = \mathrm{Fix}(h)$ and $(\hat{D}_0^1 f)(X) = \mathrm{Fix}(\hat{D}_0^1 f) = \mathrm{Fix}(\hat{D}_0^1 h)$; c) If one of the two sets in b) is $\{0\}$, so is the other one.*

The proof below of a) *and* b) *is due to Mazet-Vigué.*

Proof. a) We again set $h_0 = id_\Omega$, $h_n = h_{n-1} \circ h$ for $n \in \mathbb{N}^*$, and moreover $M_n h = \dfrac{1}{n}(h_0 + h_1 + \cdots + h_{n-1})$: $M_n h(0) = 0$, $M_n h \in \mathscr{A}(\Omega, \Omega)$ since Ω is convex; then, for each $x \in \Omega$, the set $\{M_n h(x): n \in \mathbb{N}^*\}$ is relatively compact for the weakened topology and, given an ultrafilter Φ on $\mathbb{N}^*$ finer than the Fréchet filter, $M_n h(x) \underset{\Phi}{\rightarrow} g(x)$ for this topology. This means that, for all $x' \in X'$, $x' \circ M_n h \underset{\Phi}{\rightarrow} x' \circ g$ pointwise, hence uniformly on compact sets since the $x' \circ M_n h$ are equicontinuous: $g \in \mathcal{G}(\Omega, X)$; but $g(\Omega) \subset \bar{\Omega}$ (Prop. 1.1.1.b) implies $g(\Omega)$ bounded, finally $g \in \mathcal{A}(\Omega, \Omega)$ since $g(0) = 0$ (See the proof of Th. 6.2.4); moreover, for any $x_0 \in X$, $k \in \mathbb{N}^*$: $x' \circ (\hat{D}_0^k M_n h)(x_0) \underset{\Phi}{\rightarrow} x' \circ (\hat{D}_0^k g)(x_0)$ or $(\hat{D}_0^k M_n h)(x_0) \underset{\Phi}{\rightarrow} (\hat{D}_0^k g)(x_0)$ for the weakened topology. Now

$$(\hat{D}_0^1 h) \circ (\hat{D}_0^1 M_n h)(x_0) = \frac{1}{n}[(\hat{D}_0^1 h_1) + (\hat{D}_0^1 h_2) + \cdots + (\hat{D}_0^1 h_n)](x_0)$$

$$= (\hat{D}_0^1 M_n h)(x_0) + \frac{1}{n}[(\hat{D}_0^1 h_n)(x_0) - x_0],$$

from which follows that

$$x' \circ (\hat{D}_0^1 h) \circ (\hat{D}_0^1 M_n h)(x_0) - x' \circ (\hat{D}_0^1 M_n h)(x_0) \underset{\Phi}{\rightarrow} 0$$

for all $x' \in X'$ and, since $x' \circ (\hat{D}_0^1 h) \in X'$: $(\hat{D}_0^1 h) \circ (\hat{D}_0^1 g) = \hat{D}_0^1 g$.

By a similar argument, from $(M_n h) \circ h = (M_n h) + \frac{1}{n}(h_n - h_0)$ follows $g \circ h = g$. Now we iterate g: $g_0 = id_\Omega$, $g_n = g_{n-1} \circ g$ for $n \in \mathbb{N}^*$: since each set $\{g_n(x): n \in \mathbb{N}^*\}$ is relatively compact for the weakened topology, again $g_n(x) \underset{\Phi}{\rightarrow} f(x)$ for this topology, $f \in \mathcal{A}(\Omega, \Omega)$, and $g \circ h = g$ implies $g_n \circ h = g_n$ for all n, $f \circ h = f$. The following steps of the proof aim at the other relation $h \circ f = f$.

1st step. Let u, $u' \in \mathcal{A}(\Omega, X)$, $u(0) = u'(0) = 0$: if u and u' have the same homogeneous polynomial expansion (h.p.e.) up to the degree $n - 1$ ($n \geqslant 2$), then $h \circ u$ and $h \circ u'$ (defined on some neighbourhood of the origin) also have the same h.p.e. up to the degree $n - 1$, and actually up to the degree n if $\hat{D}_0^1 h \equiv 0$; this common h.p.e. up to the degree n vanishes if $\hat{D}_0^k h \equiv 0$ for all $k \leqslant n$.

In fact, the assumption means that, for any $x \in X$, ζ^n divides the Taylor expansion of $u(\zeta x) - u'(\zeta x)$; then, if λ is an m-linear continuous map $X^m \to X$, ζ^n if $m = 1$, ζ^{n+1} if $m \geqslant 2$, divides the Taylor expansion of $\lambda[u(\zeta x), \ldots, u(\zeta x)] - \lambda[u'(\zeta x), \ldots, u'(\zeta x)]$ and (Th. 2.2.9c) this remains true for $\varphi \circ u(\zeta x) - \varphi \circ u'(\zeta x)$ if φ is an m-homogeneous continuous polynomial map. So the statements follow from the uniform summability, for sufficiently small $|\zeta|$, of the series

$$h \circ u(\zeta x) = \sum_{m \in \mathbb{N}^*} \frac{1}{m!}(\hat{D}_0^m h) \circ u(\zeta x),$$

$$h \circ u'(\zeta x) = \sum_{m \in \mathbb{N}*} \frac{1}{m!} (\hat{D}_0^m h) \circ u'(\zeta x)$$

(Th. 3.1.5.a).

2nd *step*. For each $n \in \mathbb{N}*$, $h \circ g_n$ and g_n have the same h.p.e. up to the degree n.

This is true for $n = 1$, since $\hat{D}_0^1(h \circ g) = \hat{D}_0^1 g$; so we proceed by an induction on n, assuming that $n \geqslant 2$ is given, that $h \circ g_{n-1}$ and g_{n-1} have the same h.p.e. up to the degree $n - 1$. By the 1st step (with h_{p-1} instead of h), this implies that the terms of degree $\leqslant n - 1$ in the h.p.e. of $h_p \circ g_{n-1}$ do not depend on $p \in \mathbb{N}$; their sum u is also the sum of the same terms in the expansion of $(M_p h) \circ g_{n-1}$, hence also in the expansion of $g \circ g_{n-1} = g_n$ since, for all $x' \in X'$, $x' \circ (M_p h) \circ g_{n-1} \underset{\Phi}{\rightrightarrows} x' \circ g_n$ uniformly on compact sets; but $u_p = \dfrac{1}{n!} \hat{D}_0^n(h_p \circ g_{n-1})$ does depend on $p \in \mathbb{N}$ and

$$\frac{1}{p}(u_0 + u_1 + \cdots + u_{p-1})(x_0) \underset{\Phi}{\rightrightarrows} \frac{1}{n!} \hat{D}_0^n g_n(x_0) = v(x_0), \; x_0 \in X,$$

for the weakened topology.

Now we set $\dfrac{1}{k!} \hat{D}_0^k h = \varphi_k$, $k \in \mathbb{N}*$. By the 1st step (with $u + u_p$ and $h_p \circ g_{n-1}$ instead of u and u'), $h_{p+1} \circ g_{n-1}$ and $\varphi_1 \circ u_p + \varphi_1 \circ u + \cdots + \varphi_n \circ u$ have the same h.p.e. up to the degree n: this means that $u_{p+1} - \varphi_1 \circ u_p$ is the sum of the terms of degree $\leqslant n$ in the h.p.e. of $\varphi_1 \circ u + \cdots + \varphi_n \circ u - u$, therefore does not depend on $p \in \mathbb{N}$, say $u_{p+1} - \varphi_1 \circ u_p = w$, hence

$$u_0 + \cdots + u_{p-1} = pw + (u_0 - u_p) + \varphi_1 \circ (u_0 + \cdots + u_{p-1})$$

and

$$v = w + \varphi_1 \circ v$$

by going over to the Φ-limit for the weakened topology.

Similarly, by the 1st step (with $u + v$ and g_n instead of u and u'), $h \circ g_n$ and $\varphi_1 \circ v + \varphi_1 \circ u + \cdots + \varphi_n \circ u$ have the same h.p.e. up to the degree n; by the definition of w, so have $h \circ g_n$ and $\varphi_1 \circ v + w + u = u + v$. In other words, $h \circ g_n$ and g_n have the same h.p.e. up to the degree n, q.e.d. in this 2nd step.

3rd *step*. Since the assumption of the 2nd step is proved, we may use a result obtained in the course of the argument: the sum of the terms of degree $\leqslant n - 1$ is the same, namely u, in the expansions of $g_{n-1}, g_n, g_{n+1}, \ldots$, in the expansion of f. Finally, by the 1st step, $h \circ f$ and $h \circ g_n$ have the same h.p.e.

up to the degree $n - 1$; since this is also true for $h \circ g_n$ and g_n, for g_n and f, it is true for $h \circ f$ and f.

b) $f(\Omega) \subset \text{Fix}(h)$ follows from $h \circ f = f$ and the inclusions $\text{Fix}(h) \subset \text{Fix}(g) \subset \text{Fix}(f) \subset f(\Omega)$ are obvious; $(\hat{D}_0^1 f)(X) \subset \text{Fix}(\hat{D}_0^1 h)$ follows from $(\hat{D}_0^1 h) \circ (\hat{D}_0^1 f) = \hat{D}_0^1 f$ and the inclusions $\text{Fix}(\hat{D}_0^1 h) \subset \text{Fix}(\hat{D}_0^1 g) \subset \text{Fix}(\hat{D}_0^1 f) \subset (\hat{D}_0^1 f)(X)$ are obvious. Moreover $(\hat{D}_0^1 g)(X) \subset \text{Fix}(\hat{D}_0^1 h)$ follows from $(\hat{D}_0^1 h) \circ (\hat{D}_0^1 g) = \hat{D}_0^1 g$ and the inclusions $\text{Fix}(\hat{D}_0^1 h) \subset \text{Fix}(\hat{D}_0^1 g) \subset (\hat{D}_0^1 g)(X)$ are obvious: the second set in b) is also $(\hat{D}_0^1 g)(X)$.

c) $f \equiv 0$ implies $\hat{D}_0^1 f \equiv 0$. Conversely, let $\hat{D}_0^1 g \equiv 0$: if g or some iterate g_n is identically 0, the proof is over since $\text{Fix}(h) \subset \text{Fix}(g_n)$. If not, let k_0 be the smallest integer k such that $\hat{D}_0^k g \not\equiv 0$: the density number at the origin is k_0 for $\ln \|g\|$ (Ex. 5.5.4(B)), at least k_0^2 for $\ln \|g_2\| = (\ln \|g\|) \circ g$ (Th. 5.5.6) and, by induction, at least k_0^n for $\ln \|g_n\| = (\ln \|g_{n-1}\|) \circ g$, which means $\hat{D}_0^k g_n \equiv 0$ for all $k < k_0^n$, and $k_0 \geq 2$: then $\hat{D}_0^k f \equiv 0$ for all k, $f \equiv 0$. $\square$

Exercises

6.2.1. Let Ω be the open unit ball in the Banach space $\mathscr{C}(T)$, where T is a compact space: show that $c_\Omega(a; c) = \sup\limits_{T} \dfrac{|c|}{1 - |a|^2}$ for all $a \in \Omega$, $c \in \mathscr{C}(T)$. Compute $C_\Omega(a, b)$, $(a, b) \in \Omega^2$.

6.2.4. a) Show that a continuous map $x: [-1, 1] \to X$ is the uniform limit on $[-1, 1]$ of a sequence of polynomials $\mathbb{C} \to X$.

Hint. Set $L_n(t) = c_n(1 - t^2)^n$, $f_n(t) = \int\limits_{-1}^{1} x(u) L_n(t - u)\, du$, where the coefficient c_n is such that $\int\limits_{-1}^{1} L_n(t)\, dt = 1$.

b) Show that $d_\Omega(a, b) < \infty$ for any Ω and $(a, b) \in \Omega^2$.

6.1, 6.2, 6.3. Using Theorem 5.5.1., show that, if $G_\Omega \not\equiv +\infty$, the p.s.h. function $- G_\Omega(a, .)$ has, at the point a, a density number $\in \mathbb{N}^*$, which is 1:
a) for all $a \in \Omega$ if there is an $x' \in X'$ such that $x'(\Omega)$ is a Green open set in $\mathbb{C}$, in particular if the convex hull of Ω is not the whole space;
b) if a lies on some complex geodesic in Ω.

6.4.8. Let Ω be the open unit ball in the product, endowed with the norm $\|(x_1, x_2)\| = \sup(\|x_1\|, \|x_2\|)$, of two Hilbert spaces X_1, X_2.
1) Let the linear map $(x_1, x_2) \to (Ax_1 + Bx_2, Cx_1 + Dx_2)$ belong to $\mathrm{Aut}(\Omega)$. Then:
a) $\|x_1\| = \|x_2\| = 1$ implies $\|Ax_1 + Bx_2\| = \|Cx_1 + Dx_2\| = 1$;
b) $(Ax_1|Bx_2) = (Cx_1|Dx_2) = 0$ for all $x_1 \in X_1$, $x_2 \in X_2$;
c) either B and C or A and D are identically 0.
2) Find all elements of $\mathrm{Aut}(\Omega)$.

Bibliography

The list below does not at all claim to be exhaustive as regards analyticity in infinite dimensional spaces: it contains only the titles mentioned in the text, whether they regard this domain or not. A quasi exhaustive list (725 titles) of papers appeared on the subject up to 1980 can be found in the book $[Di]_5$.

[A1]   H. Alexander: Analytic functions on a Banach space (Univ. of California at Berkeley, Thesis, 1968).

[Ba Ma Na]   J.A. Barroso, M. Matos, L. Nachbin: On holomorphy versus linearity in classifying locally convex spaces (Infinite dimensional holomorphy and applications, North Holland math. studies, **12**, 1977, p. 31–74).

[Ba]   T.J. Barth: The Kobayashi indicatrix at the center of a circular domain (Proc. amer. math. Soc., **88**, 1983, p. 527–530).

[Boc Sic]   J. Bochnak, J. Siciak: Analytic functions in topological vector spaces (Studia math., **39**, 1971, p. 77–112).

$[Bo]_1$ N. Bourbaki: Topologie générale (Hermann, Paris, 1951).

$[Bo]_2$ N. Bourbaki: Espaces vectoriels topologiques (Hermann, Paris, 1953).

$[Brel]_1$ M. Brelot: Sur la théorie autonome des fonctions sousharmoniques (Bull. Sc. math., **65**, 1941, p. 72–98).

$[Brel]_2$ M. Brelot: Sur l'allure des fonctions harmoniques et sousharmoniques à la frontière (Math. Nachr., **4**, 1950, p. 298–307).

$[Brel]_3$ M. Brelot: Eléments de la théorie classique du potentiel (Centre de documentation universitaire, Paris, 1959).

[Br Ch]   M. Brelot and G. Choquet: Le théorème de convergence en théorie du potentiel (Journal Madras Univ., **27**, 1957, p. 277–286).

[Brem]   H. Bremermann: Ueber die Aequivalenz der pseudokonvexen Gebiete und der Holomorphiegebiete im Raum von n komplexen Veränderlichen (Math. Ann., **128**, 1954, p. 63–91).

[Brez]   H. Brezis: Analyse fonctionnelle (Masson, Paris, 1983), especially §IV.3.

[Bu]   J.D. Buckholtz: A characterization of the exponential series (Amer. math. Monthly, **73**, 1966, p. 121–123).

[Ca]   H. Cartan: Sur les fonctions de plusieurs variables complexes (Math. Zeit., **35**, 1932, p. 760–773).

[Ca Th]   H. Cartan and P. Thullen: Regularitäts- und Konvergenzbereiche (Math. Ann., **106**, 1932, p. 617–647).

[Ch]   G. Choquet: Theory of capacities (Ann. Inst. Fourier, **5**, 1955, p. 131–295), especially §26.12.

[Coe]   G. Coeuré: Fonctions plurisousharmoniques sur les espaces vectoriels topologiques et applications à l'étude des fonctions analytiques (Ann. Inst. Fourier, **20-1**, 1970, p. 361–432).

[Col]   J.F. Colombeau: Sur les applications $\mathscr{G}$-analytiques et analytiques en dimension infinie (Sém. Lelong, 1971/72, Lecture notes, **332**, p. 48–58).

[Di]$_1$ S. Dineen: Bounding subsets of a Banach space (Math. Ann., **192**, 1971, p. 61–70).

[Di]$_2$ S. Dineen: Unbounded holomorphic functions on a Banach space (J. London math. Soc., **4**, 1972, p. 461–465).

[Di]$_3$ S. Dineen: Holomorphically complete locally convex topological vector spaces (Sém. Lelong, 1971/72, Lecture notes, **332**, p. 77–111).

[Di]$_4$ S. Dineen: Holomorphic functions on locally convex topological vector spaces. II. Pseudoconvex domains (Ann. Inst. Fourier, **23-3**, 1973, p. 155–185).

[Di]$_5$ S. Dineen: Complex analysis in locally convex spaces (North Holland math. studies **57**, 1981).

[Di No Sch]   S. Dineen, P. Noverraz, M. Schottenloher: Le problème de Levi dans certains espaces vectoriels topologiques localement convexes (Bull. Soc. math. France, **104**, 1976, p. 87–97).

[Di Ti Vi]   S. Dineen, R. Timoney, J.P. Vigué: Pseudodistances invariantes sur les domaines d'un espace localement convexe (Ann. Sc. Norm. Sup. Pisa, **12**, 1985, p. 515–529).

[Du Sc]   N. Dunford and J. Schwartz: Linear operators (Wiley, New York, 1958), especially Chapter II, §3.

[Ea Ha]   C. Earle and R. Hamilton: A fixed point theorem for holomorphic mappings (Proc. Symp. in pure Math., A.M.S., **16**, 1970, p. 61–65).

[Ev]   G.C. Evans: On potentials of positive mass (Trans. amer. math. Soc., **37**, 1935, p. 226–253).

[Fa]   L. Fantappiè: I funzionali analitici (Mem. R. Accad. Lincei, **3**, 1930, p. 453–683).

[Fe Si]   J.P. Ferrier and N. Sibony: Approximation pondérée sur une variété totalement réelle de $\mathbb{C}^n$ (Ann. Inst. Fourier, **26-2**, 1976, p. 101–115).

[Gr]   L. Gruman: The Levi problem in certain infinite dimensional vector spaces (Ill. J. Math., **18**, 1974, p. 20–26).

[Gr Ki]   L. Gruman and C. Kiselman: Le problème de Levi dans les espaces de Banach à base (C.R. Acad. Sc. Paris, **274**, 1972, p. 1296–1299).

[Har]$_1$ L. Harris: A continuous form of Schwarz's lemma in normed linear spaces (Pacific J. Math., **38**, 1971, p. 635–639).

[Har]$_2$ L. Harris: Schwarz-Pick systems of pseudometrics for domains in normed linear spaces (North Holland math. studies, **34**, 1979, p. 345–406).

[Ha Su]   T. Hayden and T. Suffridge: Biholomorphic maps in Hilbert space have a fixed point (Pacific J. Math., **38**, 1971, p. 419–422).

[He]$_1$ M. Hervé: Analytic and plurisubharmonic functions in finite and infinite dimensional spaces (Lecture notes, **198**, 1971).

[He]$_2$ M. Hervé: Lindelöf's principle in infinite dimensions (Proc. on infinite dimensional holomorphy, Lecture notes, **364**, 1974, p. 41–57).

[He]$_3$ M. Hervé: Les fonctions analytiques (Presses universitaires de France, Paris, 1982).

[He]$_R$ R.M. Hervé: Recherches axiomatiques sur la théorie des fonctions surharmoniques et du potentiel (Ann. Inst. Fourier, **12**, 1962, p. 415–571).

[Hi Ph]   E. Hille and R. Phillips: Functional analysis and semigroups (Amer. math. Soc. Colloquium, **31**, 1957).

[Hir]   A. Hirschowitz: Ouverts d'analyticité en dimension infinie (Sém. Lelong, 1970, Lecture notes, **205**, p. 11–20).

[Hö]   L. Hörmander: An introduction to complex analysis in several variables (Van Nostrand, 1966).

[Ho]   J. Horváth: Topological vector spaces and distributions (Addison-Wesley, 1966).

[Jo]$_1$ B. Josefson: A counterexample in the Levi problem (Proc. on infinite dimensional holomorphy, Lecture notes, **364**, 1974, p. 168–177).

[Jo]$_2$ B. Josefson: Weak sequential convergence in the dual of a Banach space does not imply norm convergence (Arkiv för Math., **13**, 1975, p. 79–89).

[Kad]   M.I. Kadec: Spaces isomorphic to a locally uniformly convex space (Izv. Vysš. Zaved. Mat., n°6, **13**, 1959, p. 51–57 and **25**, 1961, p. 186–187).

[Ka Kl]   S. Kakutani and V. Klee: The finite topology of a linear space (Archiv der Math., **14**, 1963, p. 55–58).

[Kau]   L. and B. Kaup: Holomorphic functions of several variables (de Gruyter studies in Math., **3**, 1983).

[Kau Up]   W. Kaup and H. Upmeier: Banach spaces with biholo- morphically equivalent unit balls are isomorphic (Proc. amer. math. Soc., **58**, 1976, p. 129–133).

[Ki]$_1$ C. Kiselman: Geometric aspects of the theory of bounds for entire functions in normed spaces (Infinite dimensional holomorphy and applications, North Holland math. studies, **12**, 1977, p. 249–275).

[Ki]$_2$ C. Kiselman: The partial Legendre transformation for plurisub harmonic functions (Invent. math., **49**, 1978, p. 137–148).

[Ki]$_3$ C. Kiselman: Stabilité du nombre de Lelong par restriction à une sous-variété (Colloque de Wimereux, Lecture notes, **919**, 1982, p. 324–336).

[Lel]$_1$ P. Lelong: Fonctions plurisousharmoniques et fonctions analytiques réelles (Ann. Inst. Fourier, **11**, 1961, p. 515–562).

[Lel]$_2$ P. Lelong: Topologies semi-vectorielles; applications à l'analyse complexe (Ann. Inst. Fourier, **25-3**, 1975, p. 381–407).

[Lel]$_3$ P. Lelong: Fonctions plurisousharmoniques dans les espaces vectoriels topologiques (Sém. Lelong, 1967/68, Lecture notes, **71**, p. 167–190).

[Lel]$_4$ P. Lelong: Fonctions plurisousharmoniques et ensembles polaires (Sém. Lelong, 1969, Lecture notes, **116**, p. 1–20).

[Lel]$_5$ P. Lelong: Théorème de Banach-Steinhaus pour les polynomes; applications entières d'espaces vectoriels complexes (Sém. Lelong, 1970, Lecture notes, **205**, p. 87–112).

[Lel]$_6$ P. Lelong: Un théorème de fonctions inverses dans les espaces vectoriels topologiques complexes et ses applications à des problèmes de croissance en analyse complexe (Sém. Lelong-Skoda, 1976/77, Lecture notes, **694**, p. 172–195).

[Lel]$_7$ P. Lelong: Calcul du nombre densité et lemme de Schwarz pour les fonctions plurisousharmoniques dans un espace vectoriel topologique (Sém. Lelong-Skoda, 1980/81, Lecture notes, **919**, p. 167–176).

[Lel]$_8$ P. Lelong: Notions de croissance pour les fonctions holomorphes sur un espace vectoriel topologique (Aspects of math. and its applications, North Holland, 1986, p. 551–572).

[Lem]$_1$ L. Lempert: La métrique de Kobayashi et la représentation des domaines sur la boule (Bull. Soc. math. France, **109**, 1981, p. 427–474).

[Lem]$_2$ L. Lempert: Holomorphic retracts and intrinsic metrics in convex domains (Anal. math., **8**, 1982, p. 257–261).

[Lev]   E.E. Levi: Sulle ipersuperficie dello spazio a 4 dimensioni che possono essere frontiera del campo di esistenza di una funzione analitica di 2 variabili complesse (Ann. di Mat., **18**, 1911, p. 69–79).

[Mu]   J. Mujica: Complex analysis in Banach spaces (North Holland math. studies, **120**, 1986).

[Ng]   T.V. Nguyen: Fonctions séparément analytiques et prolonge-
            ment analytique faible en dimension infinie (Ann. Pol. math., **33**,
            1976, p. 71–83).

[Ni]   A. Nissenzweig: Weak* sequential convergence (Israel J. Math., **22**,
            1975, p. 79–89).

[Nor]   F. Norguet: Sur les domaines d'holomorphie des fonctions uni-
            formes de plusieurs variables complexes (Bull. Soc. math. France,
            82, 1954, p. 137–159).

[Nos]   K. Noshiro: Cluster sets (Ergebnisse der Math., Heft 28, 1960).

[Nov]$_1$ P. Noverraz: Fonctions plurisousharmoniques et analytiques dans
            les espaces vectoriels topologiques complexes (Ann. Inst. Fourier,
            19-2, 1969, p. 419–493).

[Nov]$_2$ P. Noverraz: Pseudoconvexité et bases de Schauder dans les
            espaces localement convexes (Sém. Lelong, 1973/74, Lecture notes,
            474, p. 63–82).

[Nov]$_3$ P. Noverraz: Sur la mesure gaussienne des ensembles polaires en
            dimension infinie (Journées de fonctions analytiques, Toulouse
            1976, Lecture notes, **578**, p. 265–268).

[Pa]   G. Patrizio: On holomorphic maps between domains in $\mathbb{C}^n$ (Ann.
            Sc. Norm. Sup. Pisa, **13**, 1986, p. 267–279).

[Re]   A. Renaud: Quelques propriétés des applications analytiques d'une
            boule de dimension infinie dans une autre (Bull. Sc. math., **97**, 1973,
            p. 129–159).

[Ro]   H. Royden: Remarks on the Kobayashi metric (Several complex
            variables II, Lecture notes **185**, p. 125–137).

[Sch]   M. Schottenloher: Das Leviproblem in unendlichdimensionalen
            Räumen mit Schauderzerlegung (Habilitationsschrift, München,
            1974).

[Sic]   J. Siciak: A polynomial lemma and analytic mappings in topo-
            logical vector spaces (Sém. Lelong, 1970/71, Lecture notes **275**,
            p. 131–142).

[Sil]   J.S. Silva: Conceitos de função differenciavel em espaços local-
            mente convexos (Centro de estudos matematicos de Lisboa, 1957).

[Sz]   E. Szpilrajn: Remarques sur les fonctions sous-harmoniques (Ann.
            of Math., **34**, 1933, p. 588–594).

[Ta]$_1$ A. Taylor: On the properties of analytic functions in abstract
            spaces (Math. Ann., **115**, 1938, p. 466–484).

[Ta]$_2$ A. Taylor: Introduction to functional analysis (Wiley, New
            York).

[Te]   O. Teichmüller: Über die Stetigkeit linearer analytischer Funck-
            tionale (Deutsche Math., **1**, 1936, p. 350–352).

[Th Wh]   E. Thorp and R. Whitley: The strong maximum modulus theorem for analytic functions into a Banach space (Proc. amer. math. Soc., **18**, 1967, p. 640–646).

[Ts]   M. Tsuji: Potential theory in modern function theory (Tokyo, 1959).

[Va]   F. Vasilesco: Sur la continuité du potentiel à travers les masses (Comptes rendus Acad. Sc. Paris, **200**, 1935, p. 1173–1174).

[Ve]$_1$ E. Vesentini: Invariant distances and invariant differential metrics in locally convex spaces (Proc. Stefan Banach intern. centre, **8**, 1982, p. 493–512).

[Ve]$_2$ E. Vesentini: Variations on a theme of Caratheodory (Ann. Sc. Norm. Sup. Pisa, **7**, 1979, p. 39–68).

[Ve]$_3$ E. Vesentini: Complex geodesics (Comp. math., **44**, 1981, p. 375–394).

[Ve]$_4$ E. Vesentini: Complex geodesics and holomorphic maps (Sympos. math., **26**, 1982, p. 211–230).

[Vi]$_1$ J.P. Vigué: Sur le groupe des automorphismes analytiques d'un domaine borné d'un espace de Banach complexe (Comptes Rendus, **282** A, 1976, p. 111–114).

[Vi]$_2$ J.P. Vigué: Sur les applications holomorphes isométriques pour la distance de Caratheodory (Ann. Sc. Norm. Sup. Pisa, **9**, 1982, p. 255–261).

[Vi]$_3$ J.P. Vigué: La distance de Caratheodory n'est pas intérieure (Result. der Math., **6**, 1983, p. 100–104).

[Vi]$_4$ J.P. Vigué: Géodésiques complexes et points fixes d'applications holomorphes (Adv. in Math., **52**, 1984, p. 241–247).

[Zo]   M. Zorn: Gâteaux differentiability and essential boundedness (Duke math. J., **12**, 1945, p. 579–583).

Glossary of Notations

Roman letters

$\text{Aut}(\Omega)$	The group of analytic automorphisms of Ω.
C_n^k	The binomial coefficient $n!/k!(n-k)! (0 \leqslant k \leqslant n)$.
$\text{csn}(X)$	The set of all seminorms on the space X continuous for its topology.
$c_0(A)$ [resp. $c_{00}(A)$]	The space of maps $x: A \to \mathbb{C}$ such that $x^{-1}[C \backslash \Delta(0, \varepsilon)]$ is finite for all $\varepsilon > 0$ [resp. $x^{-1}(\mathbb{C}^*)$ is finite] with the sup norm.
$C_\Omega; c_\Omega$	The Caratheodory pseudodistance; its infinitesimal form.
dist	The distance in a metric space.
$D^k f(a) (k \in \mathbb{N}^m)$	The partial derivatives $(\in Z)$ at $a \in U$ of an analytic map $f: \mathbb{C}^m \supset U \to Z$.
$\hat{D}_a^k f [k \in \mathbb{N}^*, a \in \Omega, f \in \mathscr{A}(\Omega, Z)$	The k-homogeneous polynomial map whose value at x is the coefficient of ζ^k in the Taylor expansion of $f(a + \zeta x)$ around 0.
$D^k f(a) (k \in \mathbb{N}^*)$	The value $(\in Z)$ at a fixed point of the k-homogeneous polynomial map $\hat{D}_a^k f$.
E_u	The set where $u < u^*$ (u^* upper regularized of u).
$(e_1, \ldots, e_m)$	The canonical basis of $\mathbb{C}^m$.
$(e_n)_{n \in N^*}$	A countable Hamel basis or a Schauder basis.
$\text{Fix}(h)$	The set of fixed points of a map $h: \Omega \to \Omega$.
G_U	The Green function of an open set $U \subset \mathbb{C}$.
H_Ω	A pseudodistance on Ω^2 related to the positive pluriharmonic functions on Ω.
$\hat{K}$	The convex hull in Ω, with respect to $\mathscr{P}(\Omega)$ or $\mathscr{A}(\Omega, \mathbb{C})$, of a compact set $K \subset \Omega$.
$K(\mathscr{A})$ [resp. $K^*(\mathscr{A})$]	The set of sums $\sum_{j \in J} \zeta_j a_j$, $|\zeta_j| \leqslant 1$ (resp. $= 1$) for all $j \in J$, where $\mathscr{A} = (a_j)_{j \in J}$ is a summable family.
$K_\Omega; k_\Omega$	The Kobayashi pseudodistance; its infinitesimal form.
l.c.	locally convex (space).
$l^\infty(A)$	The Banach space of bounded maps $A \to \mathbb{C}$ with the sup norm.
$MV u(e^{i\theta})$	The mean value $\dfrac{1}{2\pi} \int\limits_{-\pi}^{\pi} u(e^{i\theta}) \, d\theta$ when $u(e^{i\theta})$ is $d\theta$-measurable.
$\overline{MV} u(e^{i\theta}) =$	$\inf\{MV v(e^{i\theta}): v \geqslant u, v(e^{i\theta}) \, d\theta\text{-measurable}\}$.

p.h.h.
  (resp. p.s.h.)   plurihypo-(resp. plurisub-) harmonic (function).
$P(\zeta, e^{i\varphi})$ The Poisson kernel for the unit disc.
q.c.   quasi-complete (space).
$R(f, p, q)$ $[p \in csn(X), q \in csn(Z)]$ The (p, q)-radius of boundedness, around a given point, of $f \in \mathscr{A}(X, Z)$.
s.c., s.r.   sequentially complete, semi-reflexive (space).
$S_\sigma(X)$ Classes of functions $\in \mathscr{P}(X)$ with logarithmic growth.
U An open set in a finite dimensional space $\mathbb{C}^m$.
u^* The lower or upper regularized of a function u with values in $\bar{R}$, namely $u^*(x) = \liminf_{y \to x}$ or $\limsup_{y \to x} u(y)$.
(X, π) A l.c. space whose topology is defined by the family π of seminorms.
X' The adjoint space to X, i.e. the space of all continuous linear maps $X \to \mathbb{C}$.
$\langle x, x' \rangle$ The value at $x \in X$ of $x' \in X'$.
$(x - a)^k =$ $\prod_{j=1}^{m} (\xi_j - \alpha_j)^{k_j}$ when $x = (\xi_1, \ldots, \xi_m)$ and $a = (\alpha_1, \ldots, \alpha_m)$ are points of $\mathbb{C}^m$.

Script letters

$\mathscr{A}(\Omega, Z)$ The class of (Fréchet-) analytic maps of an open set $\Omega \subset X$ into a s.c.l.c. space Z.
$\mathscr{A}(\Omega, \Gamma) =$ $\{f \in \mathscr{A}(\Omega, Z): f(\Omega) \subset \Gamma\}$.
$\mathscr{G}(\Omega, Z)$ The class of Gâteaux-analytic maps of a finitely open set $\Omega \subset X$ into the same Z.
$\mathscr{N}(U)$ A class of nearly subharmonic functions on an open set $U \subset \mathbb{C}$.
$\mathscr{P}(\Omega)$ The class of p.s.h. functions on Ω.
$\mathscr{P}^1(\Omega)$ The class of locally bounded from above suprema of sequences $\subset \mathscr{P}(\Omega)$.
$\mathscr{P}^2(\Omega)$ The class of limits $(> -\infty$ on a dense set) of decreasing sequences $\subset \mathscr{P}^1(\Omega)$ *and so on.*
$\mathscr{Q}(U)$ The class of quasisubharmonic functions on an open set $U \subset \mathbb{C}$.
$\mathscr{R}e, \mathscr{I}m$ Real, imaginary part of a complex number.
$\mathscr{S}(U)$ The class of superharmonic (resp. positive superharmonic)
  [resp. $\mathscr{S}^+(U)$]   functions on an open set $U \subset \mathbb{C}$.

Greek letters

$\Delta(\alpha, r)$ [resp. $\bar{\Delta}(\alpha, r)$]	The open (resp. closed) disc in $\mathbb{C}$ with centre α, radius r.
$\Delta(\omega)$	The fine boundary of an open set ω.
$v_w(a)$	The density number for a funciton $w \in \mathscr{P}(\Omega)$ at a point $a \in \Omega$.
$\pi(\alpha, \beta)$	The Poincaré distance in the unit disc Δ.
$\omega(a)(a \in \Omega)$	The largest balanced subset of the translated set $\Omega - a$.

Subject Index

Abel's lemma 21
adjoint space 3
analytic maps in the sense of
  Gâteaux 35
  Fréchet 52
  Fantappiè 63
Baire space 9
balanced set 14
Banach space 11
Banach-Steinhaus theorem for
  continuous polynomial maps 56,
  146
barrel 14
barrelled space 11
Bernstein-Walsh inequality 30
bornological space 3
bounded set 2
bounded family of analytic maps 25,
  54
bounding set 73
Caratheodory pseudo-distance 105,
  168
Cauchy (inequalities of –) 23
– (generalized ineq. of –) 38
complete space 3
complex geodesic 179
composition property for
  Gâteaux-analytic maps 42
  continuous analytic maps 57
  a p.s.h. function 95
cone 149
convergence theorems 90, 111
convex functions 84
– hull 130
– sets 137
density number of a p.s.h. function
  156

domain of existence, of holomorphy
  135
entire maps and functions 65
equicontinuity of anal. maps 25, 55
equi-Schauder basis 139
exponential type 72
extremal points 182
Fantappiè (anal. in the sense of) 63
fine boundary of an open set 123, 125
fine topology 123, 124
fine maximum principle 125
finitely open set 35
fixed point Th. of Earle-Hamilton 188
– – – Hayden-Suffridge 189
Fréchet space 3, 11
– (anal. in the sense of –) 52
Gâteaux (anal. in the sense of –) 35
gauge 13
generalized derivative 38
– polydisc 97
Green function of an open set $\mathbb{C}$ 90
– – with pole at ∞ 31
Green open set 164
Harnack (inequality of –) 89
Harris (inequality of –) 175
Hartogs (property of –) for
  Gâteaux-analytic maps 41
  continuous analytic maps 58
holomorphically convex 135
homogeneous polynomial map 33
– – expansion 38
homogeneous p.s.h. function 152
inductive limit 14
– – (strict –) 15
intrinsic pseudodistances 168
inverse function theorem 116
Josefson-Nissenzweig Th. 68

Kobayashi pseudodistance 171
Legendré transform 157
Lempert's Th. for convex domains 172
Leja's polynomial lemma 31
Levi problem 135
Lindelöf principle 165
Liouville Th. for entire maps 71
– – p.s.h. functions 82
locally bounded from above family 84
locally convex space 2
maximum modulus principle 40
maximum principle for p.s.h. f^{ns} 82
– – (fine –) 125
meagre set 9
metrizable l.c. space 2
nearly subharmonic functions 91
Nissenzweig (Josefson –) Th. 68
order of growth 149
pluriharmonic function 104
plurihypoharmonic function 82
plurisubharmonic – 83
pluripolar set 113
Poincaré distance 164
Poisson kernel 88
polar set 89
polynomial 28
– lemma of Leja 31
– map 32
principle (maximum modulus –) 40

– (maximum –) 82
– of anal. continuation 22
pseudoconvexity 130
pseudodistances 105, 106, 168
quasi-complete space 3
quasisubharmonic function 92
radius of boundedness 70, 144
– – uniform convergence 70
scalarly analytic map 20
Schauder basis 139
Schwarz lemma for analytic maps 184
– – – p.s.h. f^{ns} 156
semi-reflexive space 4
sheaf property 36, 83
star-shaped 13
strict inductive limit 15
strictly pluripolar set 113
strong topology 4
submedian function 85
summable family 6
Taylor expansion 22
thin set 120
translation lemma 35
type of growth 73, 153
uniformly convex seminorm 144
– summable family 6
unipolar set 107
upper regularized 84
weakened topology 3
weak* – 4
Zorn's theorem 46

de Gruyter Studies in Mathematics

Volume 1:   **W. Klingenberg**
Riemannian Geometry
1982. X, 396 pages. Cloth   ISBN 3 11 008673 5

Volume 2:   **M. Métivier**
Semimartingales
A Course on Stochastic Processes
1982. XII, 287 pages. Cloth   ISBN 3 11 008674 3

Volume 3:   **L. Kaup / B. Kaup**
Holomorphic Functions of Several Variables
An Introduction to the Fundamental Theory
1983. XVI, 350 pages. Cloth   ISBN 3 11 004150 2

Volume 4:   **C. Constantinescu**
Spaces of Measures
1984. 444 pages. Cloth   ISBN 3 11 008784 7

Volume 5:   **G. Burde / H. Zieschang**
Knots
1984. XII, 400 pages. Cloth   ISBN 3 11 008675 1

Volume 6:   **U. Krengel**
Ergodic Theorems
1985. VIII, 357 pages. Cloth   ISBN 3 11 008478 3

Volume 7:   **H. Strasser**
Mathematical Theory of Statistics
Statistical Experiments and Asymptotic Decision Theory
1985. XII, 492 pages. Cloth   ISBN 3 11 010258 7

Volume 8:   **T. tom Dieck**
Transformation Groups
1987. X, 312 pages. Cloth   ISBN 3 11 009745 1

Volume 9:   **H.-O. Georgii**
Gibbs Measures and Phase Transitions
1988. XIV, 525 pages. Cloth   ISBN 3 11 010455 5

de Gruyter · Berlin · New York

Walter de Gruyter & Co., Genthiner Str. 13, D-1000 Berlin 30, FRG, Phone: (0 30) 2 60 05-0 · Telex 1 84 027
Walter de Gruyter, Inc., 200 Saw Mill River Road, Hawthorne, N.Y. 10532, USA
Phone (914) 747-0110 · Telex 6 46 677

Volume 1 · 1989

Forum Mathematicum

An international journal devoted to pure and applied
mathematics as well as mathematical physics

Forum Mathematicum is devoted entirely to the publication of original
research articles in all fields of pure and applied mathematics, including
mathematical physics. Expository surveys, research announcements,
book reviews, etc. will not be published.

Manuscripts should be submitted to any of the editors or to:

Forum Mathematicum
Mathematisches Institut der Universität
Bismarckstraße 1 ^{1}/2
D-8520 Erlangen, Federal Republic of Germany

de Gruyter · Berlin · New York

Walter de Gruyter & Co., Genthiner Str. 13, D-1000 Berlin 30, FRG, Phone (0 30) 2 60 05-0 · Telex 1 84 027
Walter de Gruyter, Inc., 200 Saw Mill River Road, Hawthorne, N.Y. 10532, USA,
Phone (914) 747-0110 · Telex 6 46 677